Symbolic Data Analysis

WILEY SERIES IN COMPUTATIONAL STATISTICS

Wiley Series in Computational Statistics is comprised of practical guides and cutting edge research books on new developments in computational statistics. It features quality authors with a strong applications focus. The texts in the series provide detailed coverage of statistical concepts, methods and case studies in areas at the interface of statistics, computing, and numerics.

With sound motivation and a wealth of practical examples, the books show in concrete terms how to select and to use appropriate ranges of statistical computing techniques in particular fields of study. Readers are assumed to have a basic understanding of introductory terminology.

The series concentrates on applications of computational methods in statistics to fields of bioinformatics, genomics, epidemiology, business, engineering, finance and applied statistics.

Symbolic Data Analysis

Conceptual Statistics and Data Mining

Lynne Billard

University of Georgia, USA

Edwin Diday

University of Paris, France

Telephone (+44) 1243 779777

Email (for orders and customer service enquiries): cs-books@wiley.co.uk
Visit our Home Page on www.wiley.com

Other Wiley Editorial Offices

John Wiley & Sons Inc., 111 River Street, Hoboken, NJ 07030, USA

Jossey-Bass, 989 Market Street, San Francisco, CA 94103-1741, USA

Wiley-VCH Verlag GmbH, Boschstr. 12, D-69469 Weinheim, Germany

John Wiley & Sons Australia Ltd, 42 McDougall Street, Milton, Queensland 4064, Australia

John Wiley & Sons (Asia) Pte Ltd, 2 Clementi Loop #02-01, Jin Xing Distripark, Singapore 129809

John Wiley & Sons Canada Ltd, 6045 Freemont Blvd, Mississauga, ONT, L5R 4J3 Canada

Wiley also publishes its books in a variety of electronic formats. Some content that appears in print may not be available in electronic books.

Library of Congress Cataloging in Publication Data

Billard, L. (Lynne), 1943–
Symbolic data analysis / L. Billard, E. Diday.
p. cm.
Includes bibliographical references.
ISBN-13: 978-0-470-09016-9
ISBN-10: 0-470-09016-2
1. Multivariate analysis. 2. Data mining. I. Diday, E. II. Title.
QA278.B54 2007
519.5′35—dc22

2006028488

British Library Cataloguing in Publication Data

A catalogue record for this book is available from the British Library

ISBN-13 978-0-470-09016-9 (HB)
ISBN-10 0-470-09016-2 (HB)

Typeset in 10/12pt Times by Integra Software Services Pvt. Ltd, Pondicherry, India

This book is printed on acid-free paper responsibly manufactured from sustainable forestry in which at least two trees are planted for each one used for paper production.

Contents

1

Introduction

Our aim is to provide an introduction to symbolic data and how such data can be analyzed. Classical data on p random variables are represented by a single point in p-dimensional space $\Re^p$. In contrast, symbolic data with measurements on p random variables are p-dimensional hypercubes (or hyperrectangles) in $\Re^p$, or a Cartesian product of p distributions, broadly defined. The 'hypercube' would be a familiar four-sided rectangle if, for example, $p = 2$ and the two random variables take values over an interval $[a_1, b_1]$ and $[a_2, b_2]$, say, respectively. In this case, the observed data value is the rectangle $R = [a_1, b_1] \times [a_2, b_2]$ with vertices (a_1, a_2), (a_2, b_2), (b_1, a_2), and (b_1, b_2). However, the $p = 2$ dimensional hypercube need not be a rectangle; it is simply a space in the plane. A classical value as a single point is a special case. Instead of an interval, observations can take values that are lists, e.g., {good, fair} with one or more different values in the list. Or, the observation can be a histogram. Indeed, there are many possible formats for symbolic data. Basic descriptions of symbolic data, the types of symbolic data, how they contrast with classical data, how they arise, and their inherent internal properties are covered in Chapter 2, before going on to specific analytic methodologies in the chapters that follow.

At the outset, however, it is observed that a symbolic observation in general has an internal variation. For example, an individual whose observed value of a random variable is $[a, b]$, $a \neq b$, is interpreted as taking (several) values across that interval. (This is not to be confused with uncertainty or impression when the variable takes a (single) value in that interval with some level of uncertainty.) Unless stated otherwise, this interpretation applies throughout this volume. Analogous arguments apply for other types of symbolic data. A classical observation with its single point value perforce has no internal variation, and so analyses deal with variation between observations only. In contrast, symbolic data deal with the internal variation of each observation plus the variation between observations.

Symbolic Data Analysis: Conceptual Statistics and Data Mining L. Billard and E. Diday

Symbolic data arise in a variety of different ways. Some data are inherently symbolic. For example, if the random variable of interest is 'Color' and the population is 'Species of birds', then any one realization for a given species can take several possible colors from the set of all colors, e.g., a magpie has colors {white, black}. An all-black bird is not a magpie, it is a different kind of bird. A different kind of species, not a magpie, may be such that some birds are all white, some all black, or some both black and white. In that case, the possible colors are {white, black, white and black}. Or, it may not be possible to give the exact cost of an apple (or shirt, or product, or ...) but only its cost that takes values in the range [16, 24] cents (say). We note also that an interval cost of [16, 24] differs from that of [18, 22] even though these two intervals both have the same midpoint value of 20. A classical analysis using the same midpoint (20) would lose the fact that these are two differently valued realizations with different internal variations.

In another direction, an insurance company may have a database of hundreds (or millions) of entries each relating to one transaction for an individual, with each entry recording a variety of demographic, family history, medical measurements, and the like. However, the insurer may not be interested in any one entry per se but rather is interested in a given individual (Colin, say). In this case, all those single entries relating to Colin are aggregated to produce the collective data values for Colin. The new database relating to Colin, Mary, etc., will perforce contain symbolic-valued observations. For example, it is extremely unlikely that Mary always weighed 125 pounds, but rather that her weight took values over the interval [123, 129], say. Or, in a different scenario, when a supermarket keeps track of the sales of its products, it is not really interested in the sale of any one item (e.g., my purchase of bread yesterday) but more interested in the variation of its sales of bread (or cheese, or meat, or ...) in a given period. It may also be interested in patterns of purchases, e.g., are the two products bread and butter often bought at the same time? Likewise, a drug company is interested in the overall patterns of purchases of certain medications rather than Frank's particular use; a hospital wants to track different types of patient ailments; an automotive insurance company is less interested in a single car accident but more interested in car accidents of different driver demographic categories, or of different types of vehicles. Rather than focusing on a single DNA sequence, a geneticist may be interested in tracking the occurrences of a particular amino acid in a protein chain instead of one single DNA sequence. For example, CCU, CCC, CCA, and CCG all code for proline; similarly, GCU, GCC, GCA, and GCG code for alanine.

In all these and similar situations, there is a basic concept of interest, in effect a second level of observations, whose values are obtained by aggregation over the observations for those individuals (at a first level) who constitute the descriptor of that concept. By 'aggregation', it is meant the collection of individuals who satisfy the descriptor of interest. For example, suppose the variable is farm acreage, and Drew and Calvin have acreages of 7.1 and 8.4, respectively, both in the northeast region. Then, if the concept of interest is acreage of farms in the northeast, aggregating Drew and Calvin produces an acreage of [7.1,8.4]. How aggregation is implemented is discussed in Chapter 2. We note that the original dataset may itself contain symbolic data. The aggregated dataset, however, will almost certainly contain symbolic data regardless of whether the first-level entries are classical or symbolic values.

In these two types of settings, the original database can be small or large. A third setting is when the original database is large, very large, such as can be generated by contemporary computers. Yet these same computers may not have the capacity to execute even reasonably elementary statistical analyses. For example, a computer requires more memory to invert a matrix than is needed to store that matrix. In these cases, aggregation of some kind is necessary even if only to reduce the dataset to a more manageable size for subsequent analysis. There are innumerable ways to aggregate such datasets. Clearly, it makes sense to seek answers to reasonable scientific questions (such as those illustrated in the previous paragraphs) and to aggregate accordingly. As before, any such aggregation will perforce produce a dataset of symbolic values.

Finally, we note that the aggregation procedures adopted and their emergent symbolic datasets are not necessarily obtained from clustering procedures. However, standard (classical, and indeed also symbolic) clustering methods do produce clusters of observations whose values when summarized for the respective random variables are symbolic in nature (unless the analyst chooses to use classical summaries such as average values, which values as we shall see do not retain as much information as do the symbolic counterparts).

Chapter 3 presents methodologies for obtaining basic descriptive statistics for one random variable whose values are symbolic valued such as intervals, namely, a histogram and its empirical probability distribution relative, along with the empirical mean and variance. Descriptive statistics are extended to $p = 2$ dimensions in Chapter 4. A key feature in these calculations is the need to recognize that each observation has its own internal variation, which has to be incorporated into the respective methodologies. Thus we see, for example, that a single observation (a sample of size 1) whose value is the interval $[a, b] = [2, 8]$, say, has a sample variance of 3, whereas a single classical observation (the midpoint 5, say) has sample variance of 0. That is, the classical analysis loses some of the information contained in the datum. In both Chapter 3 and Chapter 4, methodologies are developed for multi-valued, interval-valued, modal multi-valued and modal interval-valued (i.e., histogram-valued) observations.

A second distinctive feature is the presence of so-called rules. In addition to the internal variation mentioned in the previous paragraph, there can be structural variations especially after aggregation of a larger dataset. For example, in studying batting performances of teams in a baseball league, the numbers of hits and at-bats are aggregated over the players of the respective teams. While for each individual player the number of hits would not exceed the number of at-bats, it is possible that when aggregated by team the number of hits could exceed the number of at-bats for some regions of the resulting hypercube. Clearly, it is necessary to add a rule to the team data so as to retain the fact that, for the individual players, the number of hits does not exceed the number of at-bats (see Section 4.6 for a more detailed discussion of this example). This type of rule is a logical dependency rule and is designed to maintain the data integrity of the original observations. Such rules are often necessary for symbolic data but not for classical data. On the other hand, there are numerous rules (the list is endless) that apply equally to classical and symbolic data, e.g., structural rules such as taxonomy and hierarchy trees that can prevail. Special attention is paid to logical dependency rules when calculating the basic descriptive statistics of Chapters 3 and 4. Subsequent methodologies will by and large be described without specific detailed attention to these rules.

Chapters 5–7 deal with principal components, regression and clustering techniques, respectively. These methods are extensions of well-known classical theory applied or extended to symbolic data. Our approach has been to assume the reader is knowledgeable about the classical results, with the present focus on the adaptation to the symbolic data setting. Therefore, only minimal classical theory is provided. Regression methodologies are provided for multi-valued data, interval-valued data and histogram-valued data, along with methods for handling taxonomy tree structures and hierarchy tree structures.

The work on principal components considers two methods, the vertices method and the centers method. Both deal with interval-valued data, the former as its name suggests focusing on the vertices of the hypercube associated with each data value, and the latter using the midpoint center values. Principal component methods for multi-valued and modal-valued data currently do not exist.

Clustering techniques in Chapter 7 look at partitioning, hierarchies, divisive clustering, and pyramid clustering in particular. Since these methods depend on dissimilarity and distance measures, the chapter begins with developing these measures for symbolic data. Many of these measures and techniques are also extensions of their classical counterparts. Where available, methodologies are presented for multi-valued, interval-valued, and histogram-valued variables. In some instances, a mixed variable case with some variables multi-valued and other variables interval-valued is considered (usually through the well-known oils dataset).

Perhaps the most striking aspect of this catalogue of techniques is the paucity of available methodologies for symbolic data. It is very apparent that contemporary datasets will be increasingly symbolic in content. Indeed, Schweizer (1984) has declared that 'distributions are the numbers of the future', and 'distributions' are examples of what are symbolic data. Therefore, it is imperative that suitable techniques be developed to analyze the resulting data. In that sense the field is wide open providing lots of opportunities for new research developments. In another direction, while there are some methodologies presently existing, most seem intuitively correct, especially since results for the particular case of classical data do indeed emerge from the symbolic analyses. However, except in rare isolated instances, there is next to no work yet being done on establishing the mathematical underpinning necessary to achieve rigor in these results. Again, opportunity knocks for those so inclined.

The attempt has been made to provide many illustrative examples throughout. In several places, datasets have been presented with more random variables than were actually used in the accompanying example. Those 'other' variables can be used by the reader as exercises. Indeed, most datasets provided (in any one example or section) can be analyzed using techniques developed in other sections; again these are left as exercises for the reader.

The reader who wants to learn how to analyze symbolic data can go directly to Chapter 3. However, Section 2.1 would serve as a introduction to the types of data and the notation used in subsequent chapters. Also, Section 2.3 with its comparison between classically adapted analyses of symbolic data provides some cautionary examples that can assist in understanding the importance of using symbolic analytic methods on symbolic data.

Symbolic data appear in numerous settings, all avenues of the sciences and social sciences, from medical, industry and government experiments and data collection pursuits. An extensive array of examples has been included herein drawing upon datasets from as wide a range of applications as possible. Having said that, some datasets are used a number of times in order to provide a coherency as well as a comparative dialog of different methodologies eliciting different information from the same data. In this way, a richer knowledge base of what is contained within a dataset can be exposed, providing for a broader set of interpretations and scientific conclusions.

The datasets used are publicly available (at http://www.stat.uga.edu/faculty/LYNNE/Lynne.html) source references as to where they can be found are provided (usually when they first appear in the text). All can be obtained from the authors. As is evident to the reader/user, some examples use the entire dataset presented in the text's illustrations, while others use a portion of the data, be that some but not all of the observations or some but not all of the random variables. Therefore, the user/reader can take other portions, or indeed the complete dataset, and thence analyze the data independently as exercises. Some are provided at the end of each chapter. Since several of the methodologies presented can be applied to such data, these alternative analyses can also be performed as yet another extensive range of potential exercises. The reader might be encouraged to work through the details of the methodology independently even if only to persuade him/herself that the technique has been understood. The SODAS software package (available free at http://www.ceremade.dauphine.fr/%7Etouati/sodas-pagegarde.htm) provides the computational power for most of the methodology presented in this text.

In some sense this volume owes its origins to the review paper by Billard and Diday (2003) and the edited volume by Bock and Diday (2000). Those publications form initial attempts to bring together the essential essence of what symbolic data are. The forthcoming edited volume by Diday and Noirhomme (2006) complements the present volume in so far as it provides algorithmic and computational details of some of the software routines in the SODAS software package.

References

Billard, L. and Diday, E. (2003). From the Statistics of Data to the Statistics of Knowledge: Symbolic Data Analysis. *Journal of the American Statistical Association* 98, 470–487.

Bock, H.-H. and Diday, E. (eds.) (2000). *Analysis of Symbolic Data: Exploratory Methods for Extracting Statistical Information from Complex Data*. Springer-Verlag, Berlin.

Diday, E. and Noirhomme, M. (eds.) (2006). *Symbolic Data and the SODAS Software*. John Wiley & Sons, Ltd, Chichester.

Schweizer, B. (1984). Distributions are the Numbers of the Future. In: *Proceedings of The Mathematics of Fuzzy Systems Meeting* (eds. A. di Nola and A. Ventes). University of Naples, Naples, Italy, 137–149.

2

Symbolic Data

Suppose we want to describe the fruit (e.g., apricots) produced by a certain orchard. We might say 'The weight is between 30 and 50 grams and its color is orange or green, but when the color is green then the weight is less than 35 grams (i.e., it's not yet ripe for eating).' It is not possible to put this kind of information in a standard classical data table, since such tables consist of cells with a single point value. Nor is it easy to represent rules in a classical data table. This apricot is more easily represented by the description

$$\text{'[Weight} = [30, 50]], \text{[Color} = \{\text{orange, green}\}] \text{ and [If } \{\text{Color} = \text{green}\} \text{ then } \{\text{Weight} < 35\}].\text{'}$$

This is a description associated with the concept of an apricot. The random variable values

$$(\text{Weight, Color}) = ([30, 50], \{\text{orange, green}\})$$

is an example of a (two-dimensional) symbolic-valued random variable. If we have a set (a sample, or a population) of orchards (or of different varieties of apricots), one question is: How do we construct a histogram or find an appropriate classification of the orchards/apricots? One aim of symbolic data analysis is to develop tools to provide such answers.

In general, a symbolic data 'point' assumes values as a hypercube (or hyper-rectangle, broadly defined) or a Cartesian product of distributions in p-dimensional

Symbolic Data Analysis: Conceptual Statistics and Data Mining L. Billard and E. Diday

space or a mixture of both, in contrast to a classical data 'point' taking a single point value in p-dimensional space. That is, symbolic values may be lists, intervals, and so on. Symbolic data may exist in their own right. Or, they result after aggregating a base dataset over individual entries that together constitute a second higher level (or category) of interest to the researcher. Or, they may emerge as a result of aggregating (usually) very large datasets (especially those so large that standard analyses cannot be performed) into smaller more manageably sized datasets, or aggregating into datasets that provide information about categories (for example) of interest (which datasets perforce are smaller in size). For datasets consisting of symbolic data, the question is how the analyses of symbolic data proceeds. For datasets that are too large to be analysed directly, the first question concerns the criteria to be adopted for the data aggregation to take place – there are numerous possible criteria with the impetus behind the criteria often driven by some scientific question. Such summarization methods necessarily involve symbolic data and symbolic analyses in some format (though some need not). Behind the aggregation method lies the notion of a symbolic object. There are many possible aggregations associated with any particular dataset with a specific concept being necessarily tied to a specific aim of an ensuing analysis.

After describing what symbolic data are in Section 2.1, we discuss the underlying concepts and how they relate to categories and symbolic objects in Section 2.2. Classical data and symbolic data and their analyses are contrasted in Section 2.3. While we shall see that there are several distinctions, a major difference is that a symbolic data point has internal variation whereas a classical data point does not.

2.1 Symbolic and Classical Data

2.1.1 Types of data

Suppose we have a dataset consisting of the medical records of individuals such as might be retained by a health insurance company. Suppose that for each individual, there is a record of geographical location variables, such as region (north, northeast, south, etc.), city (Boston, Atlanta, etc.), urban/rural (Yes, No), and so on. There can be a record of type of service utilized (dental, medical, optical). There can be demographic variables such as gender, marital status, age, information on parents (such as the number alive), siblings, number of children, employer, health provider, etc. Basic medical variables could include weight, pulse rate, blood pressure, etc. Other health variables (for which the list of possible variables is endless) would include incidences of certain ailments and diseases; likewise, for a given incidence or prognosis, treatments and other related variables associated with the disease are recorded. Such a typical dataset may follow the lines of Table 2.1a; Table 2.1b provides the descriptors of the random variables $Y_1, Y_2, \ldots$ used in this table.

The entries in Tables 2.1 are classical data values with each line referring to the observed values, or realizations, for the random variable $Y = (Y_1, \ldots, Y_p)$ for a single individual. Here, there are only $n = 51$ observations and $p = 30$ variables. At this size, the data can be analyzed using classical techniques. When the size n is large, very large (e.g., $n = 100$ million, $p = 200$), standard classical analyses can be problematic. Any analysis of such a table produces results dealing with individuals. There are two possible issues here. One is simply the question of how the dataset can be reformulated to a size that allows analyses to proceed. In order to do this in a meaningful way, it is first necessary to consider what we want to learn from the data. For example, are we interested in studying what happens to those who engage medical services? This question relates to the second issue, one that is in fact of primary importance. Therefore, regardless of the size of the dataset, rather than the individuals, it may be of greater interest to the insurance company to determine what happens to certain categories, such as women who used medical services denoted by 'medical women', or 20 year olds who went to the dentist denoted by 'dental 20 year olds' rather than the individuals who make up those categories. Since each of these categories consists of several individuals, the observed 'value' is no longer a single point. For example, the weight of the 'medical women' (in Tables 2.1) is the list {73, 75, 113, 124, 124, 129, 138, 141, 152, 157, 161, 161, 165, 167}. These values may instead be represented as realizations in the interval [73, 167]; or the weight could be written as the histogram {[78, 110), 3/14; [110, 160), 7/14; [160, 170], 4/14}, i.e., 7/14 or 50% weigh between 110 and 160 pounds, 3/14 or 21.4% weigh less than 110 pounds, and 2/7 or 28.6% weigh 160 pounds or more. The weight variable for the category 'medical women' now has assumed a value which is a list, or an interval, or a histogram, respectively; and as such, each is an example of symbolic data.

This category 'medical women' is an example of a symbolic concept (or simply, concept), whose definition, construction, and properties will be explored in Section 2.2. Indeed, rather than any of the specific individuals in Tables 2.1, interest may center on the type $\times$ gender categories (or, concept). Here, there are three types (dental, medical, optical) and two genders (male, female) to give a total of $3 \times 2 = 6$ possible categories. In this case, by using the methods described in Section 2.2, Tables 2.2 can result when the data of Table 2.1 are aggregated according to these categories, for the variables $(Y_1, Y_3, Y_5, Y_7, Y_9, Y_{10})$, say; we leave the aggregation for the remaining variables as an exercise. We shall refer to categories $\omega_u \in E = \{w_1, \ldots, w_m\}$; sometimes these will be represented as classes $E = \{C_1, \ldots, C_m\}$ where each class consists of the set of individuals which satisfy the description of a category. Note, in particular, that a particular category w_u may have been formed from a single classical observation.

Table 2.3 is a partial list of credit card expenditures by a coterie of individuals over a 12-month period. The card issuer is interested in the range of each individual's itemized expenses (in dollars) for food, social entertainment, travel, gas, and clothes ($Y_1, \ldots, Y_5$, respectively) per person per month. Thus, rather than

Table 2.1a Classical data.

i	Y_1	Y_2	Y_3	Y_4	Y_5	Y_6	Y_7	Y_8	Y_9	Y_{10}	Y_{11}	Y_{12}	Y_{13}	Y_{14}	Y_{15}
1	Boston	M	24	M	S	2	2	0	165	68	120	79	183	83	86
2	Boston	M	56	M	M	1	2	2	186	84	130	90	164	64	60
3	Chicago	D	48	M	M	1	3	2	175	73	126	82	229	109	122
4	El Paso	M	47	F	M	0	1	1	141	78	121	86	239	69	74
5	Byron	D	79	F	M	0	3	4	152	84	150	88	187	67	64
6	Concord	M	12	M	S	2	1	0	73	69	126	85	109	98	107
7	Atlanta	M	67	F	M	1	6	0	166	81	134	89	190	90	96
8	Boston	O	73	F	M	0	2	4	164	77	121	81	181	81	84
9	Lindfield	D	29	M	M	2	0	2	227	62	124	81	214	94	101
10	Lindfield	D	44	M	M	1	3	3	216	71	125	79	218	98	107
11	Boston	D	54	M	S	1	5	0	213	57	118	88	189	69	66
12	Chicago	M	12	F	S	2	2	0	75	69	115	81	153	54	45
13	Macon	M	73	F	M	0	3	1	152	58	123	82	188	87	93
14	Boston	D	48	M	M	0	2	4	206	73	113	72	264	72	62
15	Peoria	O	79	F	M	0	3	3	153	72	106	78	118	40	35
16	Concord	D	20	M	S	2	0	1	268	79	123	80	205	85	89
17	Boston	D	20	F	S	2	4	0	157	75	116	87	180	60	52
18	Chicago	D	17	M	S	2	2	0	161	69	114	78	169	49	39
19	Stowe	D	31	M	M	1	3	2	183	81	118	84	185	66	62
20	Tara	M	83	M	M	0	3	1	128	80	108	80	224	48	65
21	Akron	M	20	M	S	1	3	0	182	68	114	76	150	51	40
22	Detroit	M	85	F	M	0	3	2	161	73	122	76	185	83	89
23	Ila	D	66	F	S	0	4	3	166	66	126	87	218	98	108
24	Marion	M	6	M	S	2	1	0	35	72	114	76	136	52	28
25	Albany	M	24	M	M	2	1	1	177	81	111	82	149	51	39
26	Salem	D	76	M	M	0	5	2	192	77	115	73	173	53	44
27	Quincy	O	57	M	S	1	3	2	159	72	114	75	234	131	157
28	Yuma	M	11	F	S	2	2	0	73	62	118	80	96	56	43
29	Fargo	M	27	F	M	2	2	1	124	70	114	72	167	67	63
30	Reno	D	43	F	M	2	4	4	148	66	135	97	172	52	43
31	Amherst	M	53	F	S	1	0	3	165	65	135	96	236	134	161
32	Boston	M	14	M	S	2	1	0	132	66	125	87	149	51	39
33	New Haven	D	29	F	M	1	0	1	153	70	133	92	217	97	106
34	Kent	M	84	M	M	0	4	1	239	85	114	75	229	126	150
35	Shelton	M	52	M	M	0	4	1	206	63	125	86	236	134	161
36	Atlanta	O	86	M	M	0	3	3	184	72	114	72	152	53	42
37	Medford	M	23	F	S	2	1	0	138	71	125	85	197	96	105
38	Bangor	M	51	M	M	2	2	2	172	81	119	78	172	73	71
39	Boston	M	70	M	M	1	6	3	183	75	114	74	151	52	42
40	Barry	M	65	M	M	0	4	2	191	84	120	80	175	75	75
41	Trenton	M	82	M	M	0	3	4	201	79	123	84	188	87	93
42	Concord	M	60	M	S	0	4	0	175	74	117	76	163	63	58
43	Chicago	M	48	M	M	1	4	1	187	88	132	98	182	82	86
44	Omaha	M	29	M	M	1	1	1	166	59	122	82	178	78	79
45	Tampa	M	21	F	M	2	2	1	124	72	119	79	169	70	67
46	Lynn	M	81	F	M	0	5	3	161	79	128	89	210	109	124
47	Quincy	D	70	F	M	0	3	2	178	72	119	78	230	110	124
48	Wells	M	27	F	M	2	0	0	113	77	121	80	179	79	80
49	Buffalo	M	56	F	M	2	4	1	129	76	119	81	172	72	71
50	Quincy	M	64	M	S	1	2	0	194	81	128	89	210	109	124
51	Boston	M	87	F	M	0	5	2	157	88	128	88	171	71	70
⋮															

[a] See Table 2.1b for description of Y_i variables.

Y_{16}	Y_{17}	Y_{18}	Y_{19}	Y_{20}	Y_{21}	Y_{22}	Y_{23}	Y_{24}	Y_{25}	Y_{26}	Y_{27}	Y_{28}	Y_{29}	Y_{30}
2.21	88	92	16	1.4	12	21	6.9	5.2	14.2	43.6	2.32	N	N	0
2.55	69	101	16	0.8	20	22	7.0	4.6	13.5	39.9	2.44	N	N	1
2.10	114	80	17	1.4	13	24	7.7	4.9	14.1	44.2	2.73	Y	N	1
3.45	44	90	15	1.1	14	20	6.7	4.6	13.9	40.7	2.17	Y	0	0
2.79	72	103	18	0.9	20	27	7.3	4.8	11.6	36.1	3.05	N	0	0
1.11	105	108	14	0.8	18	17	6.2	4.3	12.2	36.0	1.79	N	0	0
2.12	95	91	17	1.0	17	24	7.2	4.6	13.4	42.3	2.65	Y	6	0
2.24	86	112	19	0.9	22	29	8.0	4.0	14.9	43.6	3.32	N	0	0
2.28	99	89	18	1.0	18	27	7.8	4.7	15.0	43.4	3.13	N	0	0
2.23	103	83	18	1.0	18	27	7.8	4.5	12.4	37.1	3.12	Y	N	2
2.75	74	100	28	0.3	90	53	11.5	4.3	14.8	42.6	6.32	N	N	0
2.83	58	119	20	1.0	19	31	8.3	4.4	14.3	40.7	3.59	N	0	0
2.15	93	69	16	1.2	13	21	6.9	4.6	12.9	37.1	2.35	N	0	0
3.69	49	91	14	1.2	11	16	6.1	5.0	12.9	40.5	1.67	N	N	0
2.95	23	82	19	0.9	20	30	8.1	4.1	13.6	43.3	3.40	N	0	0
2.40	90	71	19	1.3	14	28	7.9	4.2	13.5	39.4	3.21	N	N	0
3.01	65	101	17	1.0	16	23	7.2	5.1	13.0	40.8	2.61	N	0	0
3.45	54	96	17	1.0	16	23	7.2	4.2	13.1	40.7	2.61	N	N	0
2.82	71	146	18	0.7	24	28	7.8	4.8	13.2	38.2	3.14	N	N	0
4.66	38	111	15	1.0	14	18	6.4	4.7	13.6	41.7	1.94	Y	N	3
2.94	55	58	13	1.2	11	14	5.8	4.0	13.7	40.7	1.43	N	N	0
2.19	90	96	8	0.9	10	17	3.9	4.2	13.1	36.8	7.19	N	0	0
2.22	103	85	18	1.4	13	26	7.6	4.2	13.4	38.0	2.98	N	4	0
2.60	41	96	16	1.2	13	20	6.8	4.5	15.4	45.2	2.25	N	N	0
2.96	55	72	19	0.8	24	30	8.2	3.8	14.6	45.4	3.48	N	N	0
3.27	58	97	17	1.3	13	23	7.2	4.6	12.4	37.1	2.60	N	N	0
1.78	139	88	17	0.8	22	24	7.2	5.0	13.0	37.5	2.65	N	N	0
1.71	56	136	20	0.8	25	31	7.4	4.6	13.6	41.0	3.59	N	0	0
2.49	72	104	13	0.7	20	15	5.9	4.7	13.7	41.4	1.53	N	0	0
3.31	57	82	17	0.8	21	25	8.3	4.2	11.7	34.0	2.83	Y	3	0
1.76	141	102	9	1.0	9	4	7.0	4.6	11.2	33.9	2.18	N	1	0
2.96	54	120	20	1.1	18	32	8.4	5.3	12.1	38.7	3.67	N	N	0
2.23	102	99	25	1.1	23	45	10.3	4.6	14.3	41.4	5.29	N	0	0
1.81	134	113	9	1.1	8	4	4.3	4.8	12.3	37.6	2.43	Y	N	5
1.77	14	114	16	1.2	13	22	4.4	4.4	12.9	37.0	0.15	N	N	0
2.88	57	92	18	1.2	15	27	6.6	4.2	15.7	49.9	3.11	N	N	0
2.05	102	70	20	1.4	15	33	8.6	4.1	14.4	40.7	3.78	N	0	0
2.38	77	105	17	1.3	13	23	7.2	4.8	13.5	39.3	2.60	N	N	0
2.90	56	79	14	0.7	21	17	6.3	5.2	12.3	36.5	1.86	Y	N	2
2.34	80	139	21	1.3	16	34	8.8	4.5	14.6	43.8	3.97	N	N	0
2.15	93	111	18	0.8	21	26	7.5	4.6	14.5	44.6	2.91	N	N	0
2.58	68	112	12	1.2	9	10	5.2	4.7	13.4	38.7	0.91	N	N	0
2.23	87	95	18	0.8	23	27	7.8	4.3	12.4	37.3	3.13	N	N	0
2.29	83	77	22	1.5	15	38	9.3	4.3	13.5	38.9	4.43	N	N	0
2.43	74	103	14	1.0	14	15	6.0	4.9	13.5	43.1	1.60	N	0	0
1.93	115	86	15	0.6	23	19	7.8	4.5	11.3	32.9	2.06	Y	3	0
2.09	115	87	12	1.2	10	10	5.3	4.4	12.5	39.1	8.88	Y	1	1
2.27	84	101	6	1.0	6	8	3.1	4.2	14.6	40.9	2.99	N	0	0
2.38	77	99	14	1.4	10	16	6.1	4.7	13.9	40.3	1.69	Y	0	4
1.93	115	88	19	1.1	17	28	7.9	5.0	15.3	49.1	3.23	N	N	0
2.40	76	87	19	1.0	19	28	7.9	4.6	13.9	44.4	7.14	N	0	0

Table 2.1b Variable identifications.

Y_i	Description: range
Y_1	City/Town of Residence
Y_2	Type: Dental(D), Medical(M), Optical(O)
Y_3	Age (in years): ≥ 0
Y_4	Gender: Male (M), Female (F)
Y_5	Marital Status: Single (S), Married (M)
Y_6	Number of Parents Alive: 0, 1, 2
Y_7	Number of Siblings: 0, 1,. . .
Y_8	Number of Children: 0, 1,. . .
Y_9	Weight (in pounds): > 0
Y_{10}	Pulse Rate: > 0
Y_{11}	Systolic Blood Pressure: > 0
Y_{12}	Diastolic Blood Pressure: > 0
Y_{13}	Cholesterol Total: > 0
Y_{14}	HDL Cholesterol Level: > 0
Y_{15}	LDL Cholesterol Level: > 0
Y_{16}	Ratio = Cholesterol Total/HDL Level: > 0
Y_{17}	Triglyceride Level: > 0
Y_{18}	Glucose Level: > 0
Y_{19}	Urea Level: > 0
Y_{20}	Creatinine Level: > 0
Y_{21}	Ratio = Urea/Creatinine: > 0
Y_{22}	ALT Level: > 0
Y_{23}	White Blood Cell Measure: > 0
Y_{24}	Red Blood Cell Measure: > 0
Y_{25}	Hemoglobin Level: > 0
Y_{26}	Hemocrit Level: > 0
Y_{27}	Thyroid TSH: > 0
Y_{28}	Cancer Diagnosed: Yes (Y), No (N)
Y_{29}	Breast Cancer # Treatments: 0, 1, . . . , Not applicable (N)
Y_{30}	Lung Cancer # Treatments: 0, 1, . . .

Table 2.2a Symbolic data.

w_u	Type × Gender	Cities Y_1	n_u
w_1	Dental Males	{Boston, Chicago, Concord, Lindfield, Salem, Stowe}	9
w_2	Dental Females	{Boston, Byron, Ila, New Haven, Quincy, Reno}	6
w_3	Medical Males	{Akron, Albany, Bangor, Barry, Boston, Chicago, Concord, Kent, Marion, Omaha, Quincy, Shelton, Tara, Trenton}	18
w_4	Medical Females	{Amherst, Atlanta, Boston, Buffalo, Chicago, Detroit, El Paso, Fargo, Lynn, Macon, Medford, Tampa, Wells, Yuma}	14
w_5	Optical Males	{Atlanta, Quincy}	2
w_6	Optical Females	{Boston, Peoria}	2

Table 2.2b Symbolic data.

w_u	Type × Gender	Age Y_3	Marital Status Y_5	Parents Living Y_6	Weight Y_9	Pulse Rate Y_{10}	Systolic Pressure Y_{11}	Diastolic Pressure Y_{12}	Cholesterol Y_{13}
w_1	Dental Males	[17, 76]	{M, S}	{0,1,2}	[161, 268]	[57, 81]	[113, 126]	[72, 88]	[179, 264]
w_2	Dental Females	[20, 70]	{M, S}	{0,1,2}	[148, 178]	[66, 84]	[116, 150]	[78, 97]	[172, 230]
w_3	Medical Males	[6, 84]	{M, S}	{0,1,2}	[35, 239]	[59, 88]	[108, 132]	[74, 98]	[109, 236]
w_4	Medical Females	[11, 87]	{M, S}	{0,1,2}	[73, 166]	[58, 88]	[114, 135]	[72, 96]	[96, 239]
w_5	Optical Males	[57, 86]	{M, S}	{0, 1}	[159, 184]	[72, 72]	[114, 114]	[72, 78]	[152, 234]
w_6	Optical Females	[73, 79]	{M}	{0}	[153, 164]	[72, 77]	[106, 121]	[78, 81]	[118, 181]

Table 2.3 Credit card dataset.

i	Name	Month	Food	Social	Travel	Gas	Clothes
1	Jon	February	23.65	14.56	218.02	16.79	45.61
2	Leigh	May	28.47	8.99	141.60	21.74	86.04
3	Leigh	July	30.86	9.55	193.14	24.26	95.68
4	Tom	July	24.13	15.97	190.40	35.71	20.02
5	Jon	April	23.40	11.61	179.38	23.73	48.89
6	Jon	November	23.11	16.71	178.78	20.55	47.96
7	Leigh	September	32.14	12.34	165.65	17.62	66.40
8	Leigh	August	25.92	20.78	201.18	32.97	70.96
9	Leigh	November	31.52	16.62	177.50	20.95	71.18
10	Jon	November	23.11	14.41	179.86	20.53	51.49
11	Jon	November	22.80	11.35	184.55	20.94	50.36
12	Leigh	September	32.83	13.93	158.65	17.04	69.41
13	Leigh	November	31.13	12.82	179.57	20.67	69.01
14	Tom	August	23.01	13.20	220.52	29.44	18.09
15	Jon	December	21.09	9.90	180.66	22.95	47.87
16	Leigh	August	30.90	13.29	202.22	32.29	68.71
17	Leigh	December	37.36	15.63	184.22	20.32	71.74
18	Tom	July	24.25	15.71	149.01	30.68	21.75
19	Tom	April	21.83	14.95	154.43	30.48	21.09
20	Jon	January	25.94	12.38	197.90	20.06	47.09
⋮	⋮	⋮	⋮	⋮	⋮	⋮	⋮

each separate use of the credit card, the issuer's interest lies in the category of person-months and the related spending patterns. Table 2.4 summarizes the information of Table 2.3 according to this criterion of interest. Notice that the value for each observation is now in an interval-valued format. For example, in January, Jon's spending on food takes values in the interval [20.81, 29.38], Tom's is [23.28, 30.00] and Leigh spent [25.59, 35.33].

In general then, unlike classical data for which each data point consists of a single (categorical or quantitative) value, symbolic data can contain internal variation and can be structured. It is the presence of this internal variation which necessitates the need for new techniques for analysis which in general will differ from those for classical data. Note, however, that classical data represent a special case, e.g., the classical point $x = a$ is equivalent to the symbolic interval $\xi = [a, a]$.

Notation 2.1: We shall refer to the ith individual as belonging to the set $i \in \Omega = \{1, \ldots, n\}$ of n individuals in the dataset; and to the uth concept/category w_u

Table 2.4 Credit card use by person-months.

Name – Month	Food	Social	Travel	Gas	Clothes
Jon – January	[20.81, 29.38]	[9.74, 18.86]	[192.33, 205.23]	[13.01, 24.42]	[44.28, 53.82]
Jon – February	[21.44, 27.58]	[10.86, 18.01]	[214.98, 229.63]	[16.08, 22.86]	[50.51, 63.57]
⋮ ⋮	⋮	⋮	⋮	⋮	⋮
Tom – January	[23.28, 30.00]	[8.67, 18.31]	[193.53, 206.53]	[26.28, 35.61]	[15.51, 25.66]
Tom – February	[20.61, 28.66]	[10.66, 17.20]	[195.53, 203.83]	[25.43, 34.18]	[12.99, 24.88]
⋮ ⋮	⋮	⋮	⋮	⋮	⋮
Leigh – January	[25.59, 35.33]	[7.07, 19.00]	[194.12, 207.05]	[17.75, 23.07]	[61.47, 75.43]
Leigh – February	[31.30, 40.80]	[9.05, 24.44]	[212.76, 227.43]	[13.81, 25.08]	[71.63, 85.58]
⋮ ⋮	⋮	⋮	⋮	⋮	⋮

taking values $w_u \in E = \{w_1, \ldots, w_m\}$, the set of m symbolic concepts. A symbolic observation typically represents the set of individuals who satisfy the description of the associated symbolic concept, or category, w_u; see Section 2.2. When it is known, we denote the number of individuals who formed the concept w_u by n_u. The dataset Ω may be a sample space of size n. Or, as is frequently the case when n is extremely large, Ω may be the entire population. □

Notation 2.2: A classical value or realization for the random variable Y_j, $j = 1, \ldots, p$, on the individual $i = 1, \ldots, n$, will be denoted by x_{ij}, and a symbolic value or realization will be denoted by ξ_{ij}. That is, $Y_j(i) = x_{ij}$ if it is a classical variable, and $Y_j(i) = \xi_{ij}$ if it is a symbolic variable. A classical value is a particular case of a corresponding symbolic value. □

More formally, let Y_j be the jth random variable, $j = 1, \ldots, p$, measured on an individual $i \in \Omega$, where $\Omega = \{1, \ldots, n\}$, and where n and p can be extremely large (but need not be). Let $Y_j(i) = x_{ij}$ be the particular value for the individual i on the jth variable Y_j in the classical setting; and let the $n \times p$ matrix $\boldsymbol{X} = (x_{ij})$ represent the entire dataset. Analogously, if Y_j is measured on a category $w_u \in E$ where $E = \{w_1, \ldots, w_m\}$, then we write its particular value as $Y_j(w_u) = \xi_{uj}$, and the complete set as $\boldsymbol{\xi} = (\xi_{uj})$. Then, if the domain of Y_j is $\mathcal{Y}_j$, $j = 1, \ldots, p$, the matrix $\boldsymbol{X}$ takes values in $\mathcal{X} = \mathcal{Y}_1 \times \ldots \times \mathcal{Y}_p = X_{i=1}^{p} \mathcal{Y}_j$ and the matrix $\boldsymbol{\xi}$ takes values in $\mathcal{Y} = \mathcal{Y}_1 \times \ldots \times \mathcal{Y}_p$. What is crucial in the classical setting is that each x_{ij} in $\mathcal{X}$ takes precisely one possible realized value, whereas in the symbolic setting ξ_{uj} takes several values. (This includes the possibility of a missing value denoted by '.'.) (It will become clear later that an individual i can have a realization that is classical x, or symbolic ξ, and likewise for a category w_u. However, for the present we assume realizations of a random variable Y on an individual are classically valued and on a category are symbolically valued.)

Example 2.1. Refer to Table 2.1. If the random variable Y_3 = Age, then possible classical values are $Y_3 = 24 (\equiv x_{31})$, or $Y_3 = 56, \ldots,$ where the domain of $Y_3 = \mathcal{Y}_3 = \Re^+$. If the random variable Y_1 = City, then Y_1 = Boston, or Y_1 = Quincy, ..., where the domain $\mathcal{Y}_1$ = {Boston, Quincy, ..., (list of cities), ...}.

Now, refer to Tables 2.2. Here, for Y_3 = Age, we have $Y_3(w_1) = [17, 76] \equiv \xi_{31}$, or $Y_2(w_2) = [20, 70]$. For Y_1 = City, we have $Y_1(1)$ = {Boston, ..., Stowe}, $Y_1(2)$ = {Boston, ..., Reno}. The entries in Tables 2.1 are classically valued while those in Tables 2.2 are symbolic values. □

Example 2.2. Consider the random variable Y = Type of Cancer with $\mathcal{Y}$ = {lung, liver, bone, ...}; then a classical response can be Y = lung, or Y = liver, or, ... A symbolic response could be Y = {lung, liver}. □

Notice that each x_{ij} observation consists of one and only one value, whereas ξ_{uj} consists of several values. This includes the variable $Y = Y_{cancer}$ in Example 2.2.

If an individual has two cancers, then the classical Y_{cancer} cannot accommodate this on its own. It becomes necessary to define several variables, each relating to a particular cancer. For example, Y_1 = lung cancer, Y_2 = brain cancer, Y_3 = liver cancer, . . . , each with possible values in $\mathcal{Y}_j = \{\text{No, Yes}\}$ (or, $\mathcal{Y}_j = \{0, 1\}$, or some other code). In contrast, the symbolic variable Y_{cancer} does permit more than one possible cancer, e.g., an individual for which $Y_{cancer} = \{\text{lung, liver}\}$ is an individual with lung, liver, or both lung and liver cancer. This is an example of a multi-valued symbolic variable.

Classical values can be qualitative (e.g., x_{ij} = Boston) or quantitative (e.g., x_{ij} = 17). Symbolic values can be multi-valued, interval valued, or modal valued with or without logical, taxonomy, or hierarchy dependency rules. We describe these in turn.

Definitions 2.1: A **categorical** variable is one whose values are names; also called **qualitative** variable. A **quantitative** variable is one whose values are subsets of the real line $\Re^1$. Note, however, that sometimes qualitative values can be recoded into apparent quantitative values. □

Example 2.3. In the health data of Table 2.1, the random variable Y_1 = City takes particular classical categorical values, e.g., for the first individual $Y_1(1) = x_{11}$ = Boston, and for the third individual $Y_1(3) = x_{31}$ = Chicago. The random variable Y_{28} = Diagnosis of Cancer takes classical categorical values, e.g., $Y_{28}(1) = x_{1,28}$ = No, and $Y_{28}(3) = x_{3,28}$ = Yes. These values can be recoded to {No, Yes} ≡ {0,1}, so that $x_{1,28} = 0$ and $x_{3,28} = 1$, as quantitative values. The random variable Y_{10} = Pulse Rate takes quantitative values, e.g., $Y_{10}(1) = x_{1,10} = 68$ and $Y_{10}(3) = x_{3,10}$ = 73, for the first and third individuals, respectively. □

Definition 2.2: A **multi-valued** symbolic random variable Y is one whose possible value takes one or more values from the list of values in its domain $\mathcal{Y}$. The complete list of possible values in $\mathcal{Y}$ is finite, and values may be well-defined categorical or quantitative values. □

Example 2.4. For the health-demographic data summarized according to the Type × Gender categories in Tables 2.2, the values for the variable Y_1 = City for the category Dental Males ($u = 1$) are

$$Y_1(w_1) = Y_1(\text{Dental Males}) = \{\text{Boston, Chicago, Concord, Lindfield, Salem, Stowe}\}.$$

That is, the individuals who make up the category of Dental Males live in any one of these listed cities. Likewise, those for Optical Females ($u = 6$) live in Boston or Peoria since

$$Y_1(w_6) = Y_1(\text{Optical Females}) = \{\text{Boston, Peoria}\}.$$

If we look at the random variable Y_5 = Marital Status, we see that for Dental Males and Optical Females, respectively,

$$Y_5(w_1) = Y_5(\text{Dental Males}) = \xi_{15} = \{\text{married, single}\},$$

$$Y_5(w_6) = Y_5(\text{Optical Females}) = \xi_{65} = \{\text{married}\}.$$

That is, there are married and single individuals in the category of Dental Males, while all those in the Optical Females category are married.

The random variables Y_1 = City and Y_5 = Marital Status are categorical-valued variables. The random variable Y_6 = Number of Parents Living is a quantitative multi-valued variable. Here, the domain is $\mathcal{Y}_6 = \{0, 1, 2\}$. Then, we see from Tables 2.2 that for Dental Males

$$Y_6(w_6) = Y_6(\text{Dental Males}) = \{0, 1, 2\};$$

while for Optical Males,

$$Y_6(w_5) = Y_6(\text{Optical Males}) = \{0, 1\}.$$

That is, there are no individuals in the ($u = 5$) Optical Males category that have both parents alive. □

Example 2.5. A dataset may consist of the types and/or make of cars (Y) within households. Here, the domain $\mathcal{Y}$ = {Chevrolet, Ford, Volvo, Honda, . . . , (list of car makes/models), . . . }. Then, one household (w_1) may have

$$Y(w_1) = \xi_1 = \{\text{Honda Accord, Thunderbird}\};$$

another (w_2) might have

$$Y(w_2) = \xi_2 = \{\text{Toyota, Volvo, Renault}\};$$

and yet another

$$Y(w_3) = \xi_3 = \{\text{Chevy Cavalier, Chevy Alero}\};$$

and so on. □

Example 2.6. The dataset of Table 2.5 consists of types of birds and their colors (Y). These data are symbolic data in their own right, since, for example, a magpie always has both white and black colors. An all-black bird would not be a magpie, it would be another type of bird (such as a crow). Thus, if $Y(w_u)$ is the color associated with the bird w_u, we have, for example,

$$Y(w_1) = \{\text{black, white}\} = \text{color of magpie},$$

$$Y(w_4) = \{\text{red, black}\} = \text{color of cardinal},$$

and so on. In each case, the actual realization is a sublist of possible colors from the domain $\mathcal{Y}$ = {list of colors}. □

Table 2.5 Bird colors.

u	Bird	Major Colors
w_1	Magpie	{black, white}
w_2	Kookaburra	{brown, black, white, blue}
w_3	Galah	{pink, gray}
w_4	Cardinal	{red, black}
w_5	Goldfinch	{black, yellow}
w_6	Quetzal	{red, green, white}
w_7	Toucan	{black, yellow, red, green}
w_8	Rainbow Lorikeet	{blue, yellow, green, red, violet, orange}

A particular case of the symbolic multi-valued observation value is a classical categorical observation where the 'list' of values in ξ contains a single value.

Example 2.7. In Tables 2.2, the symbolic random variable Y_5 = Marital Status takes values in $\mathcal{Y}$ = {married, single}. For the observation $u = 6$ corresponding to the category Optical Females, we have

$$Y_5(w_6) = Y_5(\text{Optical Females}) = \xi_{65} = \{\text{married}\}. \qquad \square$$

Definition 2.3: An **interval-valued** symbolic random variable Y is one that takes values in an interval, i.e., $Y = \xi = [a, b] \subset \Re^1$, with $a \leq b,\ a,\ b \in \Re^1$. The interval can be closed or open at either end, i.e., (a, b), $[a, b]$, $[a, b)$, or $(a, b]$. □

When, as in Tables 2.2 and Table 2.4, the intervals emerge as the result of aggregating classical data, then the symbolic values a_{uj}, b_{uj} for the variable j in category w_u are given by

$$a_{uj} = \min_{i \in \Omega_u} x_{ij}, \quad b_{uj} = \max_{i \in \Omega_u} x_{ij},$$

where Ω_u is the set of $i \in \Omega$ values which make up the category w_u.

Example 2.8. In the symbolic dataset of Table 2.2b, the random variables Y_3 = Age, Y_9 = Weight, Y_{10} = Pulse Rate, Y_{11} = Systolic Blood Pressure, Y_{12} = Diastolic Blood Pressure, and Y_{13} = Cholesterol are all interval-valued variables. Thus, for example, the age for $u = 4$,

$$Y_3(w_4) = Y_3(\text{Medical Females}) = \xi_{43} = [11, 87],$$

is the interval covering the ages of the White Females; and

$$Y_9(w_4) = Y_9(\text{Medical Females}) = \xi_{49} = [73, 166]$$

is the range of weights of the Medical Females; and so on. □

Example 2.9. Consider the credit card expenses dataset of Table 2.4. Each of the variables is interval valued. Thus, for example, the first category, corresponding to Jon's expenses in January (in dollars), ($u = 1$) is, for $j = 1, 2, 4$,

$$\boldsymbol{Y}(w_1) = (Y_1(w_1), Y_2(w_1), Y_4(w_1)) = \boldsymbol{\xi}_1 = ([20.81, 29.38], [9.74, 18.86], [13.01, 24.42]),$$

i.e., Jon's range of expenses is $[a_1, b_1] = 20.81$ to 29.38 for food, $[a_2, b_2] = 9.74$ to 18.86 for social entertainment, and $[a_4, b_4] = 13.01$ to 24.42 for gas. □

Interval-valued variables can occur naturally when, for instance, it is not possible to obtain precise measurements on individuals. Or, knowledge about some entity of interest may be relatively vague, and is only known to within some interval value. (The distinction with fuzzy data is explained in Section 2.2.5.) Here our units are concepts and therefore we focus on intervals whose observed values express the variation among the individuals who constituted the concept.

As noted earlier, a particular case of an interval-valued observation is a classical variable value, where a classical $x = a$ is equivalent to the symbolic interval value $\xi = [a, a]$.

Example 2.10. In Table 2.2b, the random variable Y_{10} = Pulse Rate, though itself a symbolic interval-valued variable, takes a classical value for the Optical Males ($u = 5$) category with

$$Y_{10}(w_5) = Y_{10}(\text{Optical Males}) = \xi_{5,10} = [72, 72] \equiv x_{5,10} = 72. \qquad \Box$$

A major class of symbolic variables falls under the rubric of modal-valued variables. Such variables take many different forms. The most common form is a histogram version of the multi-valued variable and the interval-valued variable considered thus far. However, modal variables also include distributions, models, and a variety of other entities. We give here a formal definition of a general modal-valued variable. Then, given their importance, we give definitions of histogram-valued variables separately.

Definition 2.4: Let a random variable Y take possible values $\{\eta_k; k = 1, 2, \ldots\}$ over a domain $\mathcal{Y}$. Then, a particular outcome is **modal valued** if it takes the form

$$Y(w_u) = \xi_u = \{\eta_k, \pi_k; \quad k = 1, \ldots, s_u\}$$

for an observation u, where π_k is a non-negative measure associated with η_k and where s_u is the number of values actually taken from $\mathcal{Y}$. These possible η_k can be finite or infinite in number; they may be categorical or quantitative in value. □

The measures $\{\pi_k\}$ are typically weights, probabilities, relative frequencies, and the like, corresponding to the respective outcome component η_k. However, they can also be capacities, necessities, possibilities, credibilities, and related entities (defined below in Definitions 2.7–2.10). They are the support of η_k in $\mathcal{Y}$. The $\{\eta_k\}$ outcome components can be categorical values (as in Definition 2.5), or can be subsets of the real line $\Re^1$ (as in Definition 2.6). The outcome can be a distribution function, a histogram, a model, or related stochastic entity associated with the random variable Y.

Definition 2.5: Let $\mathcal{Y}_{cat}$ be the domain of possible outcomes for a multi-valued random variable Y_{cat} with $\mathcal{Y}_{cat} = \{\eta_1, \eta_2, \ldots\}$. Then, a **modal multi-valued** variable is one whose observed outcome takes values that are a subset of $\mathcal{Y}_{cat}$ with a non-negative measure attached to each of the values in that subset. That is, a particular observation, for the category w_u, takes the form

$$Y(w_u) = \xi_u = \{\eta_{u1}, p_{u1}; \ldots; \eta_{us_u}, p_{u,s_u}\}$$

where $\{\eta_{u1}, \ldots, \eta_{us_u}\} \subseteq \mathcal{Y}_{cat}$ and where the outcome η_{uk} occurs with weight p_{uk}, $k = 1, \ldots, s_u$, and with $\sum_{k=1}^{s_u} p_{uk} = 1$. □

Example 2.11. The symbolic data of Table 2.6 is another format for the aggregated data generated from Table 2.1 for the concept or category defined by Type × Gender. Consider the random variable Y_5 = Marital Status. We see that for the Dental Males category

$$Y_5(w_1) = Y_5(\text{Dental Males}) = \xi_{15} = \{\text{married}, 2/3; \text{single}, 1/3\};$$

that is, the proportion of Dental Males that are married is 2/3 with 1/3 being single. This is potentially more informative than is the formulation of Table 2.2b; see Example 2.3. For Medical Females ($u = 4$), we have

$$Y_5(w_4) = Y_5(\text{Medical Females}) = \xi_{45} = \{\text{married}, 5/7; \text{single}, 2/7\};$$

that is, 5/7 are married and 2/7 are single. Notice that when $u = 6$, the Optical Females category gives the single classical value

$$Y_5(w_6) = Y_5(\text{Other Females}) = \xi_{65} = \{\text{married}\}$$

with the probability weight $p_1 = 1$ by implication. □

Example 2.12. For the quantitative categorical random variable Y_6 = Number of Parents Alive, in Table 2.6, we see, for example, that the possible outcomes in $\mathcal{Y} = \{0, 1, 2\}$ have associated weights (1/2, 1/6, 1/3) for Dental Females, i.e.,

$$Y_6(w_2) = Y_6(\text{Dental Females}) = \xi_{26} = \{0, 1/2; 1, 1/6; 2, 1/3\};$$

while for Optical Males, we have

$$Y_6(w_5) = Y_6(\text{Other Males}) = \xi_{56} = \{0, 1/2; 1, 1/2\}$$

with, by implication, $\eta = 2$ occurring with probability 0 in ξ_{56}. □

Table 2.6 Health insurance – symbolic modal formulation.

w_u	Type × Gender	Y_3 Age	Y_5 Marital Status	Y_6 Parents Alive	Y_9 Weight	Y_{13} Cholesterol
w_1	Dental Males	{[0, 40), 4/9; [40, 99], 4/9}	{M, 2/3; S, 1/3}	{0, 2/9; 1, 4/9; 2, 1/3}	{[150, 200), 4/9; [200, 275], 5/9}	{[<200), 4/9; [200, 300), 4/9, [≥300], 1/9]}
w_2	Dental Females	{[0, 40), 1/3; [40, 99], 2/3}	{M, 1/2; S 1/2}	{0, 1/2; 1, 1/16; 2, 1/3}	{[140, 160), 2/3; [160, 180], 1/3}	{[<200), 1/2; [200, 240), 1/2}
w_3	Medical Males	{[0, 20), 1/6; [20, 40), 2/89; [40, 60], 2/9; [60, 99], 7/18}	{M, 11/18; S 7/18}	{0, 1/3; 1, 1/3; 2, 1/3}	{[0, 120), 1/9; [120, 180), 7/8; [180, 240], 1/2}	{[<200), 7/9; [200, 240), 2/9}
w_4	Medical Females	{[0, 20), 1/7; [20, 40), 2/7; [(40, 60), 3/14; [60, 99], 5/14}	{M, 5/7; S, 2/7}	{0, 5/14; 1, 1/7; 2, 1/2}	{[50, 110), 1/7; [110, 140), 5/14; [140, 170] 1/2}	{[<200), 11/14; [200, 240), 3/14}
w_5	Optical Males	{[40, 60), 1/2; [60, 99], 1/2}	{M, 1/2; S 1/2}	{0, 1/2; 1, 1/2}	{[140, 160), 1/2; [160, 200], 1/2}	{[<200), 1/2; [200, 240), 1/2}
w_6	Optical Females	{[60, 99]}	{M}	{0}	{[140, 160), 1/2; [160, 180] 1/4}	{[<200)}

Example 2.13. A public opinion firm wished to determine the attitude of the general population towards a number of different products ($m = 5$, say). This was accomplished by providing potential clients ($n = 1000$, say) with one of the products and then asking the clients some follow-up questions ($p = 3$, say). The clients were asked to provide an answer {Strongly Agree, Agree, Neutral, Disagree, Strongly Disagree} (or, equivalently, give a coding $\{1, \ldots, 5\}$) to each question.

Then, if each product corresponds to a concept, we obtain a histogram value for the answers to each question after aggregating to form these categories. For example, for Product1 ($\equiv w_1$), we may have, for the first question (Y_1),

$$Y_1(w_1) = \{\text{Strongly Agree, 0.3; Agree, 0.4; Neutral, 0.15;} \\ \text{Disagree, 0.12; Strongly Disagree, 0.03}\};$$

while for Product2, we have for Y_1,

$$Y_1(w_2) = \{\text{Strongly Agree, 0.1; Agree, 0.2; Neutral, 0.5,} \\ \text{Disagree, 0.15; Strongly Disagree, 0.15}\}.$$

Thus, we see that 70% of those surveyed Agree or Strongly Agree that they like Product1, whereas 50% are ambivalent with no strong opinion about Product2. □

Definition 2.6: Let Y be a quantitative random variable that can take values on a finite number of non-overlapping intervals $\{[a_k, b_k),\ k = 1, 2, \ldots\}$ with $a_k \leq b_k$. Then, an outcome for observation w_u for a **histogram interval-valued random variable** takes the form

$$Y(w_u) = \xi_u = \{[a_{uk}, b_{uk}),\quad p_{uk};\quad k = 1, \ldots, s_u\},$$

where $s_u < \infty$ is the finite number of intervals forming the support for the outcome $Y(w_u)$ for observation w_u, and where p_{uk} is the weight for the particular subinterval $[a_{uk}, b_{uk}),\ k = 1, \ldots, s_u$, with $\sum_{k=1}^{s_u} p_{uk} = 1$. The intervals (a_k, b_k) can be open or closed at either end. □

Example 2.14. The dataset of Table 2.7 was extracted from a classical dataset of values for 50680 individual flights on 33 flight performance variables for the month of January 2004 on flights into JFK Airport, New York; see Falduti

Table 2.7 Flights into JFK Airport.

	Y_1 = Flight Time			Y_2 = Taxi In			Y_3 = Arrival Delay			Y_4 = Taxi Out			Y_5 = Departure Delay			Y_6 = Weather Delay	
	<120	[120, 220]	>220	<4	[4, 10]	>10	<0	[0, 60]	>60	<16	[16, 30]	>30	<0	[0, 60]	>60	No	Yes
Airline 1	0.15	0.62	0.23	0.12	0.64	0.24	0.42	0.46	0.12	0.38	0.47	0.15	0.44	0.47	0.09	0.92	0.08
Airline 2	0.89	0.11	0.00	0.21	0.65	0.14	0.52	0.39	0.09	0.45	0.40	0.15	0.32	0.60	0.08	0.90	0.10
Airline 3	0.99	0.01	0.00	0.10	0.82	0.08	0.45	0.48	0.07	0.41	0.44	0.15	0.54	0.40	0.06	0.97	0.03
Airline 4	0.22	0.60	0.18	0.27	0.67	0.06	0.47	0.50	0.03	0.55	0.37	0.08	0.37	0.60	0.03	0.99	0.01
Airline 5	0.24	0.72	0.04	0.03	0.79	0.18	0.52	0.43	0.05	0.28	0.53	0.19	0.71	0.25	0.04	0.98	0.02
Airline 6	0.95	0.05	0.00	0.31	0.57	0.12	0.51	0.37	0.12	0.51	0.37	0.12	0.46	0.42	0.12	0.96	0.04
Airline 7	0.56	0.35	0.09	0.09	0.72	0.19	0.47	0.49	0.04	0.32	0.52	0.16	0.46	0.51	0.03	0.99	0.01
Airline 8	0.81	0.19	0.00	0.07	0.72	0.21	0.49	0.44	0.07	0.50	0.41	0.09	0.21	0.73	0.06	0.99	0.01
Airline 9	0.83	0.17	0.00	0.03	0.78	0.19	0.41	0.49	0.10	0.33	0.49	0.18	0.47	0.45	0.08	1.0	0.00
Airline 10	0.00	0.01	(0.57, 0.42)	0.08	0.67	0.25	0.42	0.52	0.06	0.37	0.53	0.10	0.47	0.48	0.05	0.99	0.01
Airline 11	0.90	0.09	0.01	0.14	0.66	0.20	0.38	0.50	0.12	0.42	0.42	0.16	0.50	0.41	0.09	0.96	0.04
Airline 12	0.75	0.25	0.00	0.05	0.78	0.17	0.40	0.55	0.05	0.43	0.40	0.17	0.63	0.34	0.03	0.96	0.04
Airline 13	0.85	0.14	0.01	0.08	0.72	0.20	0.49	0.40	0.11	0.36	0.46	0.18	0.58	0.34	0.08	0.99	0.01
Airline 14	0.43	0.20	(0.21, 0.16)	0.23	0.66	0.11	0.46	0.44	0.10	0.29	0.53	0.18	0.57	0.34	0.09	0.98	0.02
Airline 15	0.98	0.02	0.00	0.24	0.70	0.06	0.63	0.35	0.02	0.58	0.33	0.09	0.69	0.30	0.01	0.99	0.01
Airline 16	0.58	0.36	0.06	0.57	0.42	0.01	0.53	0.43	0.04	0.89	0.09	0.02	0.01	0.96	0.03	0.99	0.01

and Taibaly (2004). Operating performance on any one single flight is not of great importance; rather overall performance by the various airlines is of interest. Therefore, the data were aggregated according to which of 16 airlines operated each flight. Six of the quantitative variables were selected and their values for each of these $m = 16$ airlines are displayed in Table 2.7, specifically Y_1 = Flight Time, Y_2 = Time to Taxi In to the gate after touchdown, Y_3 = Arrival Delay Time, Y_4 = Time to Taxi Out from the gate to the runway, Y_5 = Departure Delay Time, and Y_6 = Weather Delay Time; all times in minutes; a negative time means the flight left or arrived early.

As recorded, the data are histogram-valued. For example, Y_2 = Taxi In Time is given by the relative frequency of flights that take less than 4 minutes to taxi in to the gate, those that take 4 to 10 minutes inclusive, and those that take more than 10 minutes to taxi in. Likewise, the ranges for all the variables are as shown in Table 2.7. For the two airlines ($u = 10$ and $u = 14$), the Y_1 variable takes values over four ranges, $y_1 < 120$, $120 < y_1 < 220$, $220 \leq y_1 \leq 300$, and $y_1 > 300$ minutes. In the case of a weather delay Y_6, the values for Y_6 are simply No or Yes.

Consider the random variable Y_3 = Arrival Delay Time (of flights). The observed values are histogram-interval valued as shown. For example, for Airline 1 ($u = 1$), we have

$$Y_3(w_1) = Y_3(\text{Airline 1}) = \xi_{13} = \{[< 0), 0.42; [0, 60], 0.46; [> 60), 0.12\};$$

that is, for this airline 42% landed early, 46% were 0 to 60 minutes late, and 12% were more than 60 minutes late, over the study period. □

Example 2.15 Consider the health-demographic data of Table 2.6. These data resulted from an aggregation of the data of Tables 2.1 over Type × Gender categories. Unlike the aggregation of Tables 2.2 which display interval values, this present aggregation led to modal-valued observations. The random variables Y_3 = Age, Y_9 = Weight, and Y_{13} = Cholesterol are histogram interval-valued variables. Take the random variable Y_{13}. Those whose cholesterol level is 240 or greater are at risk of heart disease, those with a level $[a, b) = [200, 239)$ are borderline at risk, while those whose level is $[a, b) = [< 200)$ are not at risk. Therefore, it is of interest to study the histogram distribution of cholesterol values over these three intervals. We observe, for example, that for Dental Males ($u = 1$),

$$\begin{aligned} Y_{13}(w_1) = Y_{13}(\text{Dental Males}) &= \xi_{1,13} \\ &= \{[< 200), 4/9; [200, 300), 4/9; [\geq 300), 1/9\}; \end{aligned}$$

that is, 4/9 are borderline at risk and 1/9 are at risk of heart disease while 4/9 are not at risk. □

In the sequel, except where appropriate clarification is needed, we shall refer to both histogram multi-valued variables and histogram interval-valued variables as simply histogram-valued variables. Most of the modal-valued variables to be considered in this volume will tend to be histogram-valued variables. Sometimes, rather than the standard histogram-valued observations as shown in Examples 2.11–2.15, the measures $\{\pi_k\}$ could reflect the cumulative distribution function, or simply the distribution function.

Example 2.16. Consider the data of Table 2.7 pertaining to airline carriers flying into New York's JFK Airport; see Example 2.14. For the random variable Y_2 = Taxi In Time, rather than the histogram of Table 2.7, we may have recorded the observation values as, for $u = 1$,

$$Y_2(w_1) = Y_2(\text{Airline 1}) = \xi_{12} = \{[<4), 0.12; [\leq 10], 0.76; < \infty, 1.0\},$$

and for $u = 2$,

$$Y_2(w_2) = Y_2(\text{Airline 2}) = \xi_{22} = \{[<4), 0.21; [\leq 10], 0.86; < \infty, 1.0\},$$

and similarly for the other airlines. Here, the weights are cumulative probabilities. □

Modal-valued observations could also be specified distributions. They could be models.

Example 2.17. The distribution of different types of energy consumption across the 50 states of the USA may be given by the values, shown in Table 2.8, for petroleum products, natural gas, coal, hydroelectric power and nuclear power. (These data are based on rescaled values obtained from the US Census Bureau (2004) at www.census.gov.) In general, these could be (empirical) distributions with the parameter values estimated from the data; or they could be known distributions with known parameter values. Thus, for example, the distribution of coal consumption ($u = 3$) could follow that of a normal distribution (say) with mean $\mu = 47.0$ and standard deviation $\sigma = 43.3$. □

Table 2.8 Distribution of energy consumption.

w_u	Type	μ	σ
w_1	Petroleum	76.7	92.5
w_2	Natural gas	45.9	69.5
w_3	Coal	47.0	43.3
w_4	Hydroelectric power	6.4	14.3
w_5	Nuclear power	25.4	20.4

Example 2.18. Table 2.9 provides a time series process which models bank rates for four banks where in each case the rates $Y(t)$ at time t satisfy an autoregressive model AR(p) of order p,

$$Y(t) = \alpha_1 Y(t-1) + \ldots + \alpha_p Y(t-p) + e(t),$$

where $\{\alpha_1, \ldots, \alpha_p\}$ are the so-called autoregressive parameters for the model and where $e(t)$ is the noise component with zero mean and variance σ^2. As written, this model assumes the rates have been suitably transformed to give a stationary process; see any of numerous tests on time series analysis, e.g., Fuller (1996). Thus, from Table 2.9, we observe, for example, that the bank rates $Y(t)$ at time t for Bank2 ($u = 2$) are modeled by

$$Y(t, w_2) = Y(t, \text{Bank2}) = 0.7Y(t-1) - 0.2Y(t-2) + e(t);$$

likewise, for the other banks. □

Table 2.9 Time series models – banks.

w_u	Bank	Model
w_1	Bank1	AR(1); α_1 unknown
w_2	Bank2	AR(2); $\alpha_1 = 0.7$, $\alpha_2 = -0.2$
w_3	Bank3	AR(2); α_1, α_2 unknown
w_4	Bank4	AR(1); $\alpha_1 = -0.6$

It is clear that a $p > 1$ dimensional random variable can be mixed; in that case some Y_j may be multi-valued variables and some quantitative (either interval or histogram valued).

Example 2.19. Table 2.10 (adapted from Felsenstein, 1983) gives some observed values for three species and their corresponding DNA sequence. The variables Y_1 and Y_2 are multi-valued, and Y_3 and Y_4 are interval valued.

Table 2.10 Mixed variables: DNA sequences.

Species	DNA Sequence	Y_1	Y_2	Y_3	Y_4
D. pseudsobscura	ACCGTCCGTTA	{7, 8}	Long	[5.4, 6.7]	[0.29, 0.33]
D. obscura	ACAGGCCGTGA	{5, 7}	Medium	[9.0, 11.2]	[0.43, 0.49]
D. melanogaster	AACGTCCGTGC	{3, 4}	Short	[16.3, 21.1]	[0.50, 0.89]

□

Finally, we present some alternative measures $\{\pi\}$. These relate to capacities, possibilities, and similar entities.

Definition 2.7: Let Y be a random variable with values on the domain $\mathcal{Y}$. Then, the **capacity** of a category is the probability that at least one individual in that category has the specific value $Y = \xi_k, \ \xi_k \subset \mathcal{Y}$. □

Definition 2.8: Let Y be a random variable with values on the domain $\mathcal{Y}$. Then, the **credibility** of a category is the probability that every individual in that category displays the characteristic $\xi_k \subset \mathcal{Y}$. □

Example 2.20. A manufacturer who regularly contacts suppliers regarding a given product wants to know the smallest number of suppliers it needs to contact in order to maximize its capacity to have the product supplied. The manufacturer therefore phones three suppliers (A_1, A_2, A_3) to ascertain whether each does ($Y = 1$) or does not ($Y = 0$) have the product. Suppose the results of this telephone survey based on 10 calls gave the values shown in Table 2.11.

Table 2.11 Product availability.

Call	Supplier		
w_u	A_1	A_2	A_3
w_1	1	1	0
w_2	1	1	0
w_3	1	0	0
w_4	1	0	0
w_5	0	0	0
w_6	0	0	0
w_7	0	0	1
w_8	0	0	0
w_9	0	0	1
w_{10}	0	0	0

Then, the capacity that the first two suppliers, A_1 and A_2, have the product is $4/10 = 0.4$ since four times in the ten calls the manufacturer learns that one or both of these suppliers had the product. The credibility, of A_1 and A_2, is the probability that both suppliers have the product on hand, and is $2/10 = 0.2$. Similarly, the capacity that the suppliers A_1 and A_3 have the product is 0.6, while their credibility is 0. □

Remark 2.1: It follows that, if $K(A)$ is the capacity of A and $CR(A)$ is the credibility of A, then:

$$CR(A) = 1 - K(\bar{A}), \text{ where } \bar{A} \text{ is the complement of } A;$$
$$K(0) = \emptyset, \text{ for the empty set } \emptyset;$$

$K(A_1 \cup A_2) = P(A_1) + P(A_2) - P(A_1)P(A_2)$, if A_1 and A_2 are independent and where $P(A)$ is the probability of A;

$A_1 \subseteq A_2$ implies $K(A_1) \leq K(A_2)$;

and

$$K\left[\lim_{n\to\infty}(\cup A_n)\right] = \lim_{n\to\infty} K(\cup A_n).$$

The proof is left as an exercise.

Definition 2.9: A **possibility** measure is a mapping π from Ω to [0,1] such that

(i) $\pi(\Omega) = 1,\ \pi(\phi) = 0$, where ϕ is the empty subset; and

(ii) for all subsets $A, B \subseteq \Omega$, $\pi(A \cup B) = \max\{\pi(A), \pi(B)\}$. □

Definition 2.10: A **necessity** measure is a mapping N from Ω to [0,1] such that for all $A \subseteq \Omega$,

$$N(A) = 1 - \pi(\bar{A})$$

where $\bar{A}$ is the complement of A and $\pi(\cdot)$ is a possibility measure. □

The two measures, $\pi(\cdot)$ and $N(\cdot)$, are clearly related by definition. The theory of possibilities (necessities) models three kinds of semantics. Generally, they will evaluate vague observations of inaccessible characteristics.

Example 2.21. Suppose several experts have assessed that the possibility that an athlete can lift 300 pounds, $A = \{300\}$, is

$$\pi(A) = \pi(\{300\}) = 0.6,$$

and the possibility of lifting 350 pounds, $B = \{350\}$, is

$$\pi(B) = \pi(\{350\}) = 0.5.$$

Then, for these experts, the possibility that this athlete will lift 300 or 350 pounds is

$$\pi(A \cup B) = \max(\pi\{300\}, \pi\{350\}) = \max(0.6, 0.5) = 0.6.$$ □

The notion of possibility in Example 2.21 relates to a kind of physical possibility. Another type of possibility relates to the actual occurrence or otherwise of a statement, e.g., 'it is possible that it will rain or snow today'. Thirdly, possibility can relate to our lack of surprise when we observe that the color of a sunflower is 'yellow'.

Remark 2.2: It is easy to show that the possibility or necessity measures satisfy the following properties:

$$N(\phi) = 0; \quad N(A \cap B) = \min\{N(A), N(B)\};$$

$$\pi(\cup_{i=1}^{k} A_i) = \max_i\{\pi(A_i)\};$$

$$N(\cap_{i=1}^{k} A_i) = \min_i\{N(A_i)\};$$

$$\pi(A) \leq \pi(B) \quad \text{if} \quad A \subseteq B;$$

$$\max\{\pi(A), \pi(\bar{A})\} = 1, \quad \min\{N(A)N(\bar{A})\} = 0;$$

$$\pi(A) + \pi(\bar{A}) \geq 1, \quad N(A) + N(\bar{A}) \leq 1;$$

$$\pi(A) < 1 \quad \text{implies} \quad N(A) = 0,$$

$$N(A) > 0 \quad \text{implies} \quad \pi(A) = 1;$$

$$\pi(A) \geq N(A).$$

The definition of capacity follows the sense of Choquet (1954) and that of credibility follows the sense of Schafer (1976). The definitions for possibility and necessity follow Dubois and Prade (1988). More detailed properties of these measures can be found in Diday (1995). In subsequent sections, we shall mostly use the frequency and cumulative frequency measures. In most instances, however, the frequency measures used can be replaced by these alternative measures, but the interpretation of the results has to proceed carefully as it depends on the semantics of the weights used. For example, the meaning of a histogram using capacities differs from that using frequencies. So, for example, the histogram using capacities gives the frequency of those who have the capacity to lift 300 pounds, rather than the frequency of those who did lift 300 pounds.

2.1.2 Dependencies in the data

Variables can have dependencies inherent to the structure of data. By 'dependencies' here we refer not to the statistical dependence that may or may not exist between classical or symbolic random variables. We refer to logical, taxonomic, and hierarchical dependencies. We consider each in turn. Some can exist in both classical and symbolic data, while some are necessary by virtue of the internal structure of a symbolic value.

Definition 2.11: A **logical dependency** occurs when the outcome for one set of variables $\boldsymbol{Y}_1 = (Y_{11}, Y_{12}, \ldots)$ is conditional on the outcome of another set of variables $\boldsymbol{Y}_2 = (Y_{21}, Y_{22}, \ldots)$. This is usually expressed as a **logical dependency rule** v such as

$$v: \text{ If } \{\boldsymbol{Y}_1 \in A\} \text{ then } \{\boldsymbol{Y}_2 \in B\} \tag{2.1}$$

where A and B are subsets of the domains $\mathcal{Y}_1$ and $\mathcal{Y}_2$, respectively. □

Example 2.22. Suppose we have the random variables Y_1 = Age and Y_2 = Number of Children. Suppose we have the classical observations relating to the three individuals $i = 1, 2, 3$ shown in Table 2.12.

Table 2.12 Age and number of children.

Individual	Age	Number of Children
1	23	2
2	10	0
3	17	1

Let us suppose that the three individuals are in the same category. Then, the symbolic observation that results for this category is

$$Y = (Y_1, Y_2) = \xi = (\xi_1, \xi_2) = ([10, 23], \{0, 1, 2\}).$$

This implies that individuals in this category are aged 10 to 23 years and have 0, 1, or 2 children. This includes the possibility that a 10 year old has one (or two) children. To maintain integrity in the symbolic observation, it is necessary to add a rule v that says

$$v = \text{If } \{Y_1 < 14 \text{ (say)}\} \text{ then } \{Y_2 = 0\}. \tag{2.2}$$

□

Logical dependencies can arise in two (or three) ways, and are expressed as logical dependency rules. The first relates to underlying conditions that exist, or that are of interest to the analyst. For example, interest may center only on children, so only those data that pertain to children are used in the analysis. Or, there may be a rule that says that two variables Y_1 and Y_2 must satisfy the relation $Y_1 + \alpha Y_2 = \beta$ (say), and so on. Such rules may exist because of an inherent interest in variable outcomes that match such dependencies.

Another type of logical dependency is necessary to maintain data integrity. When aggregating individual values (which would typically be classical values at a first level of aggregation; but which can in fact be symbolic-valued variables) into the relevant symbolic-valued data that make up the categories dictating that aggregation, the very process of aggregation can produce symbolic data which perforce engage the adoption of rule(s) to maintain data integrity. Thus, Example 2.22 is one example illustrating conditions implied by the data. Another type of example is the following.

Example 2.23. Suppose we have the classical data of Table 2.13a, as shown for nine individuals, the first six of whom after appropriate aggregation will belong to the category C1 and the last three will belong to category C2. The resulting symbolic values for each of C1 and C2 are shown in Table 2.13b. Reading Table 2.13b in isolation suggests that for each of the possible Y_1 values

(a, b, c), any of the Y_2 values $(1, 2)$ can occur, including the specific combinations $(Y_1, Y_2) = (a, 1)$ and $(Y_1, Y_2) = (b, 2)$. However, closer inspection of the data tells us that whenever $Y_1 = a$, then $Y_2 = 2$ always, and whenever $Y_2 = 1$ then $Y_1 = b$ always. Therefore, to maintain this dependency in the symbolic data, we need a rule

$$v : \text{If } \{Y_1 = a\} \text{ then } \{Y_2 = 2\}, \text{ and If } \{Y_2 = 1\} \text{ then } \{Y_1 = b\}. \tag{2.3}$$

Table 2.13a Logical dependency – classical.

i	1	2	3	4	5	6	7	8	9
Category	C1	C1	C1	C1	C1	C1	C2	C2	C2
Y_1	a	a	a	b	b	c	b	b	a
Y_2	2	2	2	1	1	2	1	3	2

Table 2.13b Logical dependency – symbolic.

Category	C1	C2
Y_1	$\{a, b, c\}$	$\{a, b\}$
Y_2	$\{1,2\}$	$\{1,2,3\}$

□

These types of logical dependency rules are unique to symbolic data. They must be taken into account when analysing the data. How this is performed will be discussed and illustrated further in Section 2.2 in particular, with more specialized rules being treated as they arise in subsequent sections.

A possible third type of logical rule is one that might be viewed as a form of data cleaning. Given the huge size of many datasets encountered, some automated form of data cleaning is necessary. A rule such as $\{\text{Age} = Y_1 > 0\}$, for example, could be used to catch an observed (classical or symbolic) value of $Y_1 = -15$ (say, an obvious miskeying situation). In some circumstances, data cleaning rules can be absorbed into the two situations described above.

In contrast, taxonomy and hierarchy rules relate to the basic structure of the data and apply equally to classical and symbolic data. These structures can also include logical dependencies.

Definition 2.12: Taxonomy variables are variables organized in a tree, as nested variables, with several levels of generality. The bottom of the tree for a given variable will be referenced as the first level, with the top of the tree corresponding to the total number of levels t. □

Example 2.24. Figure 2.1 shows a taxonomy tree structure for work careers. At the top of the tree Y_1 = Type of Work takes values in $\mathcal{Y}_1$ = {white collar, blue collar, services}; let us recode this to $\mathcal{Y}_1 = \{1,2,3, \text{respectively}\}$. At the second level of the tree is the variable Y_2 = Profession. Then, if the variable $Y_1 = 1$, the next branch down the tree has possible values in $\mathcal{Y}_{21}$ = {professional, administrative}. If the variable $Y_1 = 2$, the next branch relates to the variable Y_2 = Trade with possible values in $\mathcal{Y}_{22}$ = {plumber, electrician, . . . }. If the variable $Y_1 = 3$, then Y_2 = Service Job with possible values in $\mathcal{Y}_{23}$ = {food, health, other}. □

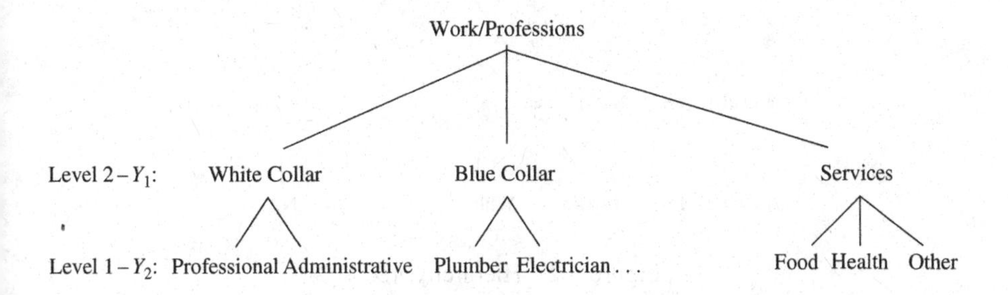

Figure 2.1 Taxonomy tree of professions.

The non-applicable values would typically be recorded as N (or NA), say. The data of Table 2.14 are a possible set of outcomes for the individuals/categories recorded. Note that there would usually be at least one other variable, beyond the structural variables, for which there is some level of interest. Thus, for the individuals on the hierarchy of Figure 2.1, there may be an annual salary Y_3 (say). For example, the third individual ($u = 3$) is a plumber with a salary of \$84,000.

Table 2.14 Taxonomy outcomes.

i/u	Y_1	Y_2	Y_3
1	White Collar	Administrator	87,000
2	White Collar	Professional	93,000
3	Blue Collar	Plumber	84,000
4	Services	Food	26,000
5	Services	Health	58,000

Definition 2.13: A **hierarchy** tree (or **Mother–Daughter** variables) is one in which whether or not a particular (Daughter) variable is operative at a given level of the tree is determined by the outcome of the (Mother) variable at the immediately preceding level of the tree. □

Example 2.25. The hierarchy tree of Figure 2.2 relates to mushroom species with the hierarchy variable at the top of the tree as Y_1 = Cap with possible values in $\mathcal{Y}_1 = \{\text{Yes, No}\}$. The next tree level gives the value for Y_2 = Color (of the cap of the mushroom), taking values in $\mathcal{Y}_2 = \{\text{brown, white,} \ldots\}$. Clearly, if Y_1 = No, then there is no value for Y_2; in this case, we write Y_2 = NA. Table 2.15 gives a possible dataset. □

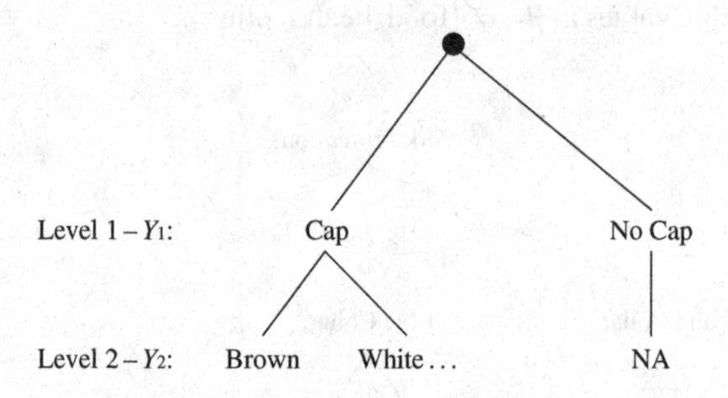

Figure 2.2 Hierarchy tree.

Table 2.15 Hierarchy outcomes.

w_u	Y_1	Y_2
w_1	No Cap	NA
w_2	No Cap	NA
w_3	Cap	Brown
w_4	Cap	White
w_5	Cap	White

2.2 Categories, Concepts, and Symbolic Objects

2.2.1 Preliminaries

At the outset, our symbolic dataset may already be in a format that allows an appropriate symbolic statistical analysis to proceed directly. An example is the data of Table 2.24 (in Section 2.3) used to illustrate a symbolic principal component analysis. More generally, however, and almost inevitably before any actual (symbolic) data analysis can be conducted especially for large datasets, there will need to be implemented various degrees of data manipulation to organize the information into classes appropriate to the specific questions at hand. In some instances, the entities in E (or Ω) are already aggregated into classes, though even here certain questions may require a reorganization into a different classification of classes regardless of whether the dataset is small or large and regardless of whether the data are classical or symbolic. For example, one set of classes $C_1, \ldots, C_m$ may represent individuals categorized according to m different types of primary diseases, while another analysis may necessitate a class structure by city, gender, age, gender and age, etc.

Another earlier stage is when initially the data are separately recorded as for classical statistical and/or computer science databases for each individual $i \in \Omega = \{1, \ldots, n\}$, with n extremely large; likewise for very large symbolic databases. This stage of the symbolic data analysis then corresponds to the aggregation of these n individuals into m classes where m is much smaller, and is designed so as to elicit more manageable formats prior to any statistical analysis with the actual format related to a scientific question. Note that this construction may, but need not, be distinct from classes that are obtained from a clustering procedure. Note also that the m aggregated classes may represent m patterns elicited from a data mining procedure.

This leads us to the notion of concepts and of symbolic objects introduced in a series of papers by Diday and his colleagues (e.g., Diday, 1987, 1989, 1990; Bock and Diday, 2000). In Section 2.1, we have alluded to the construction of symbolic datasets from classical datasets with any particular construction based on categories of individuals. While in some instances these 'categories' are true categories, in general they will be concepts. Which specific concepts are to be developed in any one construction depends on the basic underlying scientific questions whose answers the researcher seeks from the analysis of the data.

In this section, we present formal definitions of these concepts modeled by so-called symbolic objects, and how they are obtained from and relate to the individual descriptions of the original database. We concentrate on the theoretical aspects here. For illustrative clarity we shall assume that our starting database consists of classical-valued observations relating to individuals $i \in \Omega$. However, as we shall see, the basic principles developed in this section also apply if the starting dataset consists of concepts and/or if the data on the individuals are symbolic-valued observations.

2.2.2 Descriptions, assertions, extents

By way of motivation, let us consider the following two examples. First, consider the credit card dataset of Table 2.3, and suppose the concept of interest is 'Tom July'. Then, we seek all individual entries $i \in \Omega$ for which Name (i) = Tom and Month (i) = July. Let us write this as

$$d = (\text{Name} = \text{Tom}, \text{Month} = \text{July}).$$

Then, for $i = 1$,

$$Y(1) = (\text{Name}(1) = \text{Jon}, \ \text{Month}(1) = \text{February}) \neq d;$$

thus, this first observation $i = 1$ does not belong to the category of $d = (\text{Tom, July})$. However, when $i = 4$,

$$Y(4) = (\text{Name}(4) = \text{Tom}, \ \text{Month}(4) = \text{July}) = d$$

and also $Y(18)$ matches the description d. To construct the set of values associated with this category of 'Tom July', we proceed in this manner with each individual $i \in \Omega$ either matching or not matching the description d.

Secondly, let us return to the dataset of Tables 2.1. Suppose we are interested in the concept 'Northeasterner'. Thus, we have a description d representing the particular values {Boston, . . . , other NE cities, . . . } in the domain $\mathcal{Y}_1$ of Y_1 = City and we have a relation R (here $\subseteq$) linking the variable Y_1 with the particular description of interest. We write this as $[Y_1 \subseteq \{\text{Boston}, \ldots, \text{other NE cities}, \ldots\}] = a$, say. Then, each individual i in $\Omega = \{1, \ldots, n\}$ is either a Northeasterner or not. That is, a is a mapping from $\Omega \to \{0, 1\}$, where for an individual i who lives in the northeast, $a(i) = 1$; otherwise, $a(i) = 0$, $i \in \Omega$. Thus, if an individual i lives in Boston (i.e., $Y_1(i) = \{\text{Boston}\}$, then we have $a(i) = [Y_1(i) = \{\text{Boston}\} \subseteq \{\text{Boston}, \ldots, \text{other NE cities}, \ldots\}] = 1$. More generally, this expression can be written as $a(i) = [Y_1(i) \subseteq d]$ with $d = \{\text{Boston}, \ldots\}$ or $a(i) = [Y_1 R d]$ with $R \equiv \subseteq$.

The set of all $i \in \Omega$ for whom $a(i) = 1$ is called the extent of a in Ω. The triple $s = (a, R, d)$ models the concept 'Northeastener' (or 'Tom-July'). It is a symbolic object where R is a relation between the description $Y(i)$ of the (silent) variable Y and a description d and a is a mapping from Ω to L which depends on R and d. (In our two examples, $L = \{0, 1\}$.) The description d can be an intentional description, e.g., as the name suggests, we intend to find the set of individuals in Ω who live in the 'Northeast'; or we seek all those observations in Ω who are 'Tom-July'. Thus, the concept 'Northeasterner' ('Tom-July') is somewhat akin to the classical concept of population; and the extent in Ω corresponds to the sample of individuals from the northeast in the actual dataset. Recall, however, that Ω may already be the 'population' or it may be a 'sample' in the classical statistical sense of sampling.

Symbolic objects play a role in one of three major ways within the scope of symbolic data analyses. First, a symbolic object may represent a concept by its intent (e.g., its description and a way for calculating its extent) and can be used as the input of a symbolic data analysis. Thus, the concept 'Northeasterner' can be represented by a symbolic object whose intent is defined by a characteristic description and a way to find its extent which is the set of people who live in the northeast. A set of such regions and their associated symbolic descriptions can constitute the input of a symbolic data analysis. Secondly, it can be used as output from a symbolic data analysis as when a clustering analysis suggests Northeasterners belong to a particular cluster where the cluster itself can be considered as a concept and be represented by a symbolic object.

The third situation is when we have a new individual (i') who has description d', and we want to know if this individual (i') matches the symbolic object whose description is d; that is, we compare d and d' by R to give $[d' R d] \in L = \{0, 1\}$, where $[d' R d] = 1$ means that there is a connection between d' and d. This 'new' individual may be an 'old' individual but with updated data, or may be a new individual being added to the database who may or may not 'fit into' one of the classes of symbolic objects already present (e.g., should this person be provided with specific insurance coverage?).

In the context of the aggregation of our data into a smaller number of classes, were we to aggregate the individuals in Ω by city, i.e., by the value of the variable Y_1 = City, then the respective classes C_u, $u \in \{1, \ldots, m\}$ would comprise those individuals in Ω which are, in the extent of the corresponding mapping $a(u)$. Subsequent statistical analysis can take either of two broad directions. Either we analyze, separately for each class, the classical or symbolic data for the individuals in C_u as a sample of n_u observations as appropriate; or we summarize the data for each class to give a new dataset with one 'observation' per class. In this latter case, the dataset values will be symbolic data regardless of whether the original values were classical or symbolic data. For example, even though each individual in Ω is recorded as having or not having had cancer (Y_1 = Cancer with $\mathcal{Y}_1$ = {No, Yes}), i.e., as a classical data value, this variable when related to the class for city (say) will become, for example, {Yes (0.1), No (0.9)}, i.e., 10% have had cancer and 90% have not. Thus, the variable Y_1 is now a modal-valued variable.

Likewise, a class that is constructed as the extent of a concept is typically described by a symbolic dataset. For example, suppose our interest lies with 'those who live in Boston', i.e., $a = [Y_1(w) = \text{Boston}]$; and suppose the variable Y_2 = Number of children each individual $i \in \Omega$ has with possible values $\{0, 1, 2, \geq 3\}$. Suppose the data value for each i is a classical value. (The adjustment for a symbolic data value such as individual i has one or two children, i.e., $\xi_i = \{1, 2\}$, readily follows.) Then, the category representing all those who live in Boston will now have the symbolic variable Y_2 with particular value

$$Y_2(w) = \{(0, f_0), (1, f_1), (2, f_2), (\geq 3, f_3)\},$$

where f_i, $i = 0, 1, 2, \geq 3$, is the relative frequency of individuals in this class who have i children.

We need some formal definitions.

Definitions 2.14: Let the random variables Y_j, $j = 1, \ldots, p$, have domain $\mathcal{X} = X_{j=1}^{p} \mathcal{Y}_j$. Then, every point $\boldsymbol{x} = (x_1, \ldots, x_p)$ in $\mathcal{X}$ is called a **description vector**. □

Example 2.26. Consider the random variables Y_1 = City, Y_3 = Age, Y_{10} = Pulse Rate for the classical data of Table 2.1. Then, the particular values

$$x(w_1) = (\text{Boston, } 24, 68)$$

and

$$x(w_8) = (\text{Boston, } 73, 77)$$

are description vectors corresponding, respectively, to the individuals $i = 1, 8$ in Ω. The particular values

$$x(w_{27}) = (\text{Quincy, } 57, 72)$$
$$x(w_{47}) = (\text{Quincy, } 70, 72)$$
$$x(w_{50}) = (\text{Quincy, } 64, 81)$$

are the description vectors for the individuals $i = 27, 47, 50$. □

Definition 2.15: Let the random variables Y_j, $j = 1, \ldots, p$, have domain $\mathcal{X} = \mathcal{X}_{j=1}^{p} \mathcal{Y}_j$. Let D_j be a subset of $\mathcal{Y}_j$, i.e., $D_j \subseteq \mathcal{Y}_j$. Then, the p-dimensional subspace $D = (D_1, \ldots, D_p) \subseteq \mathcal{X}$ is called a **description set.** If $D = X_{j=1}^{p} D_j$ is the Cartesian product of the sets D_j, then D is called a **Cartesian description set**. □

Example 2.27. Take the random variables $Y_1 =$ City, $Y_3 =$ Age and $Y_9 =$ Weight in the symbolic dataset of Tables 2.2. Then, the observations

$$Y(w_5) = (\{\text{Atlanta, Quincy}\}, [57, 86], [159, 184])$$

$$Y(w_6) = (\{\text{Boston, Peoria}\}, [73, 79], [153, 164])$$

are description sets which describe the categories Optical Males ($u = 5$) and Optical Females ($u = 6$), respectively, and are subsets of $D' = (D_1, D_3, D_9)$.

For the two random variables Y_3 and Y_9, each of

$$d(w_5) = [57, 86] \times [159, 184]$$
$$d(w_6) = [73, 79] \times [153, 164]$$

are Cartesian description sets and are subsets of $D_3 \times D_9$. □

In general, any one observation may consist of a mixture of description vector values and of description set values. This gives us the following.

Definition 2.16: Let the random variables Y_j, $j = 1, \ldots, p$, have domain $\mathcal{X} = X_{j=1}^{p} \mathcal{Y}_j$. Let $\boldsymbol{Y} = (\boldsymbol{Y}^1, \boldsymbol{Y}^2)$ where $\boldsymbol{Y}^1 = (Y_1^1, \ldots, Y_k^1)$ takes point values $\boldsymbol{x}^1 = (x_1^1, \ldots, x_k^1) \in \mathcal{X}^1 = X_{j=1}^{k} \mathcal{Y}_j^1$ and $\boldsymbol{Y}^2 = (Y_1^2, \ldots, Y_{p-k}^2)$ takes subset values $D^2 = (D_1^2, \ldots, D_{p-k}^2) \subseteq X_{j=1}^{p-k} \mathcal{Y}_j^2$. Then, $\boldsymbol{d} = (x_1^1, \ldots, x_k^1, D_1^2, \ldots, D_{p-k}^2)$ is called a **description** d. The set of all possible descriptions is called the **description space** $\mathcal{D}$. □

Example 2.28. Suppose we aggregated the individuals $i = 27, 47, 50$ in Tables 2.1; see Example 2.26. Then, the resulting description for the random variables $Y_1 =$ City, $Y_3 =$ Age, $Y_{10} =$ Pulse Rate is

$$\boldsymbol{d} = (\{\text{Quincy}\}, [57, 70], [72, 81]).$$

Notice that we have a single point value, $x_1 =$ Quincy for the city Y_1; but subsets for age Y_3, $D_3 = [57, 70]$ and for pulse rate Y_{10}, $D_{10} = [72, 81]$, with $x_1 \in \mathcal{Y}_1$, and $D_3 \subseteq \mathcal{Y}_3$ and $D_{10} \subseteq \mathcal{Y}_{10}$.

Since these are the only three individuals in the original dataset who live in Quincy, had our primary interest been on where people live, the resulting aggregation gives us the description (for Y_3 and Y_{10}) that those who live in Quincy are aged between 57 and 70 years and have a pulse rate between 72 and 81 inclusive. □

When there are constraints on any of the variables, such as when logical dependencies exist, then the description space $\mathcal{D}$ has a 'hole' in it corresponding to those descriptions that are not permitted under the constraints.

Example 2.29. Consider the observation

$$Y = \xi = ([88, 422], [49, 149])$$

representing Y_1 = Number of At-Bats, and Y_2 = Number of Hits over a season, aggregated over the players in a baseball team. In the presence of the logical rule, which dictates that the number of hits cannot exceed the number of at-bats,

$$v : Y_1 \geq Y_2, \tag{2.4}$$

the Cartesian description set given by the rectangle [88, 422]×[49, 149] has a 'hole' in it, corresponding to the descriptions for which the number of hits exceeds the number of at-bats, an impossibility. The 'hole' here is the triangle with vertices $Y_2 = (88, 88), (88, 149), (149, 149)$; see Figure 2.3. □

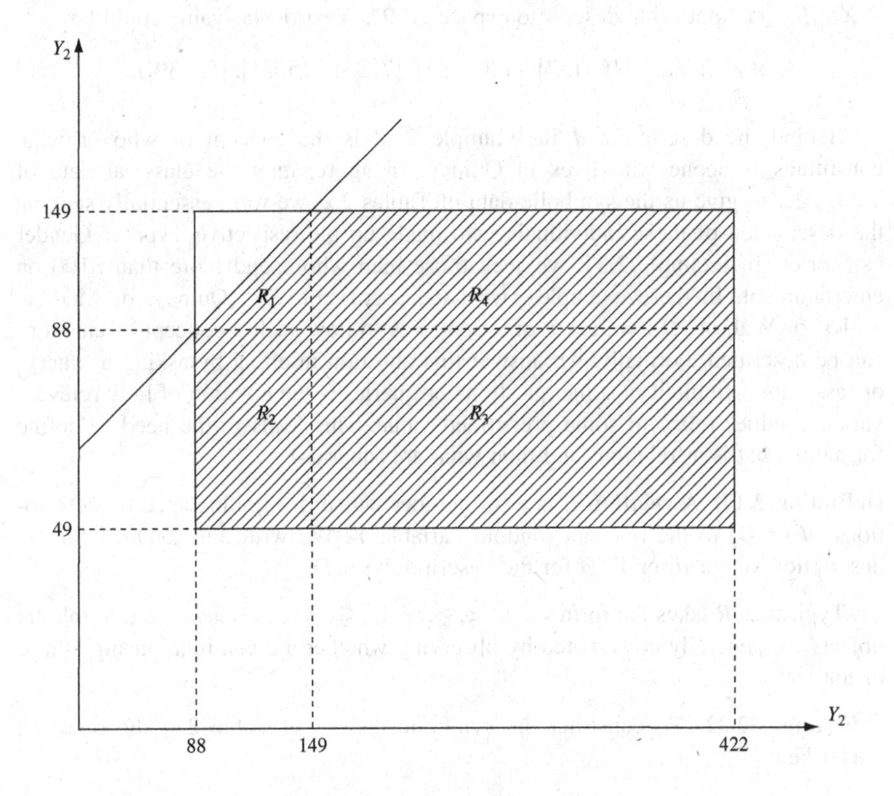

Figure 2.3 Dependency holes.

The general issue of holes in the description space can create complications in subsequent analyses. Many different examples will be considered in later sections. This baseball example is particularly interesting, especially as the seemingly innocuous but essential rule of Equation (2.4) raises a number of analytical issues; see Section 4.5 for a more complete discussion.

Another issue relates to the description space when interest (or constraints) on one or more, but not all, variables is more focused.

Example 2.30. Consider the dataset of Tables 2.1. A health insurer might be interested in individuals who are aged 50 and over but has no limits on values for the remaining variables. In this case, the description space is $\mathcal{D} = \mathcal{Y}_1 \times \mathcal{Y}_2 \times \{Y_3 \geq 50\} \times \ldots \times \mathcal{Y}_{30}$. □

Example 2.31. Consider the credit card data of Tables 2.3 and 2.4. Then, the description 'men who spend more than \$100 in a given month on entertainment' takes values in the description space

$$\mathcal{D}^* = \{\text{Men}\} \times \mathcal{Y}_1 \times D_2 \times \mathcal{Y}_3 \times \mathcal{Y}_4 \times \mathcal{Y}_5$$

where $D_2 = \{Y_2 > 100\} \subseteq \mathcal{D}_2 = \Re^+$; that is, instead of the entire domain $\mathcal{D} = X_{j=1}^5 \mathcal{Y}_j$, the applicable description space is $\mathcal{D}^*$. A particular value could be

$$\boldsymbol{d} = (\{\text{Wal}\}, [10, 157], [120, 224], [7, 23], [15, 31], [23, 39]).$$ □

Behind the description $\boldsymbol{d}$ in Example 2.28 is the concept of who or what constitutes someone who lives in Quincy. In aggregating the classical data of Tables 2.1 to give us the symbolic data of Tables 2.2, we were essentially seeking the description of those individuals who made up the respective Type × Gender categories. In Example 2.31, we seek those men who spend more than \$100 on entertainment. In these examples, we have a concept, e.g., Quincy, or Medical Males, or Wal, or . . . ; we ask a question(s) as to how such a concept or category can be described. Formally, we answer this question in effect by asking a 'query' or 'assertion'. Implicit in some questions or queries is the notation of how relevant variable values relate to some predefined value. This leads to the need to define formally what is a 'relation' and then what are 'queries'.

Definition 2.17: A **relation** R is a formal mechanism for linking specific descriptions (d or D) to the relevant random variable Y. We write this as YRd for the description vector d, or YRD for the description set D. □

Typically, R takes the form $=, \leq, <, \geq, >, \subset, \subseteq, \neq, \in, \notin$, and so on. Symbolic objects are generally constructed by observing whether a given relationship is true or not.

Example 2.32. To construct the symbolic dataset of Table 2.4, the question is asked:

$$\text{Is } Y(w) = \text{Name } (w) = \text{Jon?}$$

If yes, then the observation values (for all Y_j) are put into the category 'Jon'. Here,

$$R = \text{'='}, \text{ and } d = \text{'Jon'}.$$ □

Example 2.33. In Example 2.31, the question being asked is

$$\text{Is } Y_2 > 100?$$

Here, the relation $R =$ ' >' and $d = 100$. If this is true, then the observation is added to the category of observations satisfying this description. □

Example 2.34. Interest may center on those individuals who spend between \$100 and \$200 on entertainment. In this case, we have

$$\text{Is} Y_2 \subseteq (100, 200)?$$

Here, R is the relation '$\subseteq$', and $D = (100, 200)$. □

Example 2.35. To construct the symbolic data of Tables 2.2, two questions were asked, one about $Y_2 =$ Type and one about $Y_4 =$ Gender, i.e.,

$$\text{Is } Y_2 = \{\text{Dental}\}$$

and

$$\text{Is } Y_4 = \{\text{Male}\}, \text{ say?}$$

Here, R is the operator '='. If the answer is yes to both questions, then the individual is contained in the description space

$$\mathcal{D} = \mathcal{Y}_1 \times \{\text{Dental}\} \times \mathcal{Y}_3 \times \{\text{Male}\} \times \mathcal{Y}_5 \times \ldots \times \mathcal{Y}_{30}.$$ □

Example 2.36. It could also be that a particular focus of interest concerns those who are underweight or overweight. In this case, the relation could be, for $Y =$ Weight,

$$Y < 100, \quad Y > 180.$$

(or, equivalently, $100 \leq Y \leq 180$). Here, the relation $R = (<, >)$ and the description is $d = 100$, $d = 180$. Note that this could also be written as

$$Y \subset D_1 = [0, 100)$$

$$Y \subset D_2 = (180, \infty),$$

where now we have the relation $R =$ '$\subset$' and the sets D_1 and D_2. □

Notice in Example 2.35 that more than one relation existed; and in Example 2.36, there were two relations on the same variable. In general, we have any number of relations applying, including the possibility of several relations R on the same variable. We have:

Definition 2.18: A **product relation** is defined as the product of (any) number of single relations linking the variables Y_j, $j = 1, \ldots, p$, with specific description vectors d and/or description sets D. This is denoted by $R = X_{k=1}^{v} R_k$, with

$$yRz = \wedge_{k=1}^{v} (Y_{j_k} R_{j_k} z_{j_k})$$

where $z = d$ or $z = D$, as appropriate, and where the '$\wedge$' symbol refers to the logical conjunction 'and'. □

Example 2.37. For the random variables Y_2, Y_4, Y_9, and Y_{10} of Tables 2.2, the product relation

$$YRz = \{Y_2 = \text{Dental}\} \wedge \{Y_4 = \text{Male}\} \wedge \{Y_9 > 180\} \wedge \{Y_{10} \subseteq [60, 80]\}$$

is the relation asking that the Type (Y_2) is dental (d_2) and the Gender (Y_4) is male (d_4), the Weight (Y_9) is over 180 (d_9), and the Pulse Rate Y_{10} is in the interval $D_{10} = [60, 80]$. □

Definition 2.19: When an **assertion**, $a(i)$, also called a **query**, is asked of an individual $i \in \Omega$, it assumes a value of $a(i) = 1$ if the assertion is true, or $a(i) = 0$ if the assertion is false. We write

$$a(i) = [Y(i) \in D] = \begin{cases} 1, & Y(i) \in D, \\ 0, & Y(i) \notin D. \end{cases}$$

The function $a(i)$ is called a **truth function** and represents a mapping of Ω onto {0,1}. □

Definition 2.20: The set of all $i \in \Omega$ for which the assertion $a(i) = 1$ is called the **extension** in Ω, and is called an **extension mapping**. □

Example 2.38. For the random variables Y_1 = Cancer and Y_2 = Age, the assertions

$$a = [Y_1 = \text{Yes}], a = [Y_2 \geq 60] \tag{2.5}$$

represent all individuals with cancer, and all individuals aged 60 and over, respectively. The assertion

$$a = [Y_1 = \text{Yes}] \wedge [Y_2 \geq 60] \tag{2.6}$$

represents all cancer patients who are 60 or more years old, while the assertion

$$a = [Y_2 < 20] \wedge [Y_2 > 70] \tag{2.7}$$

seeks all individuals under 20 and over 70 years old. In each case, we are dealing with a concept 'those aged over 60', 'those over 60 with cancer', etc.; and the assertion a maps the particular individuals present in Ω onto the space $\{0,1\}$.

If instead of recording the cancer variable as a categorical {Yes, No} variable, it were recorded as a {lung, liver, breast, . . . } variable, the assertion

$$a = [Y_1 \in \{\text{lung, liver}\}] \tag{2.8}$$

is describing the class of individuals who have either lung cancer only or liver cancer only or both lung and liver cancer. Likewise, an assertion can take the form

$$a = [Y_2 \subseteq [60, 69]] \wedge [Y_1 \in \{\text{lung}\}]; \tag{2.9}$$

that is, this assertion describes those who are in their sixties and have lung cancer. □

Example 2.39. We want to construct the symbolic dataset of Tables 2.2 from the classical dataset of Tables 2.1. Our interest is in the categories Type × Gender. Thus, to build the Dental Male category, we start with the description set

$$\mathcal{D} = \mathcal{Y}_1 \times \{\text{Dental}\} \times \mathcal{Y}_3 \times \{\text{Male}\} \times \mathcal{Y}_5 \times \ldots \times \mathcal{Y}_{30};$$

or more simply

$$\mathcal{D} = \{\text{Dental}\} \times \{\text{Male}\} = \{Y_2 = \text{Dental}\} \times \{Y_4 = \text{Male}\}$$

where by implication the entire domain of the other variables $Y_1, Y_3, Y_5, \ldots, Y_{30}$ is operative. Then, the query, for each $i \in \Omega$, is

$$a(i) = [\mathbf{Y}(i) \in \mathcal{D}] = [Y_2(i) = \text{Dental}] \wedge [Y_4(i) = \text{Male}].$$

The extent consists of all those i in Ω for which this assertion is true. Thus, for example, we see that $a(1) = a(4) = 0$, since $Y_2(i) =$ medical and $Y_4(i) =$ male, for each of these two individuals ($i = 1, 2$). However, $a(3) = 1$, since $Y_2(3) =$ dental and $Y_4(3) =$ male, for the third individual ($i = 3$). Likewise, we find the $a(i)$ values for each $i \in \Omega$. Thus, we obtain the extent of the concept Dental Males as the set of individuals i with

$$i \in \{3, 5, 9, 10, 11, 14, 16, 18, 19, 26, 30, 47\}.$$

A description of this extent is shown in Tables 2.2 (for the random variables $Y_1, Y_3, Y_5, Y_6, Y_9, Y_{10}, Y_{11}, Y_{12}$, and Y_{13}). Thus, if we take just the Y_3, Y_5, Y_6, and Y_9 variables, we have

$$\mathbf{d} = ([17, 76], \{\text{M, S}\}, \{0, 1, 2\}, [161, 268]).$$

That is, the Dental Males are aged 17 to 76 years, are married or single, have no, one, or both parents still alive, and weigh 161 to 268 pounds.

If our intention is to find the individuals who are Optical Males, then our assertion is

$$a(i) = [\boldsymbol{Y}(i) \in D] = [Y_2(i) = \text{Optical}] \wedge [Y_4(i) = \text{Males}]. \tag{2.10}$$

The extent of this assertion consists of the individuals i for which

$$i \in \{27, 36\}$$

since

$$a(27) = a(36) = 1 \quad \text{and} \quad a(i) = 0 \quad \text{otherwise.}$$

The description of this extent is also shown in Tables 2.2. Proceeding in this way, we can construct the entire Tables 2.2. □

Typically, a category is constructed by identifying all the individuals (or subcategories) that matched its description, i.e., by finding all the i for which $a(i) = 1$, for some description $\boldsymbol{d}$, e.g., 'Medical Males', or '60–69 year olds who have lung cancer', and so on.

Alternatively, it may be desired to find all those (other) individuals in Ω who match the description of a specific individual i^* in Ω. In this case, the assertion becomes

$$a(i) = \wedge_{k=1}^{\nu} [Y_{j_k}(i) = Y_{j_k}(i^*)],$$

and the extension is

$$Ext(a) = \{i \in \Omega | Y(i) = Y(i^*)\}.$$

Example 2.40. An individual i^* may be a composite description of a sought-after suspect. The set Ω may consist of individuals in a particular database. For example,

$$Y(i^*) = [\{Y_1 = \text{black hair}\}] \wedge [\{Y_2 = \text{blue eyes}\}].$$

Then, the assertion

$$a(i) = [Y(i) = Y(i^*)]$$

is applied to all i in Ω, to produce the extent of this query, i.e., to find all those with black hair and blue eyes. □

Example 2.41. Unusual activity on credit card transactions can be identified by comparing the card expenditures to the usual activity as described by a person's card use description. Thus, for example, Fred may have the description

$$\boldsymbol{d}(\text{Fred}) = ([10, 65], [100, 200])$$

for two types of expenditure Y_1 and Y_2. Then, the questionable expenditures would relate to those $i \in \Omega$, or for some new i 'individuals' (i.e., expenditures) which do not match Fred's description $\boldsymbol{d}$. The assertion is

$$a(i) = [Y_1 \not\subseteq [0, 65]] \wedge [Y_2 \not\subseteq [0, 200]].$$

Clearly, attention is directed to sudden frequent excessive use of Fred's credit card when the extent of this assertion contains 'too many' individuals. □

The assertion in Definition 2.19 defines the mapping $a(i)$ as having values 0, 1. More generally, an assertion may take an intermediate value, such as a probability, i.e., $0 \leq a(i) \leq 1$, representing a degree of matching for an individual i. Flexible matching is also possible; see Esposito *et al.* (1991) for more details on flexible matching.

Definition 2.21: A **flexible mapping** occurs when the mapping a is onto the interval [0, 1]. That is, $a : \Omega \rightarrow [0, 1]$, and we say the extension of a has level α, $0 \leq \alpha \leq 1$, where now

$$Ext_\alpha(a) = Q_\alpha = \{i \in \Omega | a(i) \geq \alpha\}.$$ □

Example 2.42. Consider the assertion

$$a = [\text{Age} \subseteq [20, 30]] \wedge [\text{Weight} \subset [60, 80]] \wedge [\text{Height} \subseteq [60, 70]].$$

Suppose Joe's descriptors are

$$w = \text{Joe: Age} = (20, 25), \text{ Weight} = (60, 65), \text{ Height} = 50.$$

Then,

$$a(\text{Joe}) = 2/3,$$

since two of the three variables in the assertion 'a' match Joe's description. Hence, $\alpha = 2/3$. □

2.2.3 Concepts of concepts

In the previous section we have in effect shown how a symbolic dataset is constructed from a rational database. It is now apparent that although our examples were built from classical datasets, the original datasets could equally have been symbolic datasets, and also our initial individuals $i \in \Omega$ could have been categories or classes of individuals themselves.

Example 2.43. Consider the health dataset of Tables 2.1. We have already seen in Example 2.38 how we built the concepts of Type × Gender, with Tables 2.2 (and likewise Table 2.6) being a resulting symbolic dataset for these six concepts.

Suppose now our interest is in health (Type) categories, regardless of gender. Then, we can build the appropriate symbolic database, by finding the extent of the assertion

$$a(i) = [Y(i) \in D] = [Y_2(i) = d] \tag{2.11}$$

for each of $d =$ Dental, Medical, Optical. Thus, for the Optical category, we find, from Tables 2.2,

$$a(w_5) = a(w_6) = 1$$

with all other $a(w_u) = 0,\ u = 1, \ldots, 4$. Likewise, we can obtain the extents of the Dental, and the Medical, categories. For the $Y_3 =$ Age, $Y_5 =$ Marital Status, and $Y_9 =$ Weight random variables, the resulting descriptions are, respectively,

$$\boldsymbol{d}(\text{Dental}) = ([17, 70], \{\text{M}, \text{S}\}, [148, 268])$$

$$\boldsymbol{d}(\text{Medical}) = ([6, 87], \{\text{M}, \text{S}\}, [35, 239])$$

$$\boldsymbol{d}(\text{Optical}) = ([57, 86], \{\text{M}, \text{S}\}, [153, 184]).$$

That is, to build the concepts of Type × Gender of Tables 2.2 we considered the assertion $a(i)$ on the individuals $i \in \Omega$ (in Tables 2.1) and hence obtained the extent $E_1 = \{w_1, \ldots, w_6\}$. Then, to build the concepts Type, we treated the concepts in E_1 as though they were 'individuals'. That is, we can consider each of the $w_u \in E_1, u = \{1, \ldots, 6\}$, categories as $m = 6$ individuals or classes and build the three Type categories by applying the assertion relating to Type in Equation (2.11) to these six categories. In terms of our notation for individuals etc., we now have the new $i \equiv$ old w_u, and $\Omega \equiv E_1$. Thus, we see that for w_5 and w_6, i.e., Optical Males and Optical Females,

$$a(i) \equiv a(w_5) = [Y_2(w_5) = \text{Optical}] = 1, \quad \text{and}$$

$$a(i) \equiv a(w_6) = [Y_2(w_6) = \text{Optical}] = 1,$$

but for $u = 1, \ldots, 4$, $a(i) \equiv a(w_u) = 0$. Thus, we have obtained our concept of concepts. □

Example 2.44. A health maintenance organization (HMO) has a database of 1 million entries representing the nature of service rendered on a range of over 80000 (say) ailments to its customers. Suppose interest centers (first) on services rendered to its 15000 (say) beneficiaries rather than on any one entry. Thus, each beneficiary represents a concept, and so the data can be aggregated over its 1 million values accordingly to produce the symbolic values for the 15000 beneficiary classes.

Suppose now interest shifts to classes of beneficiaries as a function of the type (and/or number) of products prescribed as a result of the (individual) services rendered. In this case, the new concept (or concept of concepts) is defined by the product × beneficiary category. The corresponding data values are obtained by aggregating over the 15000 beneficiary values to produce the symbolic values for the 30 (say) product × beneficiary categories. □

Behind the use of assertions and relations and extents in these constructions is the somewhat vague notion of 'intent'. The intent can be loosely defined as what it is that we intend to do with the data, what we want to know. In effect, our intent is to find the extent of an assertion. Behind this is the notion of a concept, and behind that is the notion of a symbolic object. We consider each formally in turn.

Definition 2.22: A **concept** is defined by an intent and a way of finding its extent. It is modeled by a symbolic object (to be defined in Definition 2.24.) □

We are interested in a concept, e.g., Medical Females. Thus, the concept is defined by a set of properties described by the 'intent'; here the intent consists of the properties relating to type and gender, namely, Type = medical and Gender = female. The extent is the set of all individuals who are medical females. The way of finding the extent is through the assertion rule with those $i \in \Omega$ for which $a(i) = 1$ forming the extent.

Definition 2.23: The extent of a concept constitutes a **category** of units which satisfy the properties of its intent. A category is modeled by a symbolic description of this intent. □

Thus, we start with an idea of an entity of the world, i.e., a concept (e.g., Medical Females). At this stage, the entity is not yet modeled. We find its extent from Ω. This extent defines a category (e.g., of Medical Females) and this category is therefore modeled.

Example 2.45. The Type × Gender categories are modeled by the descriptions derived from their construction from the individuals in Ω. For example, from Table 2.2b,

$$d(\text{Medical Females}) = ([11, 87], \{\text{M}, \text{S}\}, [73, 166])$$

for the Y_3, Y_5, and Y_9 random variables. That is, this category is described as being of age 11 to 87 years, married or single, and with weight from 73 to 166 pounds. □

Definition 2.24: A **symbolic object** is the triple $s = (a, R, d)$ where a is a mapping $a : \Omega \rightarrow L$ which maps individuals $i \in \Omega$ onto the space L depending on the relation R between its description and the description d. When $L = \{0, 1\}$, i.e., when a is a binary mapping, then s is a **Boolean symbolic object.** When $L = [0, 1]$, then s is a **modal symbolic object**.

That is, a symbolic object is a mathematical model of a concept (see, Diday, 1995). If $(a, R, d) \in \{0, 1\}$, then s is a Boolean symbolic object and if $(a, R, d) \in [0, 1]$, then s is a modal symbolic object. For example, in Equation (2.9), we have the relations $R = (\subseteq, \in)$ and the description $d = ([60, 69], \{\text{lung}\})$. The intent is to find '60–69 year olds who have lung cancer'. The extent consists of all individuals in Ω who match this description, i.e., those i for whom $a(i) = 1$.

Example 2.46. Consider a Boolean symbolic object. Let $a = [Y R d]$ where $\mathbf{Y} = (Y_1, Y_2) =$ (Color, Height), $R = \subseteq$, and $\mathbf{d} = (\{\text{red, blue, yellow}\}, [10,15])$. That is, we are interested in those individuals (e.g., birds) that are red, blue, or yellow in Color and are 10 to 15 cm in Height. Then, we can write a as the assertion

$$a = [\text{Color} \subseteq \{\text{red, blue, yellow}\}] \wedge [\text{Height} \subseteq [10, 15]]. \quad (2.12)$$

Suppose a particular individual (bird) w has the description

$$d(w) = (\text{Color}(w) = \{\text{red, yellow}\}, \text{Height}(w) = 21).$$

Then,

$$a(w) = 0$$

since although $\text{Color}(w) \subseteq \{\text{red, blue, yellow}\}$, the variable $\text{Height}(w) = 21 \not\subseteq [10, 15]$. We conclude that the bird w does not belong to the extent, or category, which is described by Equation (2.12). □

Example 2.47. A simple example of a modal symbolic object is where a particular object w has for its age (Y) the value

$$Y(w) = \text{Age}(w) = [10, 30];$$

and suppose the assertion is

$$a = [\text{Age} \subseteq [25, 60]].$$

Then, it follows that (assuming uniformity across the interval)

$$a(w) = (30 - 25)/(30 - 10) = 0.25.$$

That is, the object w is not entirely in (or not in) the description space $D = [25, 60]$, but rather is only partially in D. □

Example 2.48. As another example of a modal symbolic object, consider the possibility functions that it will rain $d_a(t)$ and $d_b(t)$ for two assertions a and b plotted over time t; see Figure 2.4(a).

Let $a = [yRd_a]$ and $b = [yRd_b]$; hence, the assertion

$$c = a \cup b = [yR(d_a \cup d_b)].$$

The possibility function associated with c is $d_c(t)$ with

$$d_c(t) = \max_t \min\{d_a(t), d_b(t)\},$$

and is plotted in Figure 2.4(b). The values assumed by this function give the mapping probabilities α. Thus, for example, at $t = t^*$, $c(t^*) = 0.3$.

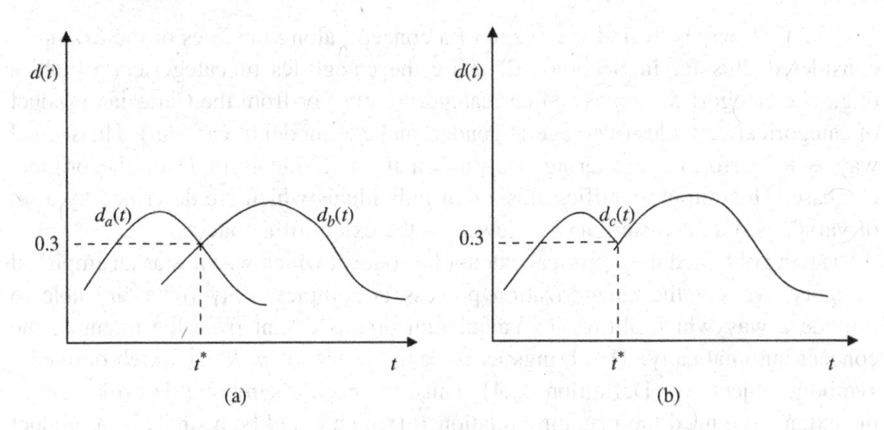

Figure 2.4 Possibility function of rain. □

Example 2.49. A different type of modal symbolic object is the following. Let a be the mapping $a = [YRd]$ where $Y = (Y_1, Y_2) =$ (Age, Marital Status) and R: $[d'RD] = \max_{k=1,2}[d'_k R_k d_k]$, where the descriptions d'_k and d_k correspond to two (s-valued) discrete probability distributions $d'_k = r$ and $d_k = q$ where the two distributions match (R) when

$$rR_k q = \sum_{j=1}^{s} r_j q_j \exp[r_j - \min(s_j, q_j)]. \tag{2.13}$$

Suppose the distribution r for Age is

$$r = \{[0, 40), 0.4; [40, 90], 0.6\}$$

and that for Marital Status q is

$$q = \{\text{married}, 0.65; \text{single}, 0.35\};$$

that is, $\boldsymbol{d} = (r, q)$ is the description that age is under 40 with probability 0.4 and age is in the interval [40, 90] with probability 0.6, and that married individuals constitute 65% and singles make up 35% of the extent. Then, an individual w belongs to this extent if there is a match where the match is defined by Equation (2.13). □

Categorization is the process of arranging individuals into categories. Here, category is taken to mean the value of a categorical variable whose extent is a class of similar units (e.g., Males, Volvo, Make of car, Leigh's credit card expenses, and so on). Since we have the concept (Males, Volvo, Leigh, . . .) and we have a way of finding its extent, it is easy to obtain the components of the category by its extent. Therefore, in practice, the question of how to categorize is transformed to the question of how to obtain the extent of the concept from the database. There are at least two ways to achieve this.

The first way is to find the extent of a concept, along the lines of the examples considered thus far in Section 2.2, using the categories of categorical variables (e.g., age categories, or make of car categories, etc.) or from the Cartesian product of categorical variables (e.g., age $\times$ gender, make $\times$ model of car, etc.). The second way is to perform a clustering analysis on the individuals in Ω of the original database. The output identifies clusters of individuals which are described by a set of variables. Each cluster can be viewed as the extent of a concept.

Having obtained the empirical extent of a concept which we view as an empirical category, we say the categorization process is complete only if we are able to provide a way which allows the calculation of this extent from the intent of the concept automatically. This brings us back to the triple (a, R, d) which defined a symbolic object (see Definition 2.24). Thus, we need a symbolic description $\boldsymbol{d}$ of the extent. We need the matching relation R (which could be a single or a product relation). We use this R to compare the description $\boldsymbol{d}(i)$ of the individual to the description $\boldsymbol{d}(w_u)$ of the concept. Thirdly, the assertion a is the mapping which determines whether the individual $i \in \Omega$ is a member of the category; this mapping a is sometimes called a membership mapping.

2.2.4 Some philosophical aspects

Whilst recognition of the need to develop methods for analyzing symbolic data and tools to describe symbolic objects is relatively new, the idea of considering higher level units as concepts is in fact ancient. Aristotle Organon in the fourth century BC (Aristotle, IVBC, 1994) clearly distinguishes first-order individuals (such as **the** horse or **the** man) which represented units in the world (the statistical population) from second-order individuals (such as **a** horse or **a** man) represented as units in a class of individuals. Later, Arnault and Nicole (1662) defined a concept by notions of an intent and an extent (whose meanings in 1662 match those herein) as:

> *Now, in these universal ideas there are two things which is important to keep quite distinct: comprehension and extension (for 'intent' and 'extent'). I call the comprehension of an idea the attributes which it contains and which cannot be taken away from it without destroying it; thus the comprehension of the idea of a triangle includes, to a superficial extent, figure, three lines, three angles, the equality of these three angles to two right angles etc. I call the extension of an idea the subjects to which it applies, which are also called the inferiors of a universal term, that being called superior to them. Thus the idea of triangle in general extends to all different kinds of triangle.*

There are essentially two kinds of concepts. The first refers to concepts of the real world such as a town, a region, a species of flower, a scenario of a road accident, a level of employment, and so on. This kind of concept is defined by an intent and an extent which exist, or have existed or will exist, in the real world.

The second kind of concept refers to concepts of our minds (one of the so-called mental objects of Changeux, 1983) which model in our mind concepts of our imagination or concepts of the real world by their properties and a way of finding their extent (by using our innate 'senses'); but the extent itself is not modeled since it is not possible for our minds to 'see' completely all the possible cases that make up that extent. Thus, we recognize a car accident, or a chair, when we see one, i.e., we have a way of finding the extent; but we cannot visualize all car accidents, or all chairs, i.e., the complete extent is beyond our reach.

A symbolic object models a concept in the same way as our mind does, by using a description $\boldsymbol{d}$ representing its properties and a mapping a which enables us to compute its extent. For example, we have a description of a car (or a chair, or a 20 year old female, or a cancer patient, etc.) and a way of recognizing that a given entity in the real world is in fact a car (or chair, etc.). Hence, whereas a concept is defined by an intent and an extent, it is modeled by an intent and a way of finding its extent by symbolic objects akin to the workings of our mind. Since it is quite impossible to obtain all the characteristic properties (the complete description $\boldsymbol{d}$) and the complete extent of a concept, a symbolic object therefore is just an approximation of a concept. This raises the question of the quality, robustness, and reliability of this approximation; we return to this issue later.

In the Aristotelian tradition, concepts are characterized by a logical conjunction of properties. In the Adansonian tradition (Adanson, 1727–1806, was a student of the French naturalist Buffon), a concept is characterized by a set of similar individuals. This contrasts with the Aristotelian tradition in which all the members of the extent of a concept are equivalent with respect to intent of that concept. A third approach, derived from the fields of psychology and cognitive science (see, e.g., Rosch, 1978), is to say that concepts must be represented by classes which 'tend to become defined in terms of prototypes or prototypical instances that contain the attributes most representative of the items inside the class'. Fourthly, Wille (1989), following Wagner (1973), says that 'in traditional philosophy things for which their intent describes all the properties valid for the individual of their extent are called "concept"'.

Symbolic objects combine the advantages of all these four approaches. The advantage of the Aristotelian tradition is that it provides the explanatory power of a logical description of concepts that are represented by Boolean symbolic objects. The Adansonian tradition provides the advantage that the members of the extent of a symbolic object are similar in the sense that they must satisfy minimally the same properties. In this sense, the concepts are polytheistic, e.g., they can be a disjunction of conjunctions. The Rosch tradition allows for their membership function to be able to provide prototypical instances characterized by the most representative attributes. Fourthly, the Wagner – Wille property is satisfied by so-called complete symbolic objects which can be shown to constitute a Galois lattice on symbolic data (see, e.g., Brito, 1994; Pollaillon 1998; Diday and Emilion 2003) and so generalize the binary model defined for example in Wille (1989).

Finally, we link the notions of the four spaces associated with individuals, concepts, descriptions, and symbolic objects by reference to Figure 2.5 (from Diday, 2005).

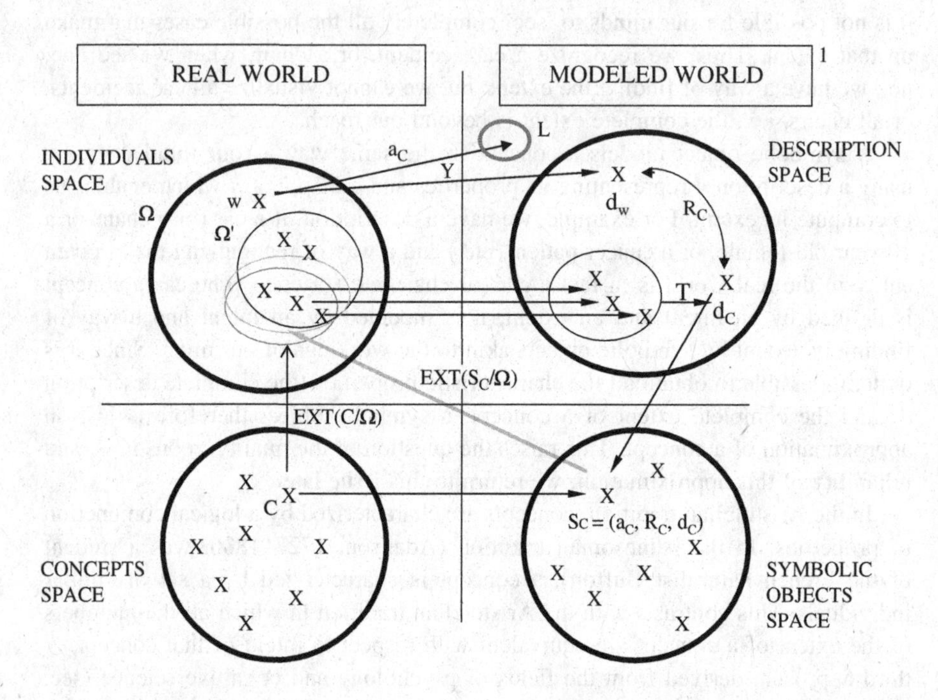

Figure 2.5 Modelization by a symbolic object of a concept known by its extent.

The real world can be viewed as consisting of individuals in the space Ω. It can also be seen as a set of concepts. Each concept C (through the assertion a) has its extent in Ω as the subset Ω' (say). For example, the concepts can relate to specific types of genetic markers. The set of individuals in Ω that have the marker C are $ExtC \equiv \Omega'$. The modeled world is the description space D; it is also the space of symbolic objects S. Individuals w in Ω' are described by a mapping $Y(w)$ which links the individual w in Ω (or Ω') to a description d_C in the description space, while the description d_C leads to the symbolic object $s \subset S$ with $s = (a_C, R, d_C)$, with the assertion $a_C = [Y(w) R_C d_C]$ measuring the fit or matching (via $0 \leq \alpha \leq 1$) between the description $Y(w)$ of the individual $w \in \Omega$ and the description of the concept d_C in D. Thus, the symbolic object models the concept C by the triple s. This model can be improved by a learning process which improves the basic choices $(a, d, R, \ldots)$ in order to reduce the difference between the extent of the concept and the extent of the symbolic object which models it.

2.2.5 Fuzzy, imprecise, and conjunctive data

Fuzzy, imprecise, and conjunctive data are different types of data, each of which relates to symbolic data in its own way. We describe each one through illustrative examples in this section. From these we see that fuzzy, imprecise, and conjunctive data are akin to symbolic data but have subtle differences. We note that the focus of this volume is on symbolic data per se.

Classical data are not fuzzy data, nor are fuzzy data symbolic data. However, fuzzy data can be transformed into symbolic data. The relationship between classical, fuzzy, and symbolic data is illustrated by the following example.

Example 2.50. Suppose four individuals, Sean, Kevin, Rob, and Jack, have Height (Y_1), Weight (Y_2), and Hair Color (Y_3) as shown in Table 2.16a. Thus, Jack is 1.95 meters in height, weighs 90 kilos, and has black hair. To transform this into fuzzy data, let us assume the random variable height takes three forms, short, average, and tall, with each form being triangularly distributed centered on 1.50, 1.80, and 1.90 meters, respectively; see Figure 2.6.

Consider Sean's height. From Figure 2.6, we observe that Sean has a fuzzy measure of 0.50 that he is of average height and 0.50 that he is tall. In contrast, Rob has a fuzzy measure of 0.50 that he is short. The heights for all four individuals expressed as fuzzy data are as shown in Table 2.16b. For the present purposes, the weight and hair color values are retained as classical numerical data.

Suppose now that we have the fuzzy data of Table 2.16b. Suppose that it is desired to aggregate the data in such a way that one category has blond hair and the second category has dark hair. The extent of the first category is Sean and Kevin, while Rob and Jack are the extent of the second category. The symbolic data that emerge from the fuzzy data are as shown in Table 2.16c. □

Table 2.16a Classical data.

Individual	Height	Weight	Hair
Sean	1.85	80	blond
Kevin	1.60	45	blond
Rob	0.65	30	black
Jack	1.95	90	black

Table 2.16b Fuzzy data.

Individual	Height			Weight	Hair
	Short	Average	Tall		
Sean	0.00	0.50	0.50	80	blond
Kevin	0.70	0.30	0.00	45	blond
Rob	0.50	0.00	0.00	30	black
Jack	0.00	0.00	0.48	90	black

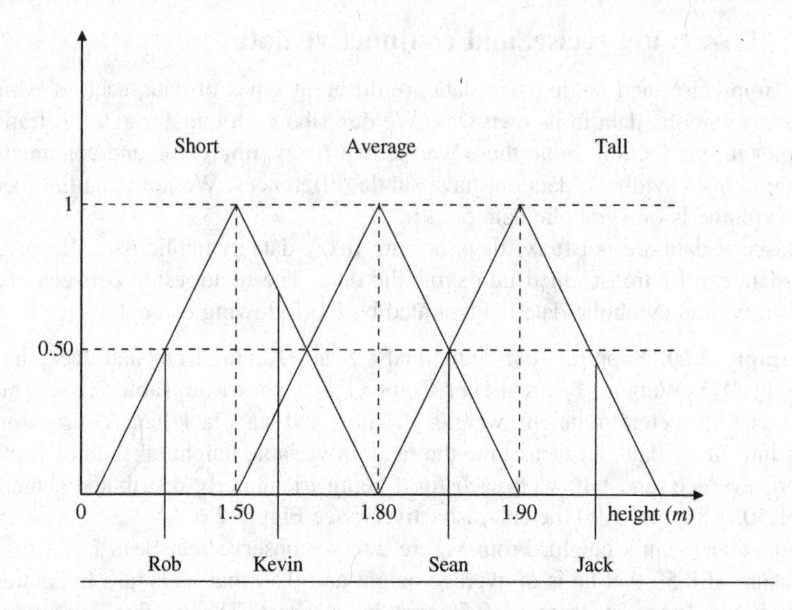

Figure 2.6 Distributions of heights.

Table 2.16c Symbolic data.

Hair Color	Height			Weight	Hair
	Short	Average	Tall		
blond	[0, 0.70]	[0.30, 0.50]	[0, 0.50]	[45, 80]	blond
black	[0, 0.50]	0	[0, 0.48]	[30, 90]	black

A related type of observation is one with (a usually unspecified) degree of imprecision and uncertainty as seen in Example 2.51.

Example 2.51. It may not be possible to record a person's pulse rate exactly. Instead, it may only be possible to record it as

$$Y = \text{Pulse Rate} = 64 \pm 1, \text{ say,}$$

i.e., we record

$$Y = \xi = [a, b] = [63, 65].$$

Or, a weighing scale can only record the weight of a person (or, parcel, or cargo, etc.) to an accuracy within 5 units, i.e.,

$$Y = \text{Weight} = 175 \pm 5$$

pounds (or ounces, or tons, etc.). Again, this observation is interval-valued, i.e., we denote this as

$$Y = \xi = [a, b] = [170, 180].$$

Or, it may be that a person's weight naturally fluctuates by 2 to 3 pounds over a weekly period, say, so that the weight is better recorded as

$$Y = \text{Weight} = [130 \pm 2] = [128, 132]. \qquad \square$$

Conjunctive data are symbolic data but care is needed in describing possible outcomes, as Example 2.52 and Example 2.53 illustrate.

Example 2.52. Consider the dataset of Table 2.5. Here, the random variable $Y =$ Color takes values from $\mathcal{Y} = \{\text{red, blue,}\ldots\}$. As presented, the species of galah has colors

$$Y(w_3) = \{\text{pink, gray}\}.$$

That is, a galah has both pink and gray colors; the species is not such that some are all pink and some all gray.

Structured differently, as conjunctive data, a galah can be represented as

$$Y(w_3) = \{\text{pink} \wedge \text{gray}\}.$$

Notice that representing a galah species by

$$Y(w_3) = \{\text{pink, gray, pink} \wedge \text{gray}\}$$

is not correct, since this representation implies galahs can be all pink or all gray or both pink and gray. Therefore, care is needed. $\square$

Example 2.53. In contrast to the bird species of Example 2.52, suppose we are dealing with species of flowers. Then, for the particular case $Y =$ Color,

$$Y(\text{sweet pea}) = \{\text{red, white, red} \wedge \text{white,}\ldots\}$$

is correct, since (unlike the galah of Example 2.52) sweet peas come in many colors including some that are 'all red', some 'all white', and some that sport both red and white colors, and so on.

Therefore, in this case, to represent the color of sweet pea as

$$Y(\text{sweat pea}) = \{\text{red, white,}\ldots\}$$

would lose the conjunctive colors (such as red and white,...). $\square$

2.3 Comparison of Symbolic and Classical Analyses

As we have indicated at the beginning of this chapter, at its simplest level classical and symbolic data differ in that a classical data point takes as its value a single point in p-dimensional space, whereas a symbolic data point takes as its value a hypercube (or hyperrectangle) in p-dimensional space or is the Cartesian product of p distributions in p-dimensional space or a mixture of both. The classical single point value is a special case of a symbolic value.

In a different direction, we shall distinguish between the classical-valued observation as being attached to an individual i in the sample (or it can be the entire population) Ω of n individuals, and the symbolic-valued observation as belonging to a category or a concept w in the sample (or population) E of m concepts. However, as we have clearly seen in Section 2.1, an observation associated with an individual can itself be symbolic valued; and likewise, a particular symbolic value associated with a 'symbolic' category or concept can be classical valued. Further, although classical datasets typically relate to individuals $i \in \Omega$, these 'individuals' can be categories or concepts (such as in Example 2.43). Also, categories themselves are Cartesian products of qualitative variables (such as age $\times$ gender, and so on) that are fully modeled (since we know precisely how to define what is the extent of the respective categories), while a concept is an 'idea', an entity of the entire world population which entity can never be defined completely. As we have seen in Section 2.2, we start with individuals and move to concepts or categories.

Example 2.54. Rather than the individual credit card expenditures (see Table 2.3 and Example 2.8), we consider the concepts of the Person (who makes those expenditures, e.g., Jon). For example, an individual use of the card for food, say, is $x = 23$, while that for Jon is $\xi = [21, 29]$, say, in January. □

Example 2.55. Instead of focusing on each individual car accident, the insurance company is interested in accidents by type and make of car, or on the age group of the driver, and so on. □

Example 2.56. Table 2.17 makes a comparison of some individuals which produce classical observations with a symbolic concept counterpart. Thus, for example, the individual 'bird' (e.g., the cardinal in your backyard) is one of the set of birds that make up the concept 'species' of birds (e.g., species of Cardinals).

Or, the Town where the individual lives is part of the concept Region (of towns in that region). Thus, the utilities commission is not interested in the electricity consumption of a particular town, but is interested in the pattern or distribution of consumption across all the towns of the various regions. That is, instead of Paris = 27 units, Lyons = 16, etc., the commission wants to know that the consumption in the north region is $\xi \sim N(21, 7)$, say. Notice too that the 'Town' can itself be a concept composed of the individual streets which in turn can be a concept which comprises the individual houses. □

Table 2.17 Individuals versus concepts.

Classical Individuals (examples)	Symbolic Concepts (examples)
Bird (this cardinal, . . .)	Species (Cardinals, Wrens, . . .)
Town (Paris, Lyons, . . .)	Region (North, South, . . .)
Football Player (Smith, . . .)	Team (Niners, . . .)
Photo (of you, . . .)	Types of Photos (Family, Vacation, . . .)
Item Sold (hammer, phone, . . .)	Types of Stores (Electrical Goods, Auto, Hardware, . . .)
WEB trace (Jones, temperature today, . . .)	WEB Usages (Medical Info, Weather, . . .)
Subscription (to Time, . . .)	Consumer Patterns (Magazines, . . .)

The description of an individual in a classical database takes a unique value, e.g.,

$$\text{Weight} = x = 60, \quad \text{Flower Color} = x = \text{red}.$$

In contrast, the description of a concept is symbolic valued, e.g.,

$$\text{Weight} = [58, 62], \quad \text{Flower Color} = \{\text{red},\ 0.3;\ \text{green},\ 0.6;\ \text{red} \wedge \text{green},\ 0.1\}$$

that is, my weight fluctuates over a weekly period from 58 to 62 kilos, and 30% of the flowers are red, 60% are green, and 10% are both red and green in color. Apart from this description which identifies a classical- or symbolic-valued variable, there are some other important distinctions that emerge especially when analyzing the datasets.

A very important distinction between classical and symbolic data is that symbolic data have internal variation and structure not present or possible in classical data.

Example 2.57. Consider the classical value $x = 5$ and the symbolic value $\xi = [2, 8]$; and consider each as a sample of size $n = m = 1$.

We know that the classical sample variance for the single observation $x = 5$ is

$$S_x^2 = 0.$$

We shall see later (from Equation (3.40)) that the symbolic sample variance for the single observation $\xi = [2, 8]$ is

$$S_\xi^2 = 3 \neq 0.$$ □

This result in Example 2.57 reflects the fact that a single symbolic data point has its own internal variation. This internal variation must be taken into account when analyzing symbolic data. We shall also see later that in the derivation of Equation (3.40), it was assumed that individual description values are uniformly distributed across each interval. Notice that a classical random variable uniformly distributed as $X \sim U(2, 8)$ has as its variance $Var(X) = 3$, exactly the value we obtained for the internal variance of the symbolic observation in Example 2.57. Along similar but different lines, two symbolic values, such as [127, 133] and [124, 136], with the same midpoint (here, 130) but different interval lengths have different interval variation (here, 3 and 12, respectively). These differences are taken into account in subsequent symbolic analyses, whereas classical analyses which are based on the midpoint cannot distinguish between these two observations.

Another distinction is that statistical analyses on classical individuals and on symbolic concepts are not necessarily the same, even though the same variable may be the object of that analysis.

Example 2.58. Table 2.18 is a partial listing of 600 birds, where Y_1 = Category (or Species) of bird with $\mathcal{Y}_1$ = {ostrich, goose, penguin}, Y_2 = Is it a Flying Species, with $\mathcal{Y}_2$ = {No, Yes}, and Y_3 = Height in cm with $\mathcal{Y}_3 = \Re^+ = \{>0\}$. Suppose there are 400 geese, 100 ostriches, and 100 penguins. Then, a frequency histogram for Y_2 gives us 400 flying birds and 200 non-flying birds; see Figure 2.7(a) for the relative frequency.

Table 2.18 Birds: individual penguins, ostriches, and geese.

Bird	Category	Flying	Height
1	Penguin	No	80
2	Goose	Yes	70
⋮	⋮	⋮	⋮
600	Ostrich	No	125

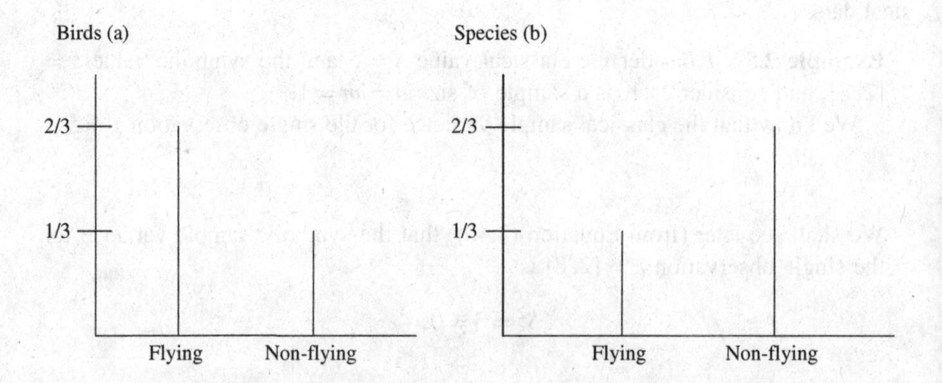

Figure 2.7 Flying and non-flying birds and species – relative frequencies.

Table 2.19 Concepts: geese, ostriches, penguins, species.

u	Species	Flying	Height	Color	Migratory
1	Geese	Yes	[60, 85]	{white, 0.3; black, 0.7}	Yes
2	Ostrich	No	[85, 160]	{white, 0.1; black, 0.9}	No
3	Penguin	No	[70, 95]	{white, 0.5, black, 0.5}	Yes

Suppose now that, rather than individual birds, it is of more interest to focus on the species of birds. The new statistical units are now geese, ostriches, and penguins, with these species constituting the concepts. Suppose the descriptions for these categories produce the symbolic data of Table 2.19. Now, a frequency histogram for Y_2 gives us one flying species and two non-flying species; see Figure 2.7(b) for the relative frequency. Notice that the relative frequency of flying and non-flying Y_2 values for the histogram of the species (concepts) is the reverse of that for the birds (individuals). □

A consequence of this distinction is that care is needed as to what entity is being analyzed. In Figure 2.7, while it was the same variable (Y_2, flying or not), the histogram differed for individuals (birds) from that for concepts (species). If, however, calculations for the histogram for species were weighted by the number of individuals that make up each concept category, then the histogram of Figure 2.7(a) would be recovered.

Therefore, it is important to be aware of the differences between classical and symbolic data, especially the presence or not of internal variation, the distinction between analyses on individuals or on concepts, and the need for logical dependency rules to maintain data integrity and structure. Thus, classical analyses and symbolic analyses are different but they are also complementary to each other.

Another distinction between classical and symbolic data is that additional variables (referred to as conceptual variables) that are symbolic valued can be added to symbolic datasets but not easily to classical datasets. This can be localized to one or two variables, or could be a result of merging two or more datasets.

Example 2.59. For the birds of Table 2.18 and Table 2.19, the random variable $Y_4 =$ Color can be easily added to Table 2.19. In this case, each species is a mixture of black and white, with, for example, penguins being 50% white and 50% black. This variable only takes symbolic values, and so cannot be added to Table 2.18 if we wish to keep it as a classical data table (except by the addition of numerous variables, one for each color etc.). On the other hand, whether a bird/species is migratory (Y_5) or not can be easily added to both datasets. □

Example 2.60. Suppose an analysis of the health data of Table 2.1 was to focus on lung cancer, with concern related to incidence by geography (represented by the random variable $Y_1 =$ City) and the extent of pollution present at these regions. Suppose interest is on the concept State (e.g., Massachusetts = {Amherst, Boston, Lindfield, Lynn, . . . }). Then, with State being the underlying concept of interest, aggregation over the dataset produces symbolic data such as that in Table 2.20. Note that instead of State, the concept could be City.

Table 2.20 Lung cancer treatments – by State

State	Y_{30} = # Lung Cancer Treatments
Massachusetts	{0, 0.77; 1, 0.08; 2, 0.15}
New York	{0, 0.50; 4, 0.50}
⋮	

Suppose further that a different dataset, such as that of Table 2.21, contains coded acid rain pollution indices for sulphates, nitrates (in kilograms/hectares). (These data are extracted and adapted from the US Environmental Protection Agency webpage: http://www.epa.gov.)

Table 2.21 Acid rain pollution measures.

State	Sulphate	Nitrate
Massachusetts	[9, 12]	[10, 12]
New York	[18, 26]	[16, 20]
⋮		

Merging the two datasets is easily accomplished (see Table 2.22), so allowing for analyses that might seek relationships between, for example, pollution and certain illnesses such as lung cancer. In this case, the pollution variable would be the added conceptual variable.

Table 2.22 Merged symbolic dataset.

State	Lung Cancer	Sulphate	Nitrate
Massachusetts	{0, 0.77; 1, 0.08; 2, 0.15}	[9, 12]	[10, 12]
New York	{0, 0.50; 4, 0.50}	[18, 26]	[16, 20]

□

There are essentially three steps in symbolic data analysis. The first is to determine the categories and/or concepts of interest and thence to build the symbolic dataset for these categories using if necessary data management algorithms (e.g., DB2S0 in SODAS) along the lines described earlier in Section 2.2. The second is to augment the symbolic dataset with additional classical- or symbolic-valued variables of interest that relate to the concepts in the symbolic database.

The third is to extract knowledge from the symbolic dataset using statistical tools, which tools minimally are extensions of classically based statistical methodologies to their symbolically based counterparts.

While there has been more than a century of effort producing an enormous range of statistical methodologies for classical data, there are comparatively few methodologies yet available for analyzing symbolic data. The following chapters will expand on many of those currently available. Two features immediately distinguish themselves. The first is that while any specific symbolic methodology seems intuitively 'correct' at least when compared to its corresponding counterpart classical method, by and large these symbolic methodologies still require rigorous mathematical proof of their validity. The second feature is an awareness of the need for many more methodologies to be developed for symbolic data. The field is open and ripe for considerable progress.

Before exploring these symbolic methodologies in detail, it is important to note that attempts to recode symbolic-valued variables into classical-valued ones to use in a classical analysis will in general result in both a loss of information and analytical solutions that are 'inferior', i.e., less informative, than would have pertained from the appropriate symbolic analysis. We close with three different examples illustrating these distinctions.

Example 2.61. Consider the symbolic data of Table 2.23 with observations representing $Y =$ Cost of articles sold at three department stores. If the interval end points were recoded to $X_1 =$ Minimum Cost and $X_2 =$ Maximum Cost, then the observed frequencies, F, for histogram intervals $I_1 = [20, 25)$, $I_2 = [25, 30)$, $I_3 = [30, 35)$, $I_4 = [35, 40]$ are, respectively,

$$X_1 : \{F[20, 25) = 2,\ F[25, 30) = 1,\ F[30, 35) = 0,\ F[35, 40] = 0\}$$

$$X_2 : \{F[20, 25) = 0,\ F[25, 30) = 1,\ F[30, 35) = 1,\ F[35, 40] = 1\}.$$

Table 2.23 Costs.

Category	Cost
Store A	[22, 27]
Store B	[26, 34]
Store C	[24, 36]

However, by constructing a symbolic histogram, we use all the values in the intervals and not just their minimum and maximum values. Therefore, utilizing the methodology described later in Section 3.3, we obtain for these symbolic data the histogram over the same histogram intervals $I_1, \ldots, I_4$ as seen in Figure 2.8 with symbolic observed frequencies F_s:

$$Y : \{F_s[20, 25) = 0.483,\ F_s[25, 30) = 1.517,\ F_s[30, 35) = 0.917,\ F_s[35, 40] = 0.083\}.$$

Notice that the symbolic observed frequencies F_s are not necessarily integer values, unlike classical observed frequencies which are always integers, but that they do sum correctly to the total number of observations (here, 3).

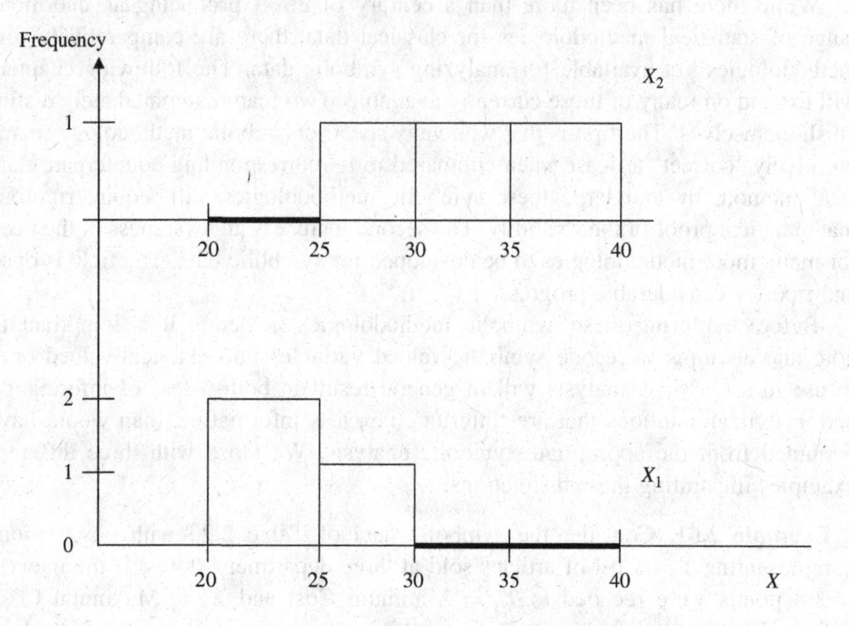

Figure 2.8(a) Classical histograms.

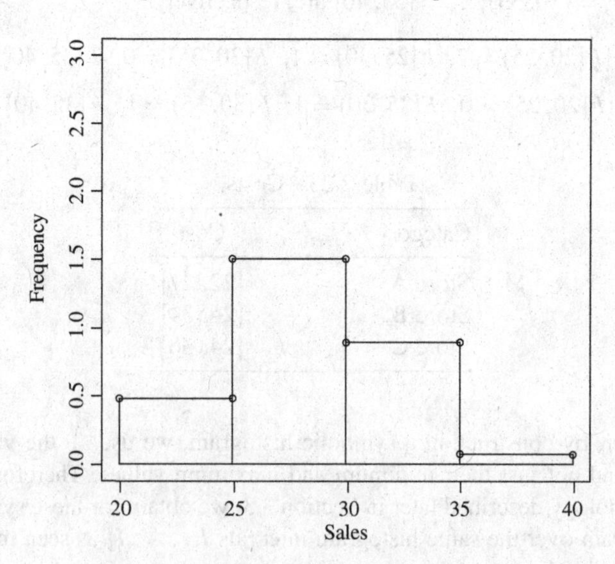

Figure 2.8(b) Symbolic histogram.

□

Example 2.62. The data in Table 2.24 represent the ranges of the random variables Y_1 = Age, Y_2 = Weight (in pounds), and Y_3 = Speed (time to run 100 yards) for members of teams (the concepts) that played in a local 'pickup' league, with team performances over a season being adjudged to be very good, good, average, fair, and poor, respectively. A classical principal component analysis on the classical midpoint values for these observations was performed. A plot of these observations with the first and second principal components as axes is shown in Figure 2.9(a). By using the methods of Chapter 5, a symbolic principal component analysis on the symbolic-valued observations gave the results shown in Figure 2.9(b). Notice that the symbolic principal component plots, being rectangle themselves, reflect the interval variation of the original symbolic values. In particular, notice that the very good team consisted of players who are more homogeneous, whereas the poor team had players who varied more widely in age, weight, and speed of movement (some young and inexperienced, some older and slower). That is, the symbolic results are more informative than are the classical results and reflect the internal variation inherent to the data.

Table 2.24 Pickup league teams.

u	Team Type	Age Y_1	Weight Y_2	Speed Y_3
w_1	Very Good	[20, 28]	[160, 175]	[10.6, 14.1]
w_2	Good	[18, 34]	[161, 189]	[10.5, 15.9]
w_3	Average	[15, 35]	[148, 200]	[10.8, 18.3]
w_4	Fair	[18, 44]	[153, 215]	[11.2, 19.7]
w_5	Poor	[16, 48]	[145, 220]	[12.3, 22.8]

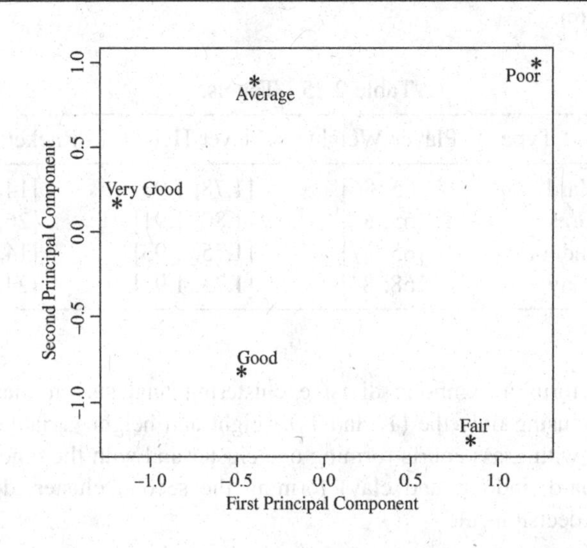

Figure 2.9(a) Classical principal component analysis on teams.

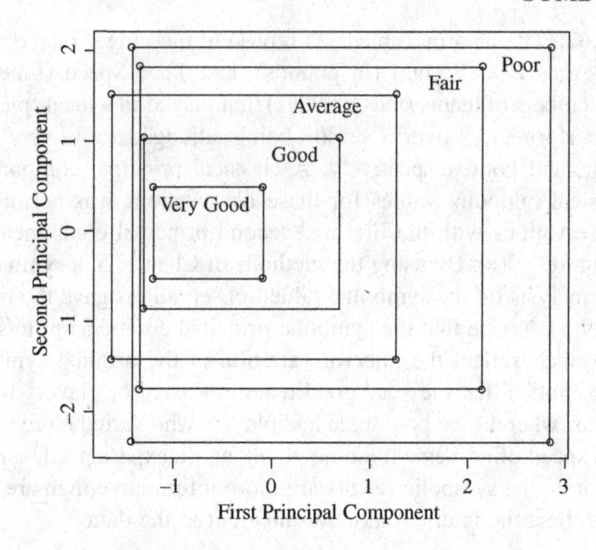

Figure 2.9(b) Symbolic principal component analysis on teams.

Example 2.63. Consider the symbolic data of Table 2.25 which gives values for Y_1 = Weight (in kilograms), Y_2 = Height (in meters) and Y_3 = Racket Tension for tennis players who participated in tournaments played on hard courts, grass courts, indoor courts, and clay courts. Here, type of court is the concept/category. These data were extracted from a larger dataset containing 22 variables and can be found on the SODAS webpage www.ceremade.dauphine.fr/%7Etouati/sodas-pagegarde.htm.

Table 2.25 Tennis.

w_u	Court Type	Player Weight	Player Height	Racket Tension
w_1	Hard	[65, 86]	[1.78, 1.93]	[14, 99]
w_2	Grass	[65, 83]	[1.80, 1.91]	[26, 99]
w_3	Indoor	[65, 87]	[1.75, 1.93]	[14, 99]
w_4	Clay	[68, 84]	[1.75, 1.93]	[24, 99]

If we perform a symbolic divisive clustering analysis on these data (see Section 7.4), using only the (Y_1 and Y_2) weight and height variables, we obtain two clusters with grass courts forming one cluster and with the other three types of courts (hard, indoor, and clay) forming the second cluster, depending on whether the decision rule

$$\text{Weight} = Y_1 \leq 74.75$$

is, or is not, satisfied. That is, grass court players tend to weigh less than 74.75 kilograms; see Figure 2.10(a).

Suppose instead that the symbolic variables were replaced by the classical variables

$$X_1 = \text{Min Weight}, \quad X_2 = \text{Max Weight}$$

$$X_3 = \text{Min Height}, \quad X_4 = \text{Max Height}$$

$$X_5 = \text{Min Tension}, \quad X_6 = \text{Max Tension}.$$

Then, a classical analysis on the $X_1, \ldots, X_4$ variables produces two clusters, the first cluster containing grass and clay courts and the second containing hard and indoor courts, depending on whether the decision rule

$$\text{Max Weight} = X_2 \leq 85$$

is, or is not, satisfied; see Figure 2.10(b).

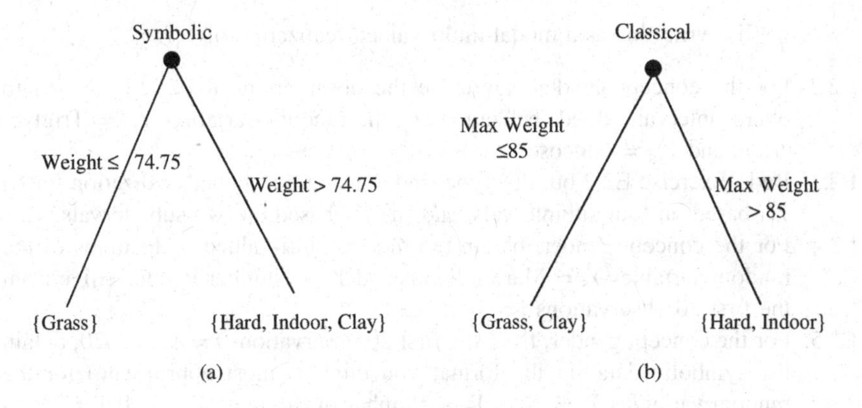

Figure 2.10 Clustering methods.

Not only do different clusters emerge, but we see that the classical analysis can only use the interval end point values, whereas the symbolic analysis is able to utilize the full range of variable values. Therefore, we can obtain the decision rule with weight at 74.75, a value not possible by the classical recoding approach.

On the other hand, when all three variables are used, both analyses produce two clusters each with the same two types of courts, but now the explanatory variable is Y_3 = racket tension. For the symbolic analysis, hard and indoor courts form one category, with the decision rule

$$\text{Tension} = Y_3 \leq 59.00$$

holding, while, when $Y_3 > 59.00$, we have grass and clay courts forming the second cluster. In contrast, the classical analysis produces the decision rule

$$X_3 = \text{Min Tension} \leq 19.00.$$

In this case, the same clusters are formed, but again the decision rule for the classical analysis is limited to only the minimum and maximum values, and does not use the full range of possible values. □

Exercises

Exercises E2.1–E2.15 relate to the data of Tables 2.1.

E2.1. Consider the concept gender represented by the random variable Y_4. For the observations $i = 1, \ldots, 15$, find the list of cities (from realizations of the random variable Y_1) associated with each gender.

(i) Write this list as an unweighted multi-valued realization of Y_4; then

(ii) write this as a modal multi-valued realization of Y_4.

E2.2. For the concept gender, aggregate the observations for $i = 1, \ldots, 15$ to obtain interval-valued realizations of the random variables Y_{17} = Triglyceride, and Y_{18} = Glucose Level.

E2.3. Redo Exercise E2.2 but this time find a histogram-valued realization for (i) Y_{17} based on four subintervals, and (ii) Y_{18} based on two subintervals.

E2.4. For the concept gender, obtain the modal multi-valued realizations of the random variables Y_5 = Marital Status and Y_6 = Number Parents Alive from the first 20 observations $i = 1, \ldots, 20$.

E2.5. For the concept gender, from the first 20 observations $i = 1, \ldots, 20$, obtain the symbolic data (in the format you think is most appropriate) for the random variables Y_3 = Age, Y_7 = Number of Siblings, Y_{14} = HDL Cholesterol Level, and Y_{24} = Red Blood Cell Measure.

E2.6. Repeat Exercise E2.5 using all 51 observations.

E2.7. Consider the concept age group with categories: age < 20, 20–29, 30–39, 40–49, 50–59, 60–69, 70–79, age ≥ 80. Using the entire dataset, obtain the multi-valued realizations associated with each concept for the random variables Y_{28} = Cancer Diagnosed and Y_{30} = Lung Cancer Treatments. Obtain these as (i) non-model multi-valued realizations, and (ii) modal multi-valued realizations of Y_{28} and Y_{29}, respectively.

E2.8. For the concept age group described in Exercise E2.7, build the interval-valued symbolic data table for the random variables Y_{23} = White Blood Cell Measure, Y_{24} = Red Blood Cell Measure, and Y_{25} = Hemoglobin Level, based on the observations $i = 1, \ldots, 20$.

E2.9. Redo Exercise E2.8 using all 51 observations.

E2.10. For the concept age group, obtain histogram-valued realizations for the random variables used in Exercise E2.8, using all 51 observations.

E2.11. Redo Exercise E2.7 but for the concept age group × gender.

E2.12. Redo Exercise E2.8 but for the concept age group × gender.

E2.13. Redo Exercise E2.9 but for the concept age group × gender.

E2.14. Redo Exercise E2.10 but for the concept age group × gender.

E2.15. Consider the age group × gender concepts associated with Tables 2.1, and the three random variables Y_{23}, Y_{24}, and Y_{25} used in Exercise E2.10. Instead of the relative frequencies used there to build histogram-valued realizations, build the modal-valued symbolic data where now the non-negative measure $\{\pi_k\}$ associated with each subinterval $\{\eta_k\}$ reflects the cumulative distribution associated with each random variable (recall Definition 2.4).

E2.16. Take the credit card data in Table 2.4. Suppose the month 'January' referred to the six months 'January June' and 'February' referred to 'July-December'. Based on these data, give a description of (i) Jon, (ii) Tom, (iii) Leigh.

E2.17. Refer to Exercise E2.16. The credit card company noticed that Jon's credit card was used to purchase travel in the amounts of (a) \$226, (b) \$566, and (c) \$335, all on one day. Would the credit card company be concerned? Why/why not?

E2.18. Refer to Exercise E2.16. Suppose Tom's credit card was used to purchase gas in the amounts (a) \$45, (b) \$55, and (c) \$58, all in one day. Would the credit card company be concerned? Why/why not?

E2.19. Refer to Exercise E2.16. Would the credit card company take particular interest if suddenly Leigh's credit card showed clothes purchases of (a) \$185, (b) \$153, and (c) \$80 on the same day? Why/why not?

References

Aristole (IVBC; 1994). Des Categories De l'Interpretation. Organon. *Librarie Philosophique Journal Vrin.*

Arnault, A. and Nicole, P. (1662). *LaLogique ou l'art Depensur.* Reprinted by Froman, Stuttgart (1965).

Bock, H.-H. and Diday, E. (eds.) (2000). *Analysis of Symbolic Data: Exploratory Methods for Extracting Statistical Information from Complex Data.* Springer-Verlag, Berlin.

Brito, M. P. (1994). Order Structure of Symbolic Assertion Objects. *IEEE Transactions on Knowledge and Data Engineering* 6, 830–835.

Changeux, J. P. (1983). *L'thomme Neuronal*, Collection Pluriel, Fayard France.

Choquet, G. (1954). Theory of Capacities. *Annals Institute Fourier* 5, 131–295.

Diday, E. (1987). Introduction à l'Approche Symbolique en Analyse des Données. *Premières Journées Symbolique-Numérique*, CEREMADE, Université Paris, Dauphine, 21–56.

Diday, E. (ed.) (1989). *Data Analysis, Learning Symbolic and Numerical Knowledge.* Nova Science, Antibes.

Diday, E. (1990). Knowledge Representation and Symbolic Data Analysis. In: *Knowledge Data and Computer Assisted Decisions* (eds. M. Schader and W. Gaul). Springer-Verlag, Berlin, 17–34.

Diday, E. (1995). Probabilist, Possibilist and Belief Objects for Knowledge Analysis. *Annals of Operations Research* 55, 227–276.

Diday, E. (2005). Categorization in Symbolic Data Analysis. In: *Handbook of Categorization in Cognitive Science* (eds. H. Cohen and C. Lefebvre). Elsevier, Amsterdam, 845–867.

Diday, E. and Emilion, R. (2003). Maximal Stochastic Galois Lattice. *Journal of Discrete Applied Mathematics* 127, 271–284.

Dubois, D. and Prade, H. (1988). *Possibility Theory.* Plenum, New York.

Esposito, F., Malerba, D., and Semeraro, G. (1991). Flexible Matching for Noisy Structural Descriptions. In: *Proceedings of the Conference on Artificial Intelligence* (eds. J. Mylopoulos and R. Reiter). Morgan Kaufmann, San Francisco, 658–664.

Falduti, N. and Taibaly, H. (2004). Etude des Retards sur les Vols des Compagnies Aériennes. Report CEREMADE, Université Paris, Dauphine, 63 pages.

Felsenstein, J. (1983). Statistical Inference of Phylogenies (with Discussion). *Journal of the Royal Statistical Society A* 146, 246–272.

Fuller, W. A. (1996). *Introduction to Statistical Time Series* (2nd ed.). John Wiley & Son, Inc., New York.

Pollaillon, G. (1998). Organization et Interprétation par les Treillis de Galois de Données de Type Multivalué, Intervalle ou Histogramme. Doctoral Dissertation, Université Paris, Dauphine.

Rosch, E. (1978). Principles of Categorization. In: *Cognition and Categorization* (ed. E. Rosch and B. B. Lloyd). Erlbaum Associates, Hillsdale, NJ, 27–48.

Schafer, G. (1976). *A Mathematical Theory of Evidence.* Princeton University Press, Princeton, NJ.

US Census Bureau (2004). *Statistical Abstract of the United States: 2004–2005* (124th ed.). Washington, DC, 575.

Wagner, H. (1973). Begriff. In: *Hanbuck Philosophischer Grundbegriffe* (eds. H. Krungs, H. M. Baumgartner, and C. Wild). Kosel, Munich, 191–209.

Wille, R. (1989). Knowledge Acquisition by Methods of Formal Concepts Analysis. In: *Data Analysis, Learning Symbolic and Numeric Knowledge* (ed. E. Diday). Nova Sciences Commack, NY, 365–380.

3

Basic Descriptive Statistics: One Variate

3.1 Some Preliminaries

Basic descriptive statistics for one random variable include frequency histograms and sample means and variances. We consider symbolic data analogues of these statistics, for multi-valued and interval-valued variables and modal variables. Symbolic data often exist in the presence of so-called rules. This is especially so when such data arise from the aggregation of other data, but rules are also relevant by the inherent nature of symbolic data. Therefore, descriptive statistics will be considered both in the presence of rules and without rules. As we develop these statistics, let us recall Example 2.58 which illustrated the need to distinguish between the levels (e.g., structure) in symbolic data when constructing histograms compared with that for a standard histogram.

We consider univariate statistics (for the $p \geq 1$ variate data) in this chapter and bivariate statistics in Chapter 4. For integer-valued and interval-valued variables, we follow the approach adopted by Bertrand and Goupil (2000). De Carvalho (1994, 1995) and Chouakria *et al.* (1998) have used different but equivalent methods to find the histogram and interval probabilities (see Equation (3.20) below) for interval-valued variables.

Virtual extensions

We first need to introduce the concept of virtual extensions. Recall from Chapter 2 that the symbolic description of an observation $w_u \in E$ (or equivalently, $i \in \Omega$)

Symbolic Data Analysis: Conceptual Statistics and Data Mining L. Billard and E. Diday

was given by the description vector $\boldsymbol{d}_u = (\xi_{u1}, \ldots, \xi_{up})$, $u = 1, \ldots, m$, or more generally, $\boldsymbol{d} \in (D_1, \ldots, D_p)$ in the space $\mathcal{D} = \times_{j=1}^{p} D_j$, where in any particular case the realization of Y_j may be an x_j as for classical data or an ξ_j of symbolic data.

Definition 3.1: Individual descriptions, denoted by x, are those descriptions for which each D_j is a set of one value only, i.e., $\boldsymbol{x} = (x_1, \ldots, x_p) \equiv \boldsymbol{d} = (\{x_1\}, \ldots, \{x_p\})$, $\boldsymbol{x} \in \mathcal{X} = \times_{j=1}^{p} \mathcal{Y}_j$. □

Example 3.1. Suppose the multi-valued random variable Y = bird color takes the particular symbolic value $Y = \xi$ = {red, blue, green}. Then, the individual descriptions associated with this ξ are x = red, x = blue, and x = green. □

Example 3.2. Consider the $p = 2$ dimensional variate (Y_1, Y_2) with Y_1 representing the type of car owned by a family with Y_1 taking values in $\mathcal{Y}_1$ = {Austin, Chevy, Corolla, Volvo, ...} and Y_2 being an indicator of having a garage with Y_2 taking values in $\mathcal{Y}_2$ = {No = 0, Yes = 1}. Then, the symbolic data point $(Y_1, Y_2) = \xi$ = ({Corolla, Volvo}, {1}) has associated with it the two individual descriptions x = (Corolla, 1) and x = (Volvo, 1). □

Example 3.3. Let (Y_1, Y_2) be the two-dimensional interval-valued variables where Y_1 = Height in inches and Y_2 = Weight in pounds. Then, the individual descriptions associated with the symbolic point $(Y_1, Y_2) = \xi = ([65, 70], [130, 150])$ are all the points $x = (x_1, x_2)$ in the rectangle $[65, 70] \times [130, 150]$. □

The calculation of the symbolic frequency histogram involves a measure of the number of individual descriptions that match certain implicit logical dependencies in the data. A logical dependency can be represented by a rule v,

$$v : [x \in A] \Rightarrow [x \in B] \tag{3.1}$$

for $A \subseteq D$, $B \subseteq D$, and $x \in \mathcal{X}$ and where v is a mapping of $\mathcal{X}$ onto {0,1} with $v(x) = 0$ (1) if the rule is not (is) satisfied by x. It follows that an individual description vector x satisfies the rule v if and only if $x \in A \cap B$ or $x \notin A$. Rules typically impact more than one variable.

Example 3.4. Suppose $x = (x_1, x_2) = (10, 0)$ is an individual description of Y_1 = Age and Y_2 = Number of Children for an individual $i \in \Omega$, and suppose there is a rule that says

$$v : \text{If } A = \{Y_1 \leq 12\}, \text{ then } B = \{Y_2 = 0\}. \tag{3.2}$$

Then, the rule that an individual whose age is less than 12 implies the individual has had no children is logically true, whereas an individual whose age is under 12 but has had two children is not logically true. Therefore, for a symbolic data point given as $Y(w) \equiv (Y_1(w), Y_2(w)) = \{(10, 26), 1\}$ with two individual descriptions $x_1 = (10, 1)$ and $x_2 = (26, 1)$, the first description x_1 cannot be logically true, while the second description x_2 is logically true under this rule. □

Clearly, then, this operation is mathematically cleaning the data (so to speak) by identifying only those values which make logical sense (by satisfying the logical dependency rule of Equation (3.1)). Thus, in Example 3.4 the data value $(Y_1, Y_2) = (10, 1)$, which records both the age as 10 years and the number of children as one, is identified as an invalid data value (under the prevailing circumstances as specified by the rule v of Equation (3.2)) and so would not be used in any analysis to calculate the basic descriptive statistics. (While not attempting to do so here, this identification does not preclude inclusion of other procedures which might subsequently be engaged to input what these values might have been.) For small datasets, it may be possible to 'correct' the data visually (or some such variation thereof). For very large datasets, this is not always possible; hence a logical dependency rule to do so mathematically/computationally is essential.

An equally important use of this rule operation occurs when dealing with aggregated large datasets. Even if there are no miskeyed data values in the original (classical or symbolic) dataset, aggregation of these individual observations into classes (e.g., by geographical region, work characteristics, and so forth) perforce produces symbolic variables some of which might have individual description values that are not feasible (or allowable).

Example 3.5. The classical data of Table 3.1 when aggregated produce the symbolic data point $Y(w) = (Y_1(w), Y_2(w)) = (\{10, 26\}, \{0, 1\})$. The individual descriptions associated with $Y(w)$ are $x_1 = (10, 0)$, $x_2 = (10, 1)$, $x_3 = (26, 0)$, $x_4 = (26, 1)$; all arise from the ensuing aggregation. Of these, x_2 cannot be logically true under the rule $A = \{\text{Age} \leq 12\}$ and $B = \{\text{Number of Children} = 0\}$.

Table 3.1 Age and children.

	Individual	Age	Number of Children
Classical	1	10	0
	2	26	1
	3	10	0
	4	10	0
	5	26	1
Symbolic	u	(10, 26)	(0, 1)

□

Example 3.6. Consider the interval-valued random variable (Y_1, Y_2) with $Y_1 =$ Cost of medication and $Y_2 =$ Insurance coverage, with the rule v_1 that

$$v_1 : A_1 = \{Y_1 = \text{Cost} < \$200\} \text{ and } B_1 = \{Y_2 = \$0\}; \tag{3.3}$$

that is, for medication costs under \$200, there is no insurance payment. Then, suppose a particular symbolic data point has the value $Y(w_u) = (Y_1(w_u), Y_2(w_u)) = \xi_u = ([100, 300], [0, 400])$. The individual descriptions for

$Y(w_3)$ are those points in the rectangle $[100, 300] \times [0, 400]$; see Figure 3.1(a). The rule v_1 implies that only the shaded portion of this rectangle can be logically true, i.e., those individual descriptions for which (x_1, x_2) takes values in the rectangle $[200, 300] \times [0, 400]$.

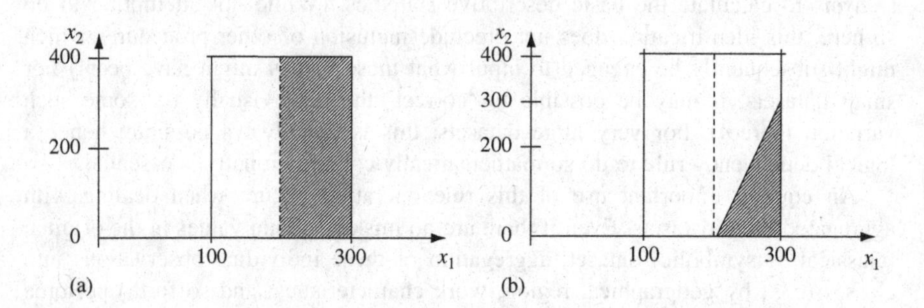

Figure 3.1 (a) Rule v_1. (b) Rule $v = (v_1, v_2)$. □

There can be more than one rule.

Example 3.7. Suppose in the medication cost–insurance of Example 3.6, a second rule limits coverage so as not to exceed actual costs, i.e., there is a v_2 which stipulates that

$$v_2 : A_2 = \{Y_1 = \text{Cost}\} \text{ and } B_2 = \{Y_2 \leq Y_1\}. \tag{3.4}$$

Under the set of rules $v = (v_1, v_2)$, the individual descriptions x that can be logically true are those in the triangle $(x_1, x_2) = \{100 \leq x_1 \leq 300, x_2 \leq x_1\}$; see Figure 3.1(b). □

This formulation of the logical dependency rule is sufficient for the purposes of establishing basic descriptive statistics. Verde and De Carvalho (1998) discuss a variety of related rules (such as logical equivalence, logical implication, multiple dependencies, hierarchical dependencies, and so on). We have the following formal definition.

Definition 3.2: The **virtual description**, $vir(d)$ of the description vector d is the set of all individual description vectors x that satisfy all the (logical dependency) rules v in $\mathcal{X}$. A typical rule can be written as in Equation (3.1). We write this as, for $V_{\mathcal{X}}$ the set of all rules v operating on $\mathcal{X}$,

$$vir(d) = \{x \in D;\ v(x) = 1, \text{ for all } v \text{ in } V_{\mathcal{X}}\}. \tag{3.5}$$

□

Therefore the symbolic value $Y(w)$ in Example 3.4 has as its virtual description the set $\{(26, 1)\}$; and for $Y(w)$ in Example 3.5, the virtual description is the set $\{(10, 0), (26, 0), (26, 1)\}$. These ideas are more fully expanded in the examples below.

Each individual $w_u \in E$ takes observation value or realization $\boldsymbol{\xi}_u = (\xi_{u1}, \ldots, \xi_{up})$ on the random variable $\boldsymbol{Y} = (Y_1, \ldots, Y_p)$. In the sequel, for ease of notation, we sometimes identify a given $w_u \in E$ by $u \in E$.

3.2 Multi-Valued Variables

Suppose we want to find the frequency distribution for the particular multi-valued symbolic random variable $Y_j = Z$ which takes possible values $\xi \in \mathcal{Z} \equiv \mathcal{Y}_j$. These can be categorical values (e.g., types of cancer), or any form of discrete random variable.

Definition 3.3: For a set of multi-valued observations $(w_1, \ldots, w_m) = E$ with realizations $\boldsymbol{\xi}_u$, the **observed frequency** of a particular $\mathcal{Z}$ is

$$O_Z(\xi) = \sum_{u \in E} \pi_Z(\xi; u) \tag{3.6}$$

where

$$\pi_Z(\xi; u) = \frac{|\{x \in vir(d_u) | x_Z = \xi\}|}{|vir(d_u)|} \tag{3.7}$$

is the percentage of the individual description vectors in $vir(d_u)$ such that $x_Z = \xi$, and where $|A|$ is the number of individual descriptions in the space A. In the summation in Equation (3.6), any u for which $vir(d_u)$ is empty is ignored. □

We note that this observed frequency $O_Z(\xi)$ is a positive real number and not necessarily an integer as for classical data; see Example 3.8. In the classical case, $|vir(d_u)| = 1$ always and so is a special case of Equation (3.7). We can easily show that

$$\sum_{\xi \in \mathcal{Z}} O_Z(\xi) = m' \tag{3.8}$$

where $m' = (m - m_0)$ with m_0 being the number of u for which $|vir(d_u)| = 0$.

Definition 3.4: For a multi-valued symbolic variable Z, taking values $\xi \in \mathcal{Z}$, the **empirical frequency distribution** is the set of pairs $[\xi, O_Z(\xi)]$ for all $\xi \in \mathcal{Z}$, and the **relative frequency distribution** or **frequency histogram** is the set for all $\xi \in \mathcal{Z}$,

$$[\xi, (m')^{-1} O_Z(\xi)]. \tag{3.9}$$

□

The following definitions follow readily.

Definition 3.5: When Z is an ordered random variable, its **empirical distribution function** is given by

$$F_Z(\xi) = \frac{1}{m'} \sum_{\xi_k \le \xi} O_Z(\xi_k). \tag{3.10}$$

□

When the possible values ξ for the multi-valued symbolic variable $Y_j = Z$ are quantitative, we can define a symbolic mean, variance, and median, as follows.

Definition 3.6: For a multi-valued quantitative variable Z, the **symbolic sample mean** is

$$\bar{Z} = \frac{1}{m'} \sum_{\xi_k \in \mathcal{Z}} \xi_k O_Z(\xi_k); \tag{3.11}$$

the **symbolic sample variance** is

$$S_Z^2 = \frac{1}{m'} \sum_{\xi_k \in \mathcal{Z}} [\xi_k - \bar{Z}]^2 O_Z(\xi_k); \tag{3.12}$$

and the **symbolic median** is the value ξ for which

$$F_Z(\xi) = 1/2. \tag{3.13}$$

□

Example 3.8. To illustrate, suppose a large dataset (consisting of patients served through a particular medical insurance company, say) was aggregated in such a way that it produced the data of Table 3.2 (from Billard and Diday, 2003, Table 3) representing the outcomes relative to the presence of Cancer (Y_1) with $\mathcal{Y}_1 = \{\text{No} = 0, \ \text{Yes} = 1\}$ and Number (Y_2) of cancer-related treatments with $\mathcal{Y}_2 = \{0, 1, 2, 3\}$ on $m = 9$ categories. Thus, for example, $d_1 = (\{0, 1\}, \{2\})$ is the description vector for the observation in row 1. Hence, for the individuals represented by this description, the observation $Y_1 = \{0, 1\}$ tells us that either

Table 3.2 Cancer incidence.

w_u	Y_1	Y_2	$vir(d_u)$	$\|vir(d_u)\|$	$\|C_u\|$
w_1	{0,1}	{2}	{(1,2)}	1	128
w_2	{0,1}	{0,1}	{(0,0), (1,0), (1,1)}	3	75
w_3	{0,1}	{3}	{(1,3)}	1	249
w_4	{0,1}	{2,3}	{(1,2), (1,3)}	2	113
w_5	{0}	{1}	ϕ	0	2
w_6	{0}	{0,1}	{(0,0)}	1	204
w_7	{1}	{2,3}	{(1,2), (1,3)}	2	87
w_8	{1}	{1,2}	{(1,1), (1,2)}	2	23
w_9	{1}	{1,3}	{(1,1), (1,3)}	2	121

some individuals have cancer and some do not or we do not know the precise Yes/No diagnosis for the individuals classified here, while the observation $Y_2 = \{2\}$ tells us that all individuals represented by d_1 have had two cancer-related treatments. In contrast, $d_7 = (\{1\}, \{2, 3\})$ represents individuals all of whom have had a cancer diagnosis ($Y_1 = 1$) and have had either two or three treatments, $Y_2 = \{2, 3\}$.

Suppose further that there is a logical dependency

$$v : Y_1 \in \{0\} \Rightarrow Y_2 = \{0\}, \tag{3.14}$$

i.e., if no cancer has been diagnosed, then there must have been no cancer treatments. Notice that

$$y_1 \in \{0\} \Rightarrow A = \{(0,0), (0,1), (0,2), (0,3)\}$$

and

$$y_2 \in \{0\} \Rightarrow B = \{(0,0), (1,0), (2,0), (3,0)\}.$$

From Equation (3.14), it follows that an individual description x which satisfies this rule is

$$x \in A \cap B = \{(0,0)\} \text{ or } x \notin A, \text{ i.e., } x \in \{(1,0), (1,1), (1,2), (1,3)\}.$$

Then all possible cases that satisfy the rule are represented by the set

$$C = \{(0,0), (1,0), (1,1), (1,2), (1,3)\}.$$

We apply Equation (3.14) to each d_u, $u = 1, \ldots, m$, in the data to find the virtual extensions $vir(d_u)$. Thus, for the first description d_1, we have

$$vir(d_1) = \{x \in \{0,1\} \times \{2\} : v(x) = 1\}.$$

The individual descriptions $x \in \{0,1\} \times \{2\}$ are (0, 2) and (1, 2), of which only one, $x = (1,2)$, is also in the space C. Therefore, $vir(d_1) = \{(1,2)\}$.

Similarly, for the second description d_2, we seek

$$vir(d_2) = \{x \in \{0,1\} \times \{0,1\} : v(x) = 1\}.$$

Here, the individual description vectors $x \in \{0,1\} \times \{0,1\}$ are (0,0), (0,1), (1,0), and (1,1) of which $x = (0,0)$, $x = (1,0)$, and $x = (1,1)$ are in C. Hence, $vir(d_2) = \{(0,0), (1,0), (1,1)\}$. The virtual extensions for all d_u, $u = 1, \ldots, 9$, are given in Table 3.2. Notice that $vir(d_5) = \phi$ is the null set, since the data value d_5 cannot be logically true in the presence of the rule v of Equation (3.14).

We can now find the frequency distribution. Suppose we first find this distribution for Y_1. By Equation (3.6), we have observed frequencies

$$O_{Y_1}(0) = \sum_{u \in E'} \frac{|\{x \in vir(d_u) | x_{Y_1} = 0\}|}{|vir(d_u)|}$$

$$= \frac{0}{1} + \frac{1}{3} + \frac{0}{1} + \frac{0}{2} + \frac{1}{1} + \frac{0}{2} + \frac{0}{2} + \frac{0}{2} = 4/3,$$

and, likewise, $O_{Y_1}(1) = 20/3$, where $E' = E - (u = 5)$ and so $|E'| = 8 = m'$. Therefore, the relative frequency distribution for Y_1 is, from Equation (3.9),

$$\text{Rel. freq. } (Y_1) : [(0, O_{Y_1}(0)/m'), \ (1, O_{Y_1}(1)/m')] = [(0, 1/6), (1, 5/6)].$$

Similarly, we have observed frequencies for the possible Y_2 values $\xi = 0, 1, 2, 3$, respectively, as

$$O_{Y_2}(0) = \sum_{u \in E'} \frac{|\{x \in vir(d_u) | x_{Y_2} = 0\}|}{|vir(d_u)|}$$
$$= \frac{0}{1} + \frac{2}{3} + \frac{0}{1} + \frac{0}{2} + \frac{1}{1} + \frac{0}{2} + \frac{0}{2} + \frac{0}{2} = 5/3,$$
$$O_{Y_2}(1) = 4/3, \qquad O_{Y_2}(2) = 2.5, \qquad O_{Y_2}(3) = 2.5;$$

hence, the relative frequency for Y_2 is, from Equation (3.9),

$$\text{Rel. freq. } (Y_2) : [(0, 5/24), (1, 1/6), (2, 5/16), (3, 5/16)].$$

The empirical distribution function for Y_2 is, from Equation (3.10),

$$F_{Y_2}(\xi) = \begin{cases} 5/24, & \xi < 1, \\ 3/8, & 1 \le \xi < 2, \\ 11/16, & 2 \le \xi < 3, \\ 1, & \xi \ge 3. \end{cases}$$

From Equation (3.11), we have the symbolic sample mean of Y_1 as

$$\bar{Y}_1 = \frac{1}{8}\left(0 \times \frac{4}{3} + 1 \times \frac{20}{3}\right) = 5/6 = 0.833,$$

and, similarly, $\bar{Y}_2 = 83/48 = 1.729$. The symbolic sample variance, from Equation (3.12), for Y_1 is given by

$$S_1^2 = \frac{1}{8}\left[(-5/6)^2 \frac{4}{3} + (1 - 5/6)^2 \frac{20}{3}\right] = 0.124,$$

and, similarly, $S_2^2 = 1.198$. The median of Y_2 is 2 from Equation (3.13). □

Finally, we observe that we can calculate weighted frequencies, weighted means and weighted variances by replacing Equation (3.6) by

$$O_Z(\xi) = \sum_{u \in E} w_u^* \pi_Z(\xi, u) \tag{3.15}$$

with weights $w_u^* \ge 0$ and $\Sigma w_u^* = 1$. For example, if the objects $w_u \in E$ are classes C_u comprising individuals from the set of individuals $\Omega = \{1, \ldots, n\}$, a possible weight is $w_u^* = |C_u|/|\Omega|$ corresponding to the relative sizes of the class C_u, $u = 1, \ldots, m$. Or, more generally, if we consider each individual description vector x as an elementary unit of the object u, we can use weights, for $u \in E$,

$$w_u^* = \frac{|vir(d_u)|}{\sum_{u \in E} |vir(d_u)|}. \tag{3.16}$$

Example 3.9. Take Example 3.8 and suppose that the observations w_u of Table 3.2 represent classes C_u of size $|C_u|$, with $|\Omega| = 1000$, as shown in Table 3.2. Then, if we use the weights $w_u^* = |C_u|/|\Omega|$, we can show that

$$O_{Y_1}(0) = \frac{1}{1000}\left[128 \times \frac{0}{1} + 75 \times \frac{1}{3} + 249 \times \frac{0}{1} + \ldots + 121 \times \frac{0}{2}\right] = 0.229;$$

likewise

$$O_{Y_1}(1) = 0.771.$$

Hence, the relative frequency of Y_1 is

$$\text{Rel. freq. } (Y_1) : [(0, 0.229), (1, 0.771)].$$

Likewise,

$$O_{Y_2}(0) = 0.254, \ O_{Y_2}(1) = 0.097, \ O_{Y_2}(2) = 0.240, \ O_{Y_2}(3) = 0.410;$$

hence the relative frequency of Y_2 is

$$\text{Rel. freq. } (Y_2) : [(0, 0.254), (1, 0.097), (2, 0.240), (3, 0.410)].$$

The empirical weighted distribution function for Y_2 becomes

$$F_{Y_2}(\xi) = \begin{cases} 0.254, & \xi < 1, \\ 0.351, & 1 \le \xi < 2, \\ 0.591, & 2 \le \xi < 3, \\ 1, & \xi \ge 3. \end{cases}$$

Similarly, we can show that the symbolic weighted sample mean of Y_1 is $\bar{Y}_1 = 0.771$ and of Y_2 is $\bar{Y}_2 = 1.805$, and the symbolic weighted sample variance of Y_1 is $S_1^2 = 0.176$ and of Y_2 is $S_2^2 = 1.484$. □

3.3 Interval-Valued Variables

The corresponding description statistics for interval-valued random variables are obtained analogously to those for multi-valued variables; see Bertrand and Goupil (2000). Let us suppose we are interested in the particular random variable $Y_j \equiv Z$, and that the realization of Z for the observation w_u is the interval $Z(w_u) = [a_u, b_u]$, for $w_u \in E$. The individual description vectors $x \in vir(d_u)$ are assumed to be uniformly distributed over the interval $Z(w_u)$. Therefore, it follows that, for each ξ,

$$P\{x \leq \xi | x \in vir(d_u)\} = \begin{cases} 0, & \xi < a_u, \\ \frac{\xi - a_u}{b_u - a_u}, & a_u \leq \xi < b_u, \\ 1, & \xi \geq b_u. \end{cases} \tag{3.17}$$

The individual description vector x takes values globally in $\bigcup_{u \in E} vir(d_u)$. Further, it is assumed that each object is equally likely to be observed with probability $1/m$. Therefore, the empirical distribution function, $F_Z(\xi)$, is the distribution function of a mixture of m uniform distributions $\{Z(w_u), u = 1, \ldots, m\}$. Therefore, from Equation (3.17),

$$\begin{aligned} F_Z(\xi) &= \frac{1}{m} \sum_{u \in E} P\{x \leq \xi | x \in vir(d_u)\} \\ &= \frac{1}{m} \left\{ \sum_{\xi \in Z(u)} \left(\frac{\xi - a_u}{b_u - a_u} \right) + |(u | \xi \geq b_u)| \right\}. \end{aligned}$$

Hence, by taking the derivative with respect to ξ, we obtain

Definition 3.7: For the interval-valued random variable Z, the **empirical density function** is

$$f(\xi) = \frac{1}{m} \sum_{u : \xi \in Z(u)} \left(\frac{1}{b_u - a_u} \right); \tag{3.18}$$

or, equivalently,

$$f(\xi) = \frac{1}{m} \sum_{u \in E} \frac{I_u(\xi)}{||Z(u)||}, \; \xi \in \Re, \tag{3.19}$$

where $I_u(.)$ is the indicator function that ξ is or is not in the interval $Z(u)$ and where $||Z(u)||$ is the length of that interval. □

Note that the summation in Equation (3.18) and Equation (3.19) is only over those observations w_u for which $\xi \in Z(u)$. The analogy with Equation (3.6) and Equation (3.7) for multi-valued variables is apparent.

To construct a histogram, let $I = [\min_{u\in E} a_u, \max_{u\in E} b_u]$ be the interval which spans all the observed values of Z in $\mathcal{X}$, and suppose we partition I into r subintervals $I_g = [\zeta_{g-1}, \zeta_g),\ g = 1, \ldots, r-1$, and $I_r = [\zeta_{r-1}, \zeta_r]$. We then can define the following.

Definitions 3.8: For the interval-valued variate Z, the **observed frequency** for the histogram subinterval $I_g = [\zeta_{g-1}, \zeta_g),\ g = 1, \ldots, r$, is

$$f_g = \sum_{u\in E} \frac{||Z(u) \cap I_g||}{||Z(u)||}, \tag{3.20}$$

and the **relative frequency** is

$$p_g = f_g/m. \tag{3.21}$$

□

Here p_g is the probability or relative frequency that an arbitrary individual description vector x lies in the interval I_g. The histogram for Z is the graphical representation of $\{(I_g, f_g), g = 1, \ldots, r\}$. If we want to plot the histogram with height f_g on the interval I_g, so that the 'area' is p_g, then

$$p_g = (\xi_g - \xi_{g-1}) \times f_g.$$

Bertrand and Goupil (2000) indicate that, using the law of large numbers, the true limiting distribution of Z as $m \to \infty$ is only approximated by the exact distribution $f(\xi)$ in Equation (3.19) since this depends on the veracity of the uniform distribution within each interval assumption. Mathematical underpinning for the histogram has been developed in Diday (1995) using the strong law of large numbers and the concepts of t-norms and t-conorms as developed by Schweizer and Sklar (1983).

Definition 3.9: For an interval-valued random variable Z, the **symbolic sample mean** is given by

$$\bar{Z} = \frac{1}{2m} \sum_{u\in E} (b_u + a_u), \tag{3.22}$$

and the **symbolic sample variance** is given by

$$S^2 = \frac{1}{3m} \sum_{u\in E} (b_u^2 + b_u a_u + a_u^2) - \frac{1}{4m^2} \left[\sum_{u\in E} (b_u + a_u) \right]^2. \tag{3.23}$$

□

To verify Equation (3.22), we recall that the empirical mean $\bar{Z}$ in terms of the empirical density function is

$$\bar{Z} = \int_{-\infty}^{\infty} \xi f(\xi) d\xi.$$

Substituting from Equation (3.19), we have

$$\begin{aligned}\bar{Z} &= \frac{1}{m}\sum_{u\in E}\int_{-\infty}^{\infty}\frac{I_u(\xi)}{||Z(u)||}\xi d\xi\\ &= \frac{1}{m}\sum_{u\in E}\frac{1}{b_u - a_u}\int_{\xi\in Z(u)}\xi d\xi\\ &= \frac{1}{2m}\sum_{u\in E}\frac{b_u^2 - a_u^2}{b_u - a_u}\\ &= \frac{1}{m}\sum_{u\in E}(b_u + a_u)/2,\end{aligned}$$

as required.

Similarly, we can derive the **symbolic sample variance**. We have

$$\begin{aligned}S^2 &= \int_{-\infty}^{\infty}(\xi - \bar{Z})^2 f(\xi)d\xi\\ &= \int_{-\infty}^{\infty}\xi^2 f(\xi)d\xi - \bar{Z}^2.\end{aligned}$$

Now, substituting for $f(\xi)$ from Equation (3.19), we have

$$\begin{aligned}\int_{-\infty}^{\infty}\xi^2 f(z)d\xi &= \frac{1}{m}\sum_{u\in E}\int_{-\infty}^{\infty}\xi^2\frac{I_u(\xi)}{||Z(u)||}d\xi\\ &= \frac{1}{m}\sum_{u\in E}\int_{a_u}^{b_u}\frac{\xi^2}{(b_u - a_u)}d\xi\\ &= \frac{1}{3m}\sum_{i\in E}\left(\frac{b_u^3 - a_u^3}{b_u - a_u}\right).\end{aligned}$$

Hence,

$$S^2 = \frac{1}{3m}\sum_{u\in E}(b_u^2 + b_u a_u + a_u^2) - \frac{1}{4m^2}\left[\sum_{u\in E}(b_u + a_u)\right]^2,$$

as required.

Notice that these formulas for the mean and variance are consistent with the assumption of uniformity within each interval, where we recall that for a classical random variable U that is uniformly distributed on (0, 1), $E(U) = 1/2$ and $Var(U) = 1/12$.

Example 3.10. The data of Table 3.3 provide measurements (in cm) of certain features of some species of mushrooms, specifically the Width of the Pileus Cap (Y_1), and the Length (Y_2) and Thickness (Y_3) of the Stipe.

These measurements are interval valued. The Edibility (Y_4) (U: unknown, Y: Yes, N: No, T: Toxic) is also shown as a multi-valued variable. Thus, for example, the width of the pileus cap for mushrooms of the *arorae* species is in the range of 3.0 to 8.0 cm, while the stipe is 4.0 to 9.0 cm in length and 0.50 to 2.50 cm thick and its edibility is unknown. These mushroom species are members of the genus *Agaricies*. The specific variables and their values displayed in Table 3.3 were extracted from the Fungi of California Species Index (http://www.mykoweb.com/CAF/species_index.html).

Table 3.3 Mushroom data.

w_u	Species	Pileus Cap Width	Stipe Length	Stipe Thickness	Edibility
w_1	*arorae*	[3.0, 8.0]	[4.0, 9.0]	[0.50, 2.50]	U
w_2	*arvenis*	[6.0, 21.0]	[4.0, 14.0]	[1.00, 3.50]	Y
w_3	*benesi*	[4.0, 8.0]	[5.0, 11.0]	[1.00, 2.00]	Y
w_4	*bernardii*	[7.0, 6.0]	[4.0, 7.0]	[3.00, 4.50]	Y
w_5	*bisporus*	[5.0, 12.0]	[2.0, 5.0]	[1.50, 2.50]	Y
w_6	*bitorquis*	[5.0, 15.0]	[4.0, 10.0]	[2.00, 4.00]	Y
w_7	*califorinus*	[4.0, 11.0]	[3.0, 7.0]	[0.40, 1.00]	T
w_8	*campestris*	[5.0, 10.0]	[3.0, 6.0]	[1.00, 2.00]	Y
w_9	*comtulus*	[2.5, 4.0]	[3.0, 5.0]	[0.40, 0.70]	Y
w_{10}	*cupreo-brunneus*	[2.5, 6.0]	[1.5, 3.5]	[1.00, 1.50]	Y
w_{11}	*diminutives*	[1.5, 2.5]	[3.0, 6.0]	[0.25, 0.35]	U
w_{12}	*fuseo-fibrillosus*	[4.0, 15.0]	[4.0, 15.0]	[1.50, 2.50]	Y
w_{13}	*fuscovelatus*	[3.5, 8.0]	[4.0, 10.0]	[1.00, 2.00]	Y
w_{14}	*hondensis*	[7.0, 14.0]	[8.0, 14.0]	[1.50, 2.50]	T
w_{15}	*lilaceps*	[8.0, 20.0]	[9.0, 19.0]	[3.00, 5.00]	Y
w_{16}	*micromegathus*	[2.5, 4.0]	[2.5, 4.5]	[0.40, 0.70]	Y
w_{17}	*praeclaresquamosus*	[7.0, 19.0]	[8.0, 15.0]	[2.00, 3.50]	T
w_{18}	*pattersonae*	[5.0, 15.0]	[6.0, 15.0]	[2.50, 3.50]	Y
w_{19}	*perobscurus*	[8.0, 12.0]	[6.0, 12.0]	[1.50, 2.00]	Y
w_{20}	*semotus*	[2.0, 6.0]	[3.0, 7.0]	[0.40, 0.80]	Y
w_{21}	*silvicola*	[6.0, 12.0]	[6.0, 12.0]	[1.50, 2.00]	Y
w_{22}	*subrutilescens*	[6.0, 12.0]	[6.0, 16.0]	[1.00, 2.00]	Y
w_{23}	*xanthodermus*	[5.0, 17.0]	[4.0, 14.0]	[1.00, 3.50]	T

Consider the pileus cap random variable Y_1. The observation intervals span the interval $I = [1.5, 21.0]$ since $\min_{u \in E} a_u = 1.5$, $\max_{u \in E} b_u = 21.0$. Let us build a histogram of the cap width on $r = 6$ intervals each of length 3.5 cm, i.e., [0, 3.5), [3.5, 7), . . . , [17.5, 21].

Then, from Equation (3.20), the observed frequency that an individual description x lies in the interval $I_1 = [0, 3.5]$ is

$$
\begin{aligned}
f_1 = & \left(\frac{3.5-3.0}{8.0-3.0}\right)_{u=1} + 0 + \ldots + 0 + \left(\frac{3.5-2.5}{4.0-2.5}\right)_{u=9} + \left(\frac{3.5-2.5}{6.0-2.5}\right)_{u=10} \\
& + \left(\frac{2.5-1.5}{2.5-1.5}\right)_{u=11} + 0 + \ldots + 0 + \left(\frac{3.5-2.5}{4.0-2.5}\right)_{u=16} + 0 + \ldots \\
& + \left(\frac{3.5-2.0}{6.0-2.0}\right)_{u=20} + 0 + \ldots + 0 \\
= & \; 3.0940.
\end{aligned}
$$

(Here, as elsewhere, the $u = \cdot$ subscript identifies the contribution of the observed value ξ_u of the observation w_u.) Notice that in effect the contribution to f_1 from the w_uth observation is the fraction of the observation interval $[a_u, b_u]$ which overlaps the histogram interval I_1. Thus, for example, the first observation ($u = 1$) on the interval [3.0, 8.0] overlaps $I_1 = [0, 3.5)$ by (3.0, 3.5) which is the fraction $\{(3.5-3.0)/(8.0-3.0)\}$, and so on. Notice too that the frequency is no longer integral (only) in value.

Then, the relative frequency for the I_1 interval is, from Equation (3.21),

$$p_1 = f_1/m = 3.0940/23 = 0.1345.$$

The frequencies f_g and the corresponding relative frequencies p_g for the remaining histogram intervals I_g, $g = 2, \ldots, r = 6$, are calculated similarly. All are shown in Table 3.4, column (a). The histogram plot is shown in Figure 3.2.

Table 3.4 Histogram frequencies – mushroom data.

(a) Pileus Cap Width			(b) Stipe Length			(c) Stipe Thickness		
I_g	f_g	p_g	I_g	f_g	p_g	I_g	f_g	p_g
[0.0, 3.5)	3.0940	0.1345	[0.0, 4.0)	4.0833	0.1775	[0.0, 1.0)	5.2500	0.2283
[3.5, 7.0)	6.5874	0.2864	[4.0, 7.0)	8.2672	0.3594	[1.0, 2.0)	9.8000	0.4261
[7.0, 10.5)	6.9960	0.3042	[7.0, 10.0)	5.1251	0.2228	[2.0, 3.0)	4.2167	0.1833
[10.5, 14.0)	4.1761	0.1816	[10.0, 13.0)	3.5680	0.1551	[3.0, 4.0)	2.9000	0.1261
[14.0, 17.5)	1.5798	0.0687	[13.0, 16.0)	1.6564	0.0720	[4.0, 5.0]	0.8333	0.0362
[17.5, 21.0]	0.5667	0.0246	[16.0, 19.0]	0.3000	0.0130			
	$\bar{X} = 8.1957$			$\bar{X} = 7.3913$			$\bar{X} = 1.8239$	
	$S^2 = 16.4871$			$S^2 = 12.6512$			$S^2 = 1.0837$	
	$S = 4.0604$			$S = 3.5569$			$S = 1.0410$	

The empirical mean cap width is found directly from Equation (3.22) as

$$\bar{X} = \frac{1}{23}\left(\frac{8.0+3.0}{2} + \ldots + \frac{17.0+5.0}{2}\right) = 8.1957$$

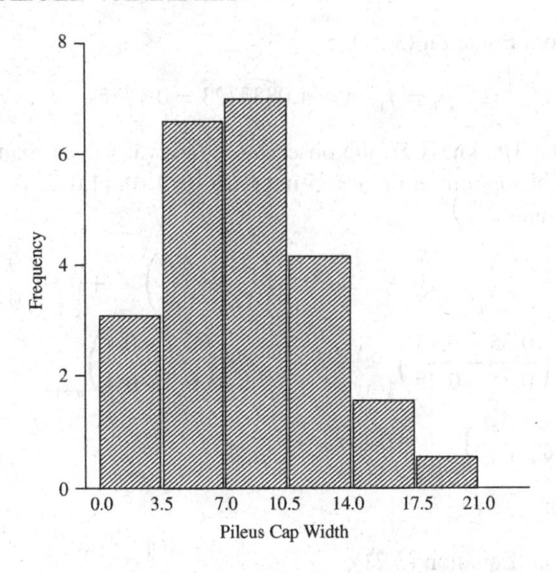

Figure 3.2 Histogram for Pileus Cap Width.

and the empirical variance from Equation (3.23) is

$$S^2 = \frac{1}{23}\left[\frac{[8.0^2 + (8.0)(3.0) + 3.0^2]}{3} + \ldots + \frac{[17.0^2 + (17.0)(5.0) + 5.0^2]}{3}\right]$$
$$- \frac{1}{23^2}\left[\left(\frac{8.0+3.0}{2}\right) + \ldots + \left(\frac{17.0+5.0}{2}\right)\right]^2$$
$$= 16.4871,$$

and hence the standard deviation is

$$S = 4.0604.$$

The histogram frequencies for the two stipe variables are also displayed in Table 3.4, column (b). The Stipe Length Y_2 values span the interval $I = [1.5, 19]$. Here, the histogram is constructed on the $r = 6$ intervals [0, 4), [4, 7), . . . , [16, 19]. Thus, for example, from Equation (3.20),

$$f_1 = 0 + \ldots + 0 + \left(\frac{4.0-2.0}{5.0-2.0}\right)_{u=5} + 0 + \left(\frac{4.0-3.0}{7.0-3.0}\right)_{u=7} + \left(\frac{4.0-3.0}{6.0-3.0}\right)_{u=8}$$
$$+ \left(\frac{4.0-3.0}{5.0-3.0}\right)_{u=9} + \left(\frac{3.5-1.5}{3.5-1.5}\right)_{u=10} + \left(\frac{4.0-3.0}{6.0-3.0}\right)_{u=11} + 0 + \ldots$$
$$+ \left(\frac{4.0-2.5}{4.5-2.5}\right)_{u=16} + 0 + 0 + 0 + \left(\frac{4.0-3.0}{7.0-3.0}\right)_{u=20} + 0 + 0 + 0$$
$$= 3.0833;$$

and hence, from Equation (3.21),

$$p_1 = f_1/m = 4.0833/23 = 0.1775.$$

For the Stipe Thickness Y_3, the observation interval values span $I = [0.25, 5]$. We build the histogram on the $r = 5$ intervals [0, 1.0), [1.0, 2.0),. . . , [4.0, 5.0]. Thus, for example,

$$\begin{aligned} f_1 = & \left(\frac{1.0-0.5}{2.5-0.5}\right)_{u=1} + 0 + \ldots + 0 + \left(\frac{1.0-0.4}{1.0-0.4}\right)_{u=7} + 0 + \left(\frac{0.7-0.4}{0.7-0.4}\right)_{u=9} \\ & + 0 + \left(\frac{0.35-0.25}{0.35-0.25}\right)_{u=11} + 0 + \ldots + \left(\frac{0.7-0.4}{0.7-0.4}\right)_{u=16} + 0 + \ldots \\ & + \left(\frac{0.8-0.4}{0.8-0.4}\right)_{u=20} + 0 + 0 + 0 \\ = & 5.2500; \end{aligned}$$

and hence, from Equation (3.21),

$$p_1 = f_1/m = 5.2500/23 = 0.2283.$$

The histogram plots are shown in Figure 3.3. The sample mean $\bar{X}$, variance S^2, and standard deviation S are easily found, from Equation (3.22) and Equation (3.23), to be, for Stipe Length, $\bar{X} = 7.3913$, $S^2 = 12.6512$, and $S = 3.5569$, and, for Stipe Thickness, $\bar{X} = 1.8239$, $S^2 = 1.0837$, and $S = 1.0410$. □

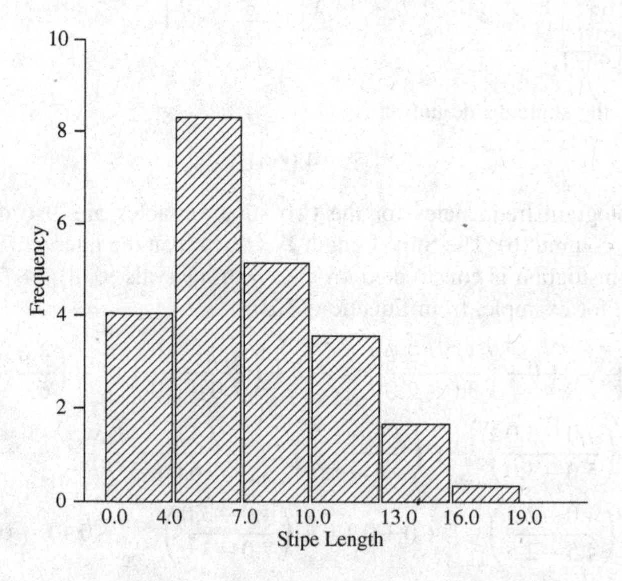

Figure 3.3 Histogram for Stipe Length.

Example 3.11. Suppose we have a dataset consisting of one observation, and suppose that observation corresponds to the $u = 1$ value of Table 3.3. Then, from Equation (3.22) the empirical mean is $\bar{X} = 5.5$, and from Equation (3.23), again with $m = 1$, the empirical variance is

$$S^2 = \frac{1}{3}(3^2 + 3 \times 8 + 8^2) - \left[\frac{(3+8)}{2}\right]^2 = 2.183.$$

Notice that this $S^2 \neq 0$, unlike the sample variance of 0 for a sample of size $m = 1$ for classical data. Recall, however, that symbolic data contain internal variation, and it is this $S^2 = 2.183$ which reflects that internal variation for this particular observation. □

As for multi-valued variables, if an observation w_u has some internal inconsistency relative to a logical rule, i.e., if w_u is such that $|vir(d_u)| = 0$, then the summations in Equations (3.19), (3.20), (3.22), and (3.23) are over only those u for which $|vir(d_u)| \neq 0$, i.e., over $u \in E'$, and m is replaced by m' (equal to the number of objects w_u in E'). Alternatively, it can be that instead of the entire observation w_u, only a portion has some internal inconsistency under a logical rule v, as in Example 3.6 and Example 3.7 and displayed in Figure 3.1. The same principles apply. In the sequel, it will be understood that m and E refer to those w_u for which these rules hold. Examples 3.12–3.15 illustrate this phenomenon further.

Example 3.12. To illustrate the application of a rule, consider the data based on Raju (1997) shown in Table 3.5, in which the Pulse Rate (Y_1), Systolic Blood Pressure (Y_2) and Diastolic Blood Pressure (Y_3) are recorded as an interval for each of the patients, $w_u, u = 1, \ldots, 15$.

Suppose there is a logical rule that says

$$v_1 : A = \{Y_1 \leq 55\} \Rightarrow B = \{Y_2 \leq 170, Y_3 \leq 100\}. \tag{3.24}$$

The set A portion of this rule includes observations $u = 5, 11, 12, 15$ only. Figure 3.4(a) displays these observations for the (Y_2, Y_3) plane. The observation w_5 and all of w_{15} impacted by A fit entirely into the valid region of the set B observations of the rule; see the shaded observations in Figure 3.4(a). However, all the (Y_2, Y_3) values of the w_{11} and w_{12} observations are not valid under this rule. Thus, this rule v_1 has the effect that the observed values for the observations w_{11} and w_{12} cannot be logically true; hence these particular observations are omitted from the calculations for the mean, variance, and histogram frequencies. Note that under this rule v_1, $m = 13$.

Let us focus on the random variable $Y_2 =$ Systolic Blood Pressure. Suppose we build a histogram on the $r = 6$ intervals [80, 100), [100, 120), . . . , [180, 200].

Table 3.5 Blood pressure interval-valued data.

w_u	Y_1 Pulse Rate	Y_2 Systolic Pressure	Y_3 Diastolic Pressure
w_1	[44, 68]	[90, 110]	[50, 70]
w_2	[60, 72]	[90, 130]	[70, 90]
w_3	[56, 90]	[140, 180]	[90, 100]
w_4	[70, 112]	[110, 142]	[80, 108]
w_5	[54, 72]	[90, 100]	[50, 70]
w_6	[70, 100]	[134, 142]	[80, 110]
w_7	[72, 100]	[130, 160]	[76, 90]
w_8	[76, 98]	[110, 190]	[70, 110]
w_9	[86, 96]	[138, 180]	[90, 110]
w_{10}	[86, 100]	[110, 150]	[78, 100]
w_{11}	[53, 55]	[160, 190]	[205, 219]
w_{12}	[50, 55]	[180, 200]	[110, 125]
w_{13}	[73, 81]	[125, 138]	[78, 99]
w_{14}	[60, 75]	[175, 194]	[90, 100]
w_{15}	[42, 52]	[105, 115]	[70, 82]

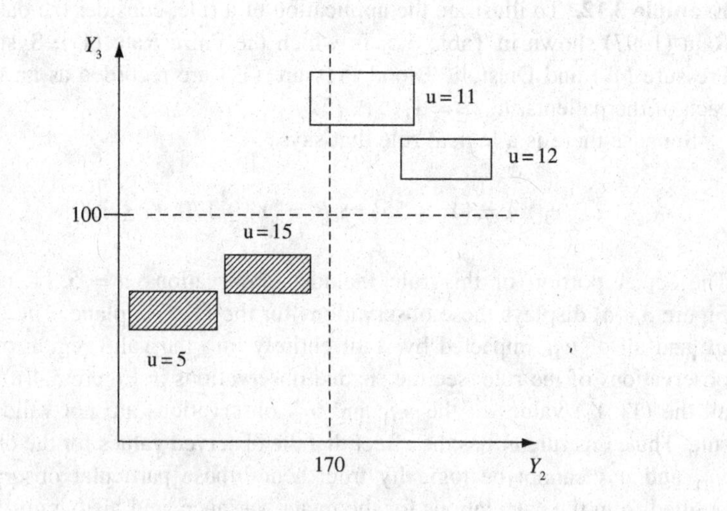

Figure 3.4a Rule v_1.

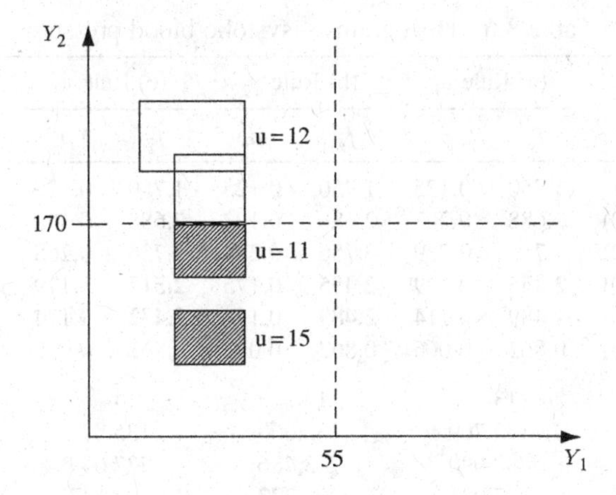

Figure 3.4b Rule v_2.

Then, we use Equation (3.20) where the set E omits the w_{11} and w_{12} values. Thus, for example, for the $g = 5$th interval $I_5 = [160, 180)$, we have

$$f_5 = 0 + 0 + \left(\frac{180 - 160}{180 - 140}\right)_{u=3} + 0 + \ldots + 0 + \left(\frac{180 - 160}{180 - 138}\right)_{u=9}$$
$$+ (0)_{u=10} + (0)_{u=13} + \left(\frac{180 - 175}{194 - 175}\right)_{u=14} + 0$$
$$= 1.489,$$

and hence

$$p_5 = f_5/m = 1.489/13 = 0.114.$$

The results for all intervals are shown in Table 3.6, column (a). The symbolic mean, variance, and standard deviation are readily calculated from Equations (3.22) and (3.23) where again the w_{11} and w_{12} values are omitted. Thus, we have

$$\bar{X} = 133.769, \quad S^2 = 730, \quad 690, \; S = 27.031. \qquad \square$$

Example 3.13. Suppose instead of the rule v_1 in Example 3.12 we have the rule

$$v_2 : A = \{Y_1 < 55\} \Rightarrow B = \{Y_2 \leq 170\}. \tag{3.25}$$

Under this rule, some of the individual description vectors x are logically valid while some are not. In particular, observations w_u with $u = 11, 12, 15$ are relevant to the set A portion of this rule; see Figure 3.4b. Of these, the observation

Table 3.6 Histograms – systolic blood pressure.

g	I_g	(a) Rule v_1		(b) Rule v_2		(c) Rule v_3		(d) Rule v_4	
		f_g	p_g	f_g	p_g	f_g	p_g	f_g	p_g
1	[80, 100)	1.750	0.135	1.750	0.125	1.750	0.125	1.750	0.135
2	[100, 120)	2.688	0.207	2.688	0.192	2.688	0.192	2.688	0.207
3	[120, 140)	3.756	0.289	3.756	0.268	3.756	0.268	3.756	0.289
4	[140, 160)	2.455	0.189	2.455	0.175	**2.512**	0.179	2.512	0.193
5	[160, 180)	1.489	0.114	**2.489**	0.178	**2.432**	0.174	**1.432**	0.110
6	[180, 200)	0.862	0.066	0.862	0.062	0.862	0.062	0.862	0.066
	m	13		14		14		13	
	$\bar{X}$	133.769		136.000		135.877		133.637	
	S^2	730.690		743.786		737.049		722.844	
	S	27.031		27.272		27.147		26.886	

w_{12} is not valid to any extent; that is, none of the description vectors satisfy Equation (3.26), so that its virtual description is zero. The observation w_{15} is completely valid, so that its virtual description contains all the description vectors of the original observation itself. A portion of the w_{11} observation is logically true and a portion cannot be valid under rule v_2. The virtual description of this w_{11} observation is the (smaller) rectangle $[53, 55] \times [160, 170]$. Only that part which is logically true under v_2, i.e., only its virtual description, is used in the calculations for the histogram, mean, and variance. The impact of this rule is seen as follows. Note that the number of observations is now $m = 14$.

Let us build the histogram on the same $r = 6$ intervals used under the first rule v_1. We note that except for the interval $I_5 = [160, 180)$, the frequencies are unchanged. For I_5, we now have, from Equation (3.20),

$$\begin{aligned} f_5 =& 0+0+\left(\frac{180-160}{180-140}\right)_{u=3} +0+\ldots+0+\left(\frac{180-160}{190-110}\right)_{u=8} \\ &+\left(\frac{180-160}{180-138}\right)_{u=9} +0+\left(\frac{170-160}{170-160}\right)_{u=11} +0+\left(\frac{180-160}{194-175}\right)_{u=14} +0 \\ =& 2.489 \end{aligned}$$

where the w_{12} observation is omitted entirely, and where the contribution of the w_{11} observation to the summation is modified according to the portion of that observation that is logically valid; that is, here $b_{11} \equiv 170$ effectively.

The relative frequencies, or probabilities, are calculated from $p_g = f_g/m$, with $m = 14$. These values under rule v_2 are shown in column (b) of Table 3.6.

To calculate the mean and variance, Equations (3.22) and (3.23) are used as before except that for the observation w_{11}, the b_u becomes $b_{11} \equiv 170$. Thus, we obtain from Equation (3.22) and Equation (3.23)

$$\begin{aligned}\bar{X} =& \frac{1}{14}[(100)_{u=1} + \ldots + (140)_{u=10} + (165)_{u=11} + (131.5)_{u=13} \\ &+ \ldots + (110)_{u=15}] \\ =& 136,\end{aligned}$$

$$\begin{aligned}S^2 =& \frac{1}{14(3)}[(90^2 + 90 \times 100 + 110^2)_{u=1} + \ldots + (160^2 + 160 \times 170 + 170^2)_{u=11} \\ &+ \ldots + (105^2 + 105 \times 115 + 115^2)_{u=15}] - 136^2 \\ =& 743.788\end{aligned}$$

and

$$S = 27.272. \quad \square$$

Example 3.14. Let us now consider the rule

$$v_3 : A = \{Y_1 \leq 70\} \Rightarrow B = \{Y_2 \leq 170\}. \tag{3.26}$$

Under the rule v_3, the observation w_{12} is not logically true at all, while portions of three observations w_3, w_{11}, and w_{14} are not valid; see Figure 3.4(c). That part of the w_{11}th observation that is valid/invalid is the same as it was under the rule v_2. Take now the w_{14} observation. Only the portion $(Y_1, Y_2) = \{(70, 75) \times (175, 194)\}$

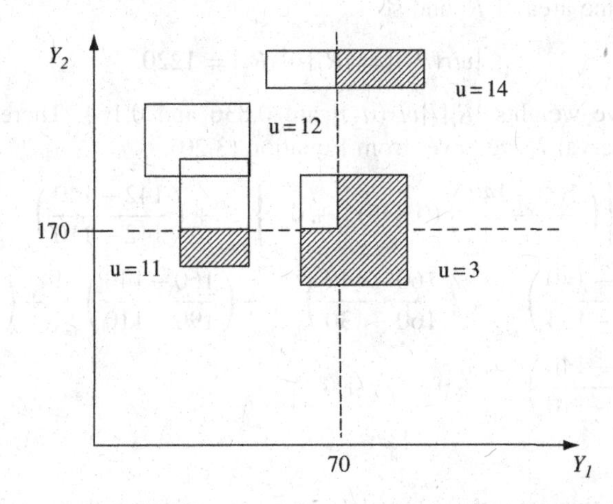

Figure 3.4c Rule v_3.

is valid with the description vectors in this rectangle being the virtual description of this observation. Notice that for this observation in calculations involving the Y_2 variable only, there is no direct numerical impact (but there is an effect for calculations involving Y_1).

In contrast, the observation w_3 is impacted in both Y_1 and Y_2 calculations. The virtual description of this observation (the shaded area in Figure 3.4(c)) is the hexagonal shape composed of the union of the two rectangles $R_1 = [56, 90] \times [140, 170]$ and $R_2 = [70, 90] \times [170, 180]$, i.e.,

$$vir(d_3) = R_1 \cup R_2$$

or equivalently,

$$vir(d_3) = R - R^*$$

where $R = (56, 90) \times (140, 180)$ is the total recorded region for the observation and $R^* = (56, 70) \times (170, 180)$ is the set of individual description x values which cannot be logically true under this rule v_3.

If we use the same I_g intervals for the histogram of systolic blood pressure values as before in Examples 3.12 and 3.13, we see that the w_3 observation is involved in the calculation of the histogram frequencies for the intervals $I_4 = [140, 160)$ and $I_5 = [160, 180)$, and the w_{11} observation involves the histogram interval I_5 under the rule v_3, while the other histogram intervals are unchanged. The contribution of this w_3 observation is determined as follows. There are two pieces, represented by the rectangles R_1 and R_2, each piece weighted proportionally to its size. In this case,

$$|R_1| = (90 - 56)(170 - 140) = 1020$$
$$|R_2| = (90 - 70)(180 - 70) = 200$$

where $|R|$ is the area of R, and so

$$|vir(d_3)| = |R_1| + |R_2| = 1220.$$

The respective weights $|R_i|/|vir(d_3)|$ are 0.836 and 0.164. Therefore, for the histogram interval I_4 we have, from Equation (3.20),

$$\begin{aligned} f_4 = & 0+0+\left\{\left(\frac{160-140}{170-140}\right)(0.836)_{R_1} + 0_{R_2}\right\}_{u=3} + \left(\frac{142-140}{142-110}\right)_{u=4} + 0 \\ & + \left(\frac{142-140}{142-134}\right)_{u=6} + \left(\frac{160-140}{160-130}\right)_{u=7} + \left(\frac{160-140}{190-110}\right)_{u=8} + \left(\frac{160-140}{180-138}\right)_{u=9} \\ & + \left(\frac{150-140}{150-110}\right)_{u=10} + 0 + \ldots + 0 \\ = & 2.512. \end{aligned}$$

Notice that in this case, the interval I_4 does not overlap with the rectangle R_2 so that there is a quantity equal to zero at this stage in the calculation of f_4.

However, in contrast, the histogram interval $I_5 = [160, 180)$ straddles both R_1 and R_2. Therefore, in this case, we have, from Equation (3.20),

$$f_5 = 0+0+\left\{\left(\frac{170-160}{170-140}\right)(0.836)_{R_1} + \left(\frac{180-170}{180-170}\right)(0.164)_{R_2}\right\}_{u=3} +0$$
$$+\ldots+\left(\frac{180-160}{190-110}\right)_{u=8} + \left(\frac{180-160}{180-138}\right)_{u=9} +0+\left(\frac{180-160}{190-160}\right)_{u=11}$$
$$+0+0+\left(\frac{180-160}{194-175}\right)_{u=14} +0$$
$$=2.432.$$

The remaining histogram intervals are not affected by this rule. All these frequencies are shown in Table 3.6, column (c), along with the corresponding relative frequencies obtained from

$$p_g = f_g/m, \quad m = 14.$$

As a cross-check, notice that the sum of the two {} terms in f_4 and f_5 corresponding to the respective contributions of the w_3 observation is one, as it should be!

To calculate the sample statistics, we use Equations (3.22) and (3.23) as before with appropriate adjustment for the observations affected by the rule. The w_{11} and w_{12} observations are handled along the lines that the w_{11} value was adjusted under rule v_2. The w_3 observation is handled similarly but with attention paid to the R_1 and R_2 pieces and with weights as appropriate. Thus, the contribution from the w_3 observation to the summation term in Equation (3.22) for the sample mean becomes

$$\left\{\left(\frac{160+170}{2}\right)(0.836) + \left(\frac{170+180}{2}\right)(0.164)\right\}_{u=3} = 166.64,$$

and the contribution to the first summation term in Equation (3.23) for the sample variance becomes

$$\{[160^2+(160)(170)+170^2](0.836)+[170^2+(170)(180)$$
$$+180^2](0.164)\}_{u=3} = 83372.8.$$

Hence, the mean is

$$\bar{X} = 135.877$$

and the variance and standard deviation are, respectively,

$$S^2 = 737.049, \quad S = 27.147. \qquad \square$$

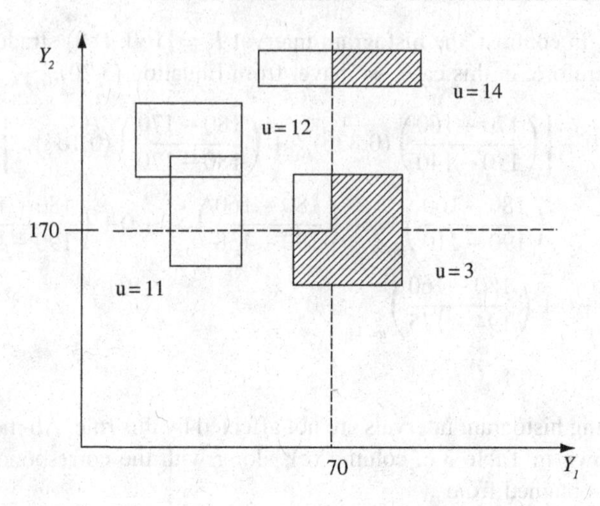

Figure 3.4d Rule v_4.

Example 3.15. Finally, suppose there are two rules in operation, i.e., rule v_4

$$v_4 = \{v_0, v_3\} \tag{3.27}$$

where v_3 is as in Equation (3.26) and v_0 is

$$v_0 : Y_3 \leq Y_2, \quad \text{i.e., } b_3 < a_2. \tag{3.28}$$

Under this rule v_4, the w_{11} observation is invalid entirely; see Figure 3.4(d). The other observations w_{12} (entirely invalid) and w_3 and w_{14} (partially valid/invalid) are as they were under rule v_3. There are now $m = 13$ observations for which their virtual descriptions $|vir(d_u)| \neq 0$. The resulting (relative) frequency distributions are shown in column (d) of Table 3.6; also shown are the $\bar{X}$, S^2, and S values. □

3.4 Modal Multi-Valued Variables

There are many types of modal-valued random variables. We consider very briefly two types only, one each of multi-valued and interval-valued random variables in this section and the next, respectively.

Let us suppose we have data from a multi-valued or categorical variable Y_j taking possible values ξ_{jk}, with relative frequencies p_{jk}, $k = 1, \ldots, s$, respectively, with $\Sigma p_{jk} = 1$. Suppose we are interested in the particular symbolic random variable $Y_j \equiv Z$, in the presence of the dependency rule v. Then, we define

Definition 3.10: The **observed frequency** that $Z = \xi_k$, $\quad k = 1, \ldots, s$, is

$$O_Z(\xi_k) = \sum_{u \in E} \pi_Z(\xi_k; u) \tag{3.29}$$

where

$$\pi_Z(\xi_k; u) = P(Z = \xi_k | x \in vir(d_u), u)$$

$$= \frac{\sum_x P(x = \xi_k | x \in vir(d_u), u)}{\sum_x \sum_{k=1}^{s} P(x = \xi_k | x \in vir(d_u), u)} \tag{3.30}$$

where, for each u, $P(x = \xi_k | x \in vir(d_u), u)$ is the probability that a particular description x ($\equiv x_j$) has the value ξ_k and that the logical dependency rule v holds. □

If for a specific observation w_u, there are no description vectors satisfying the rule v, i.e., if the denominator of Equation (3.30) is zero, then that u is omitted in the summation in Equation (3.29). We note that

$$\sum_{k=1}^{s} O_Z(\xi_k) = m. \tag{3.31}$$

Example 3.16. Suppose households are recorded as having one of the possible central heating fuel types $\mathcal{Y}_1$ taking values in $\mathcal{Y}_1 = \{\xi_0 = \text{none},\ \xi_1 = \text{gas},\ \xi_2 = \text{oil},\ \xi_3 = \text{electricity},\ \xi_4 = \text{coal}\}$ and suppose Y_2 is an indicator variable taking values in $\mathcal{Y}_2 = \{\text{No},\ \text{Yes}\}$ depending on whether a household does not (does) have central heating installed. These data are after Billard and Diday (2003, Table 4) which were themselves based on a large census dataset from the UK Office for National Statistics. Assume Y_1 and Y_2 are independent random variables.

Suppose that aggregation over thousands of households in $m = 10$ geographical regions produced the data of Table 3.7. Thus, region 1 represented by the object w_1 is such that 87% of the households are fueled by gas, 7% by oil, 5% by

Table 3.7 Modal-valued data: energy usage.

w_u	Y_1					Y_2	
	None	Gas	Oil	Elec.	Coal	No	Yes
w_1	{0.00	0.87	0.07	0.05	0.01}	{0.09	0.91}
w_2	{0.00	0.71	0.11	0.10	0.08}	{0.12	0.88}
w_3	{0.00	0.83	0.08	0.09	0.00}	{0.23	0.77}
w_4	{0.00	0.76	0.06	0.11	0.07}	{0.19	0.81}
w_5	{0.00	0.78	0.06	0.09	0.07}	{0.12	0.88}
w_6	{0.00	0.90	0.01	0.08	0.01}	{0.22	0.78}
w_7	{0.00	0.87	0.01	0.10	0.02}	{0.22	0.78}
w_8	{0.00	0.78	0.02	0.13	0.07}	{0.13	0.87}
w_9	{0.00	0.91	0.00	0.09	0.00}	{0.24	0.76}
w_{10}	{0.25	0.00	0.41	0.14	0.20}	{0.09	0.91}

electricity, 1% by coal, and 0% had none of these types of fuel; and that 9% of households did not have central heating, while 91% did.

First, let us suppose that interest centers only on the variable Y_1 (without regard to any other variable) and that there is no rule v to be satisfied. Then, from Equation (3.29), it is readily seen that the observed frequency that $Z = \xi_1 =$ gas is

$$O_{Y_1}(\xi_1) = (0.87 + \ldots + 0.91 + 0.00) = 7.41,$$

and likewise, for $\xi_0 =$ none, $\xi_2 =$ oil, $\xi_3 =$ electricity, $\xi_4 =$ coal, we have, respectively,

$$O_{Y_1}(\xi_0) = 0.25,\ O_{Y_1}(\xi_2) = 0.83,\ O_{Y_1}(\xi_3) = 0.98,\ O_{Y_1}(\xi_4) = 0.53.$$

Here, $m = 10$. Hence, the relative frequencies are

$$\text{Rel. freq. } (Y_1) : [(\xi_0, 0.025), (\xi_1, 0.741), (\xi_2, 0.083), (\xi_3, 0.098), (\xi_4, 0.053)].$$

Similarly, we can show that for the Y_2 variable,

$$\text{Rel. freq. } (Y_2) : [(\text{No}, 0.165), (\text{Yes}, 0.835)].$$ □

Example 3.17. Suppose, however, that for the situation in Example 3.16, there is the logical rule that a household must have one of ξ_k, $k = 1, \ldots, 4$, if it has central heating. Then, this logical rule can be written as the pair $v = (v_1, v_2)$

$$v = \begin{cases} v_1 : Y_2 \in \{\text{No}\} \Rightarrow Y_1 \in \{\xi_0\}, \\ v_2 : Y_2 \in \{\text{Yes}\} \Rightarrow Y_1 \in \{\xi_1, \ldots, \xi_4\}. \end{cases} \quad (3.32)$$

In the presence of the rule v, the particular descriptions $x = (Y_1, Y_2) \in \{(\xi_0, \text{No}), (\xi_1, \text{Yes}), (\xi_2, \text{Yes}), (\xi_3, \text{Yes}), (\xi_4, \text{Yes})\} = C$ can occur. Thus, the relative frequencies for each of the possible ξ_k values have to be adjusted, to give 'virtual' relative frequencies..

Take the observation w_{10}. Under the assumption that Y_1 and Y_2 are independent, we obtain the joint probability for the possible descriptions $x = (x_1, x_2)$ before any rule is invoked as shown in Table 3.8. Under the rule v of Equation (3.32) only those x in C are valid. Therefore, we standardize these probabilities by

$$P = 0.0225 + 0.0000 + \ldots + 0.1820 = 0.7050$$

where the components of P are the probabilities of those x in C. Hence, we can find the virtual probabilities for each x in C. For example, the adjusted (virtual) probability when $Y_1 = \xi_2$ is

$$\pi_{Y_1}(\xi_2; 10^*) = (0.3731)/(0.7050) = 0.5292,$$

where we have denoted the virtual w_u by u^*. The virtual probabilities for all possible (Y_1, Y_2) values are also shown in Table 3.8 for $u = 10$.

Table 3.8 Virtual relative frequencies for w_{10}.

$Y_2\backslash Y_1$	ξ_0	ξ_1	ξ_2	ξ_3	ξ_4	Total	$\pi_{Y_2}(\xi_k, u^*)$
No	0.0225	0.0000[a]	0.0369[a]	0.0126[a]	0.0180[a]	0.0900	0.0319
Yes	0.2275[a]	0.0000	0.3731	0.1274	0.1820	0.9100	0.9681
Total	0.2500	0.0000	0.4100	0.1400	0.2000		
$\pi_{Y_1}(\xi_k, u^*)$	0.0319	0.0000	0.5292	0.1807	0.2582		

[a] Invalidated when rule v invoked.

Since for all other observations, $Y_1 = \xi_0$ with probability 0, it follows that the probabilities for $k = 1, \ldots, 4$, and $u = 1, \ldots, 9$, are

$$P\{Y_1(w_u) = \xi_k,\ Y_2(w_u) = \text{Yes}\} = P\{Y_1(w_u) = \xi_k\} \times P\{Y_2(w_u) = \text{Yes}\}.$$

The complete set of virtual probabilities for the descriptions x in C is given in Table 3.9.

Table 3.9 Virtual modal-valued data.

w_u	(ξ_0, No)	(ξ_1, Yes)	(ξ_2, Yes)	(ξ_3, Yes)	(ξ_4, Yes)
w_1	0.0900	0.7917	0.0637	0.0455	0.0091
w_2	0.1200	0.6248	0.0968	0.0880	0.0704
w_3	0.2300	0.6391	0.0616	0.0693	0.0000
w_4	0.1900	0.6156	0.0486	0.0891	0.0567
w_5	0.1200	0.6864	0.0528	0.0792	0.0616
w_6	0.2200	0.7020	0.0078	0.0624	0.0078
w_7	0.2200	0.6786	0.0078	0.0780	0.0156
w_8	0.1300	0.6786	0.0174	0.1131	0.0609
w_9	0.2400	0.6916	0.0000	0.0684	0.0000
w_{10}	0.0319	0.0000	0.5292	0.1807	0.2582

We can now proceed to calculate the observed frequencies from Equation (3.29) where we use the virtual probabilities of Table 3.9. For example, the observed frequency for $Y_1 = \xi_2$ over all observations becomes under the rule v

$$O_{Y_1}(\xi_2) = 0.0637 + \ldots + 0.0000 + 0.5292 = 0.5403.$$

Similarly, we calculate that under v

$$O_{Y_1}(\xi_0) = 1.5919,\ O_{Y_1}(\xi_1) = 6.1084,\ O_{Y_1}(\xi_3) = 0.8737,\ O_{Y_1}(\xi_4) = 0.5403.$$

It is readily seen that the summation of Equation (3.31) holds, with $m = 10$. Hence, the relative frequencies for Y_1 are

$$\text{Rel. freq. } (Y_1) : [(\xi_0, 0.159), (\xi_1, 0.611), (\xi_2, 0.089), (\xi_3, 0.084), (\xi_4, 0.054)].$$

Likewise, we can show that the relative frequencies for Y_2 are

$$\text{Rel. freq. } (Y_2) : [(\text{No}, 0.159), (\text{Yes}, 0.84)].$$

Therefore, over all 10 regions, 61.1% of households use gas, 8.9% use oil, 8.4% use electricity, and 5.4% use coal to operate their central heating systems, while 15.9% do not have central heating. □

3.5 Modal Interval-Valued Variables

Let us suppose that the random variable of interest, $Y_j \equiv Z$ for observation w_u, $u = 1, \dots, m$, takes values on the intervals $\xi_{uk} = [a_{uk}, b_{uk})$ with probabilities p_{uk}, $k = 1, \dots, s_u$. One manifestation of such data would be when (alone, or after aggregation) an object w_u assumes a histogram as its observed value of Z. Such data are modal interval-valued observations, also called histogram-valued observations. One such dataset would be that of Table 3.10, in which the data (such as might be obtained after aggregation of a very large dataset compiled by, say, health insurance interests) display the histogram of weight of women by age groups (those aged in their twenties, those in their thirties, etc.). Thus, for example, we observe that for women in their thirties (observation w_2), 40% weigh between 116 and 124 pounds. Figure 3.5 displays the histograms for the respective age groups.

Table 3.10 Histogram-valued data: weight by age groups.

Age	w_u	
20s	w_1	{[70, 84), 0.02; [84, 96), 0.06; [96, 108), 0.24; [108, 120), 0.30; [120, 132), 0.24; [132, 144), 0.06; [144, 160), 0.08}
30s	w_2	{[100, 108), 0.02; [108, 116), 0.06; [116, 124), 0.40; [124, 132), 0.24; [132, 140), 0.24; [140, 150), 0.04}
40s	w_3	{[110, 125), 0.04; [125, 135), 0.14; [135, 145), 0.20; [145, 155), 0.42; [155, 165), 0.14; [165, 175), 0.04; [175, 185), 0.02}
50s	w_4	{[100, 114), 0.04; [114, 126), 0.06; [126, 138), 0.20; [138, 150), 0.26; [150, 162), 0.28; [162, 174), 0.12; [174, 190), 0.04}
60s	w_5	{[125, 136), 0.04; [136, 144), 0.14; [144, 152), 0.38; [152, 160), 0.22; [160, 168), 0.16; [168, 180), 0.06}
70s	w_6	{[135, 144), 0.04; [144, 150), 0.06; [150, 156), 0.24; [156, 162), 0.26; [162, 168), 0.22; [168, 174), 0.14; [174, 180), 0.04}
80s	w_7	{(100, 120), 0.02; [120, 135), 0.12; [135, 150), 0.16; [150, 165), 0.24; [165, 180), 0.32; [180, 195), 0.10; [195, 210), 0.04}

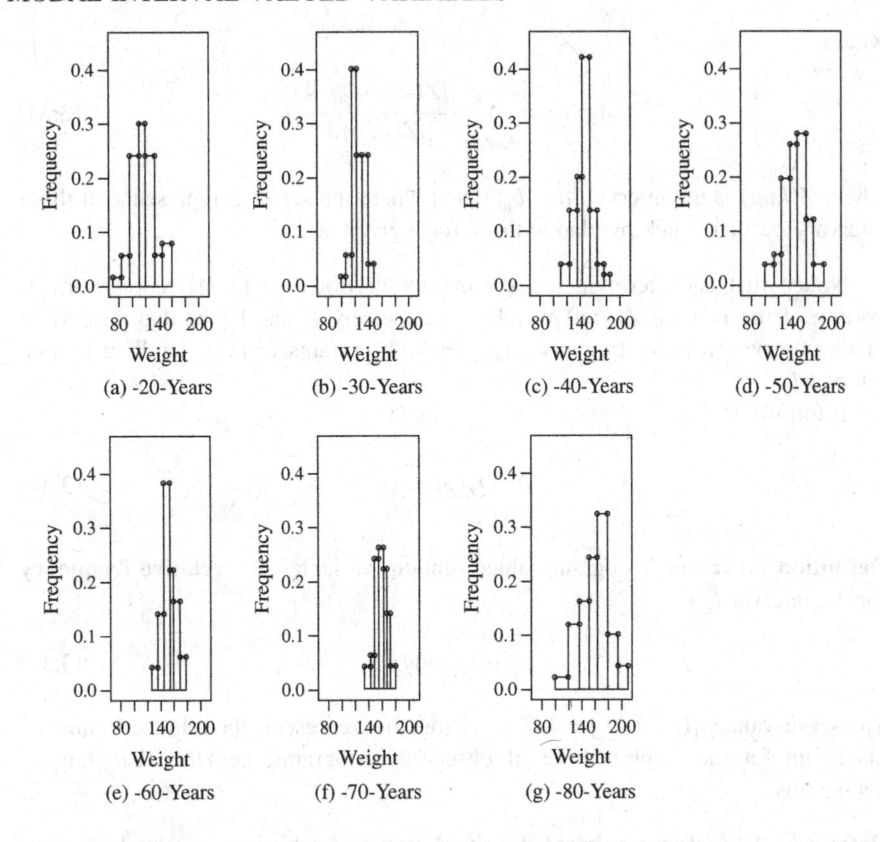

Figure 3.5 Histogram-valued data – weight by age groups.

By analogy with ordinary interval-valued variables (see Section 3.3), it is assumed that within each interval $[a_{uk}, b_{uk})$, each individual description vector $x \in vir(du)$ is uniformly distributed across that interval. Therefore, for each ξ_k,

$$P\{x \leq \xi_k | x \in vir(d_u)\} = \begin{cases} 0, & \xi_k < a_{uk}, \\ \frac{\xi_k - a_{uk}}{b_{uk} - a_{uk}}, & a_{uk} \leq \xi_k < b_{uk}, \\ 1, & \xi_k \geq b_{uk}. \end{cases} \tag{3.33}$$

A histogram of all the observed histograms can be constructed as follows. Let $I = [\min_{k,u\in E} a_{ku}, \max_{k,u\in E} b_{ku}]$ be the interval which spans all the observed values of Z in $\mathcal{Z}$, and let I be partitioned into r subintervals $I_g = [\zeta_{g-1}, \zeta_g)$, $g = 1, \ldots, r-1$, and $I_r = [\zeta_{r-1}, \zeta_r]$.

Definition 3.11: For a modal interval-valued (histogram-valued) random variable, the **observed frequency** for the interval $I_g = [\xi_{g-1}, \xi_g)$, $g = 1, \ldots, r$, is

$$O_Z(g) = \sum_{u\in E} \pi_Z(g; u) \tag{3.34}$$

where

$$\pi_Z(g; u) = \sum_{k \in Z(g)} \frac{||Z(k; u) \cap I_g||}{||Z(k; u)||} p_{uk} \tag{3.35}$$

where $Z(k; u)$ is the interval $[a_{uk}, b_{uk})$, and where the set $Z(g)$ represents all those intervals $Z(k; u)$ which overlap with I_g, for a given u. □

Notice that each term in the summation in Equation (3.34) represents that portion of the interval $Z(k; u)$ which is spanned by I_g and hence that proportion of its observed relative frequency (p_{uk}) which pertains to the overall histogram interval I_g.

It follows that

$$\sum_{g=1}^{r} O_Z(g) = m. \tag{3.36}$$

Definition 3.12: For histogram-valued random variables, the **relative frequency** for the interval I_g is

$$p_g = O_Z(g)/m. \tag{3.37}$$

The set of values $\{(p_g, I_g), g = 1, \ldots, r\}$ together represents the relative frequency histogram for the combined set of observed histograms, i.e., the histogram of histograms. □

Definition 3.13: The **empirical density function** for histogram-valued observations can be derived from

$$f(\xi) = \frac{1}{m} \sum_{u \in E} \sum_{k=1}^{s_u} \frac{I_{uk}(\xi)}{||Z(u; k)||} p_{uk}, \ \xi \in \Re, \tag{3.38}$$

where $I_{uk}(\cdot)$ is the indicator function that ξ is in the interval $Z(u; k)$. □

Definitions 3.14: For histogram-valued observations, the **symbolic sample mean** becomes

$$\bar{Z} = \frac{1}{2m} \sum_{u \in E} \left[\sum_{k=1}^{s_u} (b_{uk} + a_{uk}) p_{uk} \right], \tag{3.39}$$

and the **symbolic sample variance** is

$$S^2 = \frac{1}{3m} \sum_{u \in E} \left[\sum_{k=1}^{s_u} (b_{uk}^2 + b_{uk} a_{uk} + a_{uk}^2) p_{uk} \right] - \frac{1}{4m^2} \left[\sum_{u \in E} \sum_{k=1}^{s_u} (b_{uk} + a_{uk}) p_{uk} \right]^2. \tag{3.40}$$

□

Example 3.18. To illustrate, take the data of Table 3.10 (from Billard and Diday, 2003, Table 6). These data are observed histograms of the weights of women by age groups, such as could be obtained by aggregating a medical dataset detailing weights and age of individuals. Here the aggregation is conducted after identifying the concept age group.

Let us construct the histogram on the $r = 10$ intervals $[60, 75)$, $[75, 90), \ldots, [195, 210]$. Take the $g = 5$th interval $I_5 = [120, 135)$. Then, from Table 3.10, it follows that, from Equation (3.35),

$$\pi(5;\, 1) = \left(\frac{132-130}{132-120}\right)(0.24) + \left(\frac{135-132}{144-132}\right)(0.06) = 0.255;$$

likewise,

$$\pi(5;\, 2) = 0.5300,\ \ \pi(5;\, 3) = 0.1533,\ \ \pi(5;\, 4) = 0.1800,$$
$$\pi(5;\, 5) = 0.0364,\ \ \pi(5;\, 6) = 0.0000,\ \ \pi(5;\, 7) = 0.1200.$$

Hence, from Equation (3.34), the observed frequency is

$$O(g = 5) = 1.2747$$

and from Equation (3.37), the relative frequency is

$$p_5 = 1.2749/7 = 0.1821.$$

The complete set of histogram values $\{p_g, I_g, g = 1, \ldots, 10\}$ is given in Table 3.11 and displayed in Figure 3.6.

Table 3.11 Histogram: weights.

g	I_g	Observed frequency	Relative frequency
1	[60, 75)	0.00714	0.0010
2	[75, 90)	0.04286	0.0061
3	[90, 105)	0.24179	0.0345
4	[105, 120)	0.72488	0.1036
5	[120, 135)	1.27470	0.1821
6	[135, 150)	1.67364	0.2391
7	[150, 165)	1.97500	0.2821
8	[165, 180)	0.88500	0.1264
9	[180, 195)	0.13500	0.0193
10	[195, 210]	0.04000	0.0057

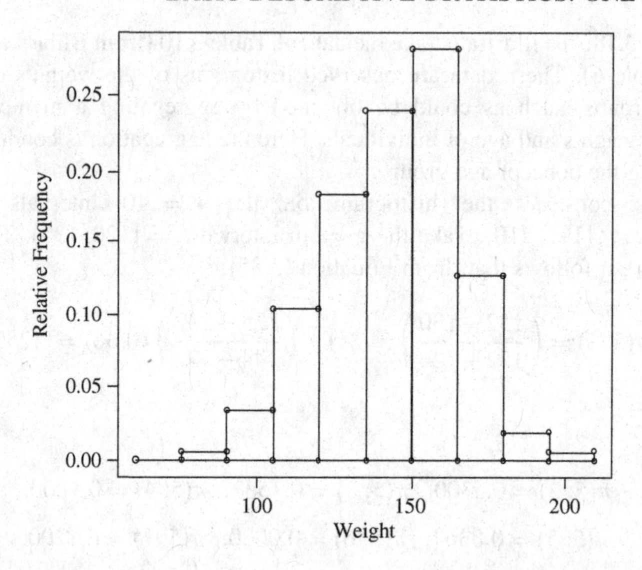

Figure 3.6 Histogram of histograms: weight × age group data.

The sample mean $\bar{X}$ is calculated from Equation (3.39) by

$$\bar{X} = \frac{1}{7}\left\{\left[\left(\frac{74+80}{2}\right)(0.02)+\ldots+\left(\frac{144+160}{2}\right)(0.08)\right]_{u=1} \right.$$
$$+\ldots$$
$$\left. +\left(\frac{100+120}{2}\right)(0.02)+\ldots+\left(\frac{195+210}{2}\right)(0.04)]_{u=7}\right\}$$
$$=143.9.$$

Likewise, the sample variance, from Equation (3.40) is

$$S^2 = \frac{1}{7}\left\{\left[\left(\frac{70^2+70\times 80+80^2}{3}\right)(0.02)+\ldots+\left(\frac{144^2+144\times 160\times 160^2}{3}\right)(0.08)\right]_{u=1}\right.$$
$$+\ldots$$
$$\left.+\left(\frac{100^2+100\times 120+120^2}{3}\right)(0.02)+\ldots\left(\frac{195^2+195\times 210+210^2}{3}\right)(0.04)\right]_{u=7}\right\}$$
$$-(143.9)^2$$
$$=447.5,$$

and hence the standard deviation is

$$S = 21.16.$$ □

Example 3.19. It is implicit in the formula of Equations (3.33)–(3.40) that the rules v hold. Suppose the data of Table 3.12 represented costs (in dollars) involved for a certain procedure at 15 different hospitals. Suppose further there is a minimum charge of \$175. This translates into a rule

$$v : \{Y < 175\} \Rightarrow \{p_k = 0\}. \tag{3.41}$$

The data for the hospitals $w_u, u = 7, \ldots, 15$, completely satisfy this rule and so their virtual description matches the actual data. However, for hospitals w_u with $u = 1, \ldots, 6$, some values satisfy this rule and some do not. Therefore, the corresponding virtual descriptions, represented by $w_u, u = u^*$, must be determined in effect by rescaling the relative frequencies over the virtual description space.

Let us take the object w_1; we find its virtual description $w_1{}^*$ as follows. The rescaling factor is

$$f_s = 1 - \left[0.006 + 0.043 + 0.100 + 0.187 + \left(\frac{175 - 170}{180 - 170}\right)(0.256)\right] = 0.536.$$

Table 3.12 Hospital histogram-valued data.

w_u	Observed cost distribution
w_1	{[130, 140), 0.006; [140, 150), 0.043; [150, 160), 0.10; [160, 170), 0.187; [170, 180), 0.256; [180, 190), 0.237; [190, 200), 0.118; [200, 210), 0.042; [210, 220), 0.012}
w_2	{[139, 153), 0.010; [153, 167), 0.044; [167, 181), 0.224; [181, 195), 0.355; [195, 209), 0.242; [209, 223), 0.105; [223, 237), 0.016; [237, 251), 0.004}
w_3	{[145, 155), 0.006; [155, 165), 0.018; [165, 175), 0.054; [175, 185), 0.163; [185, 195), 0.244; [195, 205), 0.263; [205, 215), 0.163; [215, 225), 0.069; [225, 235), 0.018; [235, 245), 0.002}
w_4	{[150, 165), 0.011; [165, 180), 0.084; [180, 195), 0.282; [195, 210), 0.375; [210, 225), 0.196; [225, 240), 0.046; [240, 255), 0.006}
w_5	{[165, 180), 0.036; [180, 195), 0.198; [195, 210), 0.387; [210, 225), 0.269; [225, 240), 0.077; [240, 255), 0.024; [255, 270), 0.009}
w_6	{[155, 170), 0.001; [170, 185), 0.030; [185, 200), 0.226; [200, 215), 0.337; [215, 230), 0.304; [230, 245), 0.086; [245, 260), 0.010; [260, 275), 0.005; [275, 290), 0.001}
w_7	{[175, 185), 0.019; [185, 195), 0.062; [195, 205), 0.144; [205, 215), 0.256; [215, 225), 0.257; [225, 235), 0.156; [235, 245), 0.081; [245, 255), 0.019; [255, 265), 0.006}
w_8	{[180, 200), 0.055; [200, 215), 0.252; [215, 230), 0.345; [230, 245), 0.272; [245, 260), 0.065; [260, 275), 0.010; [275, 290), 0.001}
w_9	{[175, 190), 0.007; [190, 205), 0.083; [205, 220), 0.289; [220, 235), 0.362; [235, 250), 0.194; [250, 265), 0.055; [265, 280), 0.010}
w_{10}	{[180, 195), 0.008; [195, 210), 0.075; [210, 225), 0.290; [225, 240), 0.356; [240, 255), 0.225; [255, 270), 0.44; [270, 285), 0.002}

Table 3.12 Continued.

w_u	Observed cost distribution
w_{11}	{[195, 210), 0.009; [210, 225), 0.169; [225, 240), 0.384; [240, 255), 0.313; [255, 270), 0.102; [270, 285), 0.021; [285, 300), 0.002}
w_{12}	{[192, 204), 0.006; [204, 216), 0.037; [216, 228), 0.138; [228, 240), 0.288; [240, 252), 0.275; [252, 264), 0.182; [264, 276), 0.064; [276, 288), 0.011}
w_{13}	{[190, 205), 0.006; [205, 220), 0.021; [220, 235), 0.187; [235, 250), 0.327; [250, 265), 0.318; [265, 280), 0.102; [280, 295), 0.034; [295, 310), 0.005}
w_{14}	{[205, 220), 0.009; [220, 235), 0.111; [235, 250), 0.333; [250, 265), 0.345; [265, 280), 0.178; [280, 295), 0.019; [295, 310), 0.005}
w_{15}	{[220, 235), 0.027; [235, 250), 0.187; [250, 265), 0.345; [265, 280), 0.327; [280, 295), 0.094; [295, 310), 0.018; [310, 325), 0.002}

Hence, the virtual frequency for the interval [175, 180) is

$$\left(\frac{180-175}{180-170}\right)\left(\frac{0.256}{0.536}\right)=0.238$$

and the virtual relative frequencies for the intervals [180, 190), [190, 200), [200, 210), [210, 220) are, respectively,

$$\left(\frac{0.237}{0.536}, \frac{0.118}{0.536}, \frac{0.041}{0.536}, \frac{0.012}{0.536}\right)=(0.220, 0.442, 0.078, 0.022).$$

The virtual relative frequencies for w_u, $u = 1^*, \ldots, 6^*$, are summarized in Table 3.13.

Then, the overall histogram is found from Equations (3.33)–(3.36) where the data are given by the virtual descriptions of Table 3.13. The resulting observed and relative frequencies are shown in Table 3.14, and plotted in Figure 3.7.

The descriptive statistics follow from Equations (3.39) and (3.40) where the $\{(a_{uk}, b_{uk}, p_{uk}),\ k = 1, \ldots, m\}$ values used are those for the virtual description histograms. Thus, from Table 3.13 and Equation (3.39), the mean cost is

$$\begin{aligned}\bar{X} &= \frac{1}{15}\left\{\left[\left(\frac{180+175}{2}\right)(0.023)+\ldots+\left(\frac{220+210}{2}\right)(0.022)\right]_{u=1}\right.\\&\quad+\ldots\\&\quad\left.+\left[\left(\frac{220+235}{2}\right)(0.027)+\ldots+\left(\frac{310+325}{2}\right)(0.002)\right]_{u=15}\right\}\\&=222.709.\end{aligned}$$

Table 3.13 Virtual hospital histograms.

u	Virtual cost distribution
1*	{[175, 180), 0.023; [180, 190), 0.442; [190, 200), 0.220; [200, 210), 0.078; [210, 220), 0.022}
2*	{[175, 181), 0.117; [181, 195), 0.434; [195, 209), 0.296; [209, 223), 0.123; [223, 237), 0.020; [237, 251), 0.005}
3*	{[175, 185), 0.177; [185, 195), 0.265; [195, 205), 0.285; [205, 215), 0.177; [215, 225), 0.075; [225, 235), 0.020; [235, 245), 0.002}
4*	{[175, 180), 0.030; [180, 195), 0.302; [195, 210), 0.402; [210, 225), 0.210; [225, 240), 0.049; [240, 255), 0.006}
5*	{[175, 180), 0.012; [180, 195), 0.203; [195, 210), 0.397; [210, 225), 0.276; [225, 240), 0.079; [240, 255), 0.025; [255, 270), 0.009}
6*	{[175, 185), 0.020; [185, 200), 0.229; [200, 215), 0.341; [215, 230), 0.307; [230, 245), 0.087; [245, 260), 0.010; [260, 275), 0.005; [275, 290), 0.001}
7*, ..., 15*	Same as for $u = 7, \ldots, 15$ in Table 3.12

Table 3.14 Frequencies: hospital data.

g	I_g	No rules		Rule v	
		Observed frequency	Relative frequency	Observed frequency	Relative frequency
1	[125, 150)	0.600	0.004	—	—
2	[150, 175)	0.766	0.051	—	—
3	[175, 200)	2.893	0.193	3.517	0.235
4	[200, 225)	4.495	0.300	4.682	0.312
5	[225, 250)	4.176	0.278	4.189	0.279
6	[250, 275)	2.193	0.146	2.193	0.146
7	[275, 300)	0.398	0.027	0.398	0.027
8	[300, 325)	0.021	0.001	0.021	0.001
	$\bar{X}$	221.182		222.709	
	S^2	843.439		716.188	
	S	29.042		26.762	

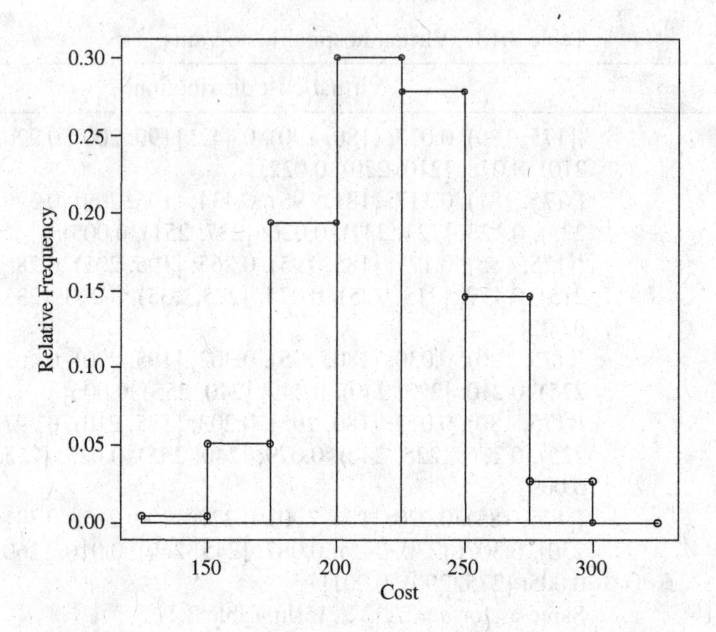

Figure 3.7 Histogram of histograms: hospital data.

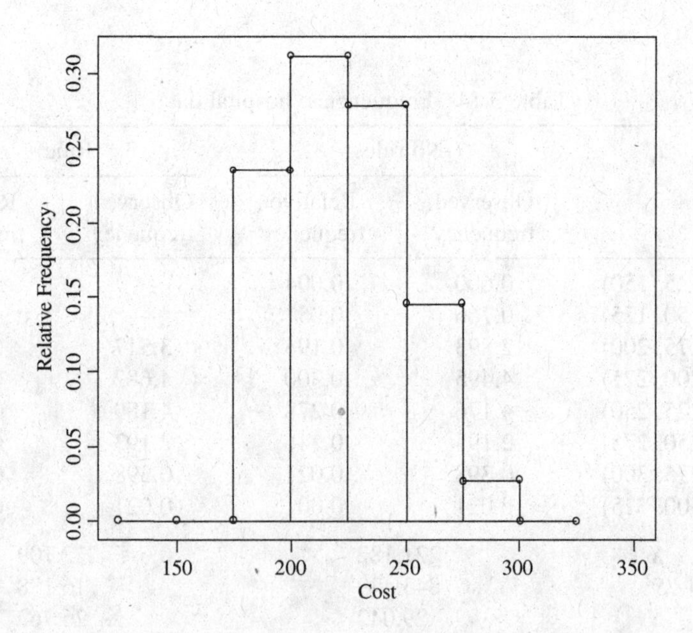

Figure 3.8 Histogram of histograms: hospital data rule $Y < 175$.

Likewise, we calculate from Table 3.13 and Equation (3.41),

$$S^2 = 719.188 \quad \text{and} \quad S = 26.762.$$

The overall empirical histogram relative frequencies along with $\bar{X}$, S^2, and S for the cost data of Table 3.12 in the absence of any rules follow easily from Equations (3.33)–(3.40); see Table 3.14 and Figure 3.8. □

Exercises

E3.1. Suppose the blood data of Table 3.5 have no restrictions/rules imposed on the realizations of the random variables. Construct the histogram for each of (i) Y_1 = Pulse Rate, Y_2 = Systolic Blood Pressure, and (iii) Diastolic Blood Pressure, respectively.

E3.2. Calculate the sample mean and sample variance for each of the Y_1, Y_2, Y_3 random variables, respectively, for the setup of Exercise E3.1.

E3.3. For the blood data of Table 3.5, suppose there is a rule on Y_1 = Pulse Rate

$$v : A = \{Y_1 \geq 55\}.$$

Calculate the sample mean and sample variance of Y_1, and build the histogram for Y_1 under this rule. How do these results compare to those for which there were no restrictions (see Exercise E3.1)?

E3.4. For the blood data of Table 3.5, suppose there is a rule

$$\text{(i)} \quad v : A = \{Y_1 \leq 75\} \Rightarrow B = \{Y_2 \leq 170\},$$

$$\text{(ii)} \quad v : A = \{Y_1 \leq 75\} \Rightarrow B = \{Y_2 \leq 160\}.$$

Calculate the sample mean and sample variance, and build the histogram for each of Y_1, Y_2, Y_3, respectively, for each separate rule (i) and (ii).

E3.5. Consider the mushroom data of Table 3.1. Suppose only edible mushrooms could be considered. Obtain the histogram for (i) Y_1 = Pileus Cap Width, (ii) Y_2 = Stipe Length, and (iii) Y_3 = Stipe Width, for the resulting data.

E3.6. For the edible mushrooms of Table 3.1 (see Exercise E3.5), find the sample mean and sample variance for each of Y_1, Y_2, and Y_3, respectively.

E3.7. Find the sample mean, sample variance, and histogram for Y_1 = Flight Time for the airline flight data of Table 2.7.

E3.8. Calculate the sample mean, sample variance, and histogram for the random variable Y_5 = Departure Delay for the airline flight data of Table 2.7.

E3.9. Redo Exercise E3.8 where now only those airline values for which the Weather Delay = $Y_6 \leq 0.03$ are used.

E3.10. Instead of ignoring these observations with $Y_6 > 0.03$ (as in Exercise E3.9), suppose now that, when weather delays are such that $Y_6 > 0.03$, the Departure Delay time (Y_5) is recorded as being [0, 60] minutes. Find the sample mean, sample variance, and histogram values for Y_5 under this rule.

E3.11. Consider the hospital data of Table 3.12. Suppose there is a minimum charge of $200. Calculate the sample mean, sample variance, and histogram for $Y =$ Costs, for this situation (see Example 3.19).

E3.12. Take the bird data of Table 2.19. Calculate the histogram for $Y_1 =$ Color. Find the sample mean and sample variance for $Y_2 =$ Height.

E3.13. For the health insurance data of Table 2.6, build the histogram for (i) $Y_5 =$ Marital Status and (ii) $Y_6 =$ Number of Parents Alive.

E3.14. Redo Exercise E3.13 but this time use the data format from Table 2.2a. Compare these results to those obtained in Exercise E3.13.

References

Bertrand, P. and Goupil, F. (2000). Descriptive Statistics for Symbolic Data. In: *Analysis of Symbolic Data: Explanatory Methods for Extracting Statistical Information from Complex Data* (eds. H.-H. Bock and E. Diday). Springer-Verlag, Berlin, 103–124.

Billard, L. and Diday, E. (2003). From the Statistics of Data to the Statistics of Knowledge: Symbolic Data Analysis. *Journal of the American Statistical Association* 98, 470–487.

Chouakria, A., Diday, E., and Cazes, P. (1998). An Improved Factorial Representation of Symbolic Objects. In: *Knowledge Extraction from Statistical Data*, European Commission Eurostat, Luxembourg, 301–305.

De Carvalho, F. A. T. (1994). Proximity Coefficients Between Boolean Symbolic Objects. In: *New Approaches in Classification and Data Analysis* (eds. E. Diday, Y. Lechevallier, M. Schader, P. Bertrand, and B. Burtschy). Springer-Verlag, Berlin, 387–394.

De Carvalho, F. A. T. (1995). Histograms in Symbolic Data Analysis. *Annals of Operations Research* 55, 299–322.

Diday, E. (1995). Probabilistic, Possibilistic and Belief Objects for Knowledge Analysis. *Annals of Operations Research* 55, 227–276.

Raju, S. R. K. (1997). Symbolic Data Analysis in Cardiology. In: *Symbolic Data Analysis and its Applications* (eds. E. Diday and K. C. Gowda). CEREMADE, Université Paris-Dauphine, 245–249.

Schweizer, B. and Sklar, A. (1983). *Probabilistic Metric Spaces*. North-Holland, New York.

Verde, R. and De Carvalho, F. A. T. (1998). Dependence Rules' Influence on Factorial Representation of Boolean Symbolic Objects. In: *Knowledge Extraction from Statistical Data*. European Commission Eurostat, Luxembourg, 287–300.

4

Descriptive Statistics: Two or More Variates

Many of the principles developed for the univariate case can be expanded to a general p-variate case, $p > 1$. We shall focus attention on obtaining joint histograms for $p = 2$; the general case follows analogously. For the sake of discussion, suppose we are interested in the specific variables Z_1 and Z_2 over the space $\mathcal{Z} = \mathcal{Z}_1 \times \mathcal{Z}_2$. We consider multi-valued, interval-valued, and modal-valued variables, in turn, starting with multi-valued variables in Section 4.1. In Section 4.2, we consider interval-valued variables. Modal-valued variables are covered in Section 4.3 and Section 4.4. Logical dependency rules, even seemingly innocuous ones, can have quite an impact on these histograms. Therefore, in Section 4.5 this aspect receives further special attention through the analysis of a baseball dataset. Then, in Section 4.6, we discuss measures of dependency between two variables.

4.1 Multi-Valued Variables

Recall that we have observations $Y(w_u) = (Y_1(w_u), \ldots, Y_p(w_u))$, $w_u \in E = \{w_1, \ldots, w_m\}$, as realizations of the random variable $Y = (Y_1, \ldots, Y_p)$. When Y_j is a multi-valued variable, it takes a list of discrete values from the space $\mathcal{Y}_j$. We are interested in determining the joint histogram for two such multi-valued variables $Y_{j_1} \equiv Z_1$ and $Y_{j_2} \equiv Z_2$, say, taking values in $\mathcal{Z}_1$ and $\mathcal{Z}_2$, respectively. As for the univariate case, a rule may be present, and also these rules may involve variables other than those for which the descriptive statistics are being calculated (here j_1 and j_2). Sometimes for descriptive simplicity, we write $u \in E$ to mean $w_u \in E$. Analogously to the definition of the observed frequency of specific values for a single variable given in Equation (3.3), we have

Symbolic Data Analysis: Conceptual Statistics and Data Mining L. Billard and E. Diday

Definition 4.1: For two multi-valued random variables Z_1 and Z_2, the **observed frequency** that $(Z_1 = \xi_1, Z_2 = \xi_2)$ is given by

$$O_{Z_1,Z_2}(\xi_1, \xi_2) = \sum_{u \in E} \pi_{Z_1,Z_2}(\xi_1, \xi_2; u) \tag{4.1}$$

where

$$\pi_{Z_1,Z_2}(\xi_1, \xi_2; u) = \frac{|\{x \in vir(d_u) | x_{Z_1} = \xi_1, x_{Z_2} = \xi_2\}|}{|vir(d_u)|} \tag{4.2}$$

is the percentage of the individual description vectors $\boldsymbol{x} = (x_1, x_2)$ in $vir(d_u)$ for which $(Z_1 = \xi_1, Z_2 = \xi_2)$, and where $vir(d_u)$ is the virtual description of the description vector d_u as defined in Definition 3.2 and Definition 3.1, respectively. □

Note that $\pi(\cdot)$ is a real number on $\Re$ in contrast to its being a positive integer for classical data. We can show that

$$\sum_{\xi_1 \in \mathcal{Z}_1, \xi_2 \in \mathcal{Z}_2} O_{Z_1,Z_2}(\xi_1, \xi_2) = m.$$

Here, the summation in Equation (4.1) is over those $w_u \in E$ for which $|vir(d_u)| \neq 0$, with m being the total of number of such objects. Then, we have the following:

Definition 4.2: For the multi-valued random variables Z_1 and Z_2, the **empirical joint frequency distribution** is the set of pairs $[\boldsymbol{\xi}, \frac{1}{m} O_{Z_1 Z_2}(\boldsymbol{\xi})]$ where $\boldsymbol{\xi} = (\xi_1, \xi_2)$ for $\boldsymbol{\xi} \in \mathcal{Z} = \mathcal{Z}_1 \times \mathcal{Z}_2$; and the **empirical joint histogram** of Z_1 and Z_2 is the graphical set $[\boldsymbol{\xi}, O_{Z_1 Z_2}(\boldsymbol{\xi})]$. □

Example 4.1. To illustrate, we find the joint histogram for the cancer data given previously in Table 3.2. We use the individual description vectors contained in the virtual descriptions $vir(d_u)$. Thus, we have that $O_{Z_1,Z_2}(\xi_1 = 0, \xi_2 = 0) \equiv O(0, 0)$, say, is, from Equation (4.1) and Equation (4.2),

$$O(0, 0) = \frac{0}{1} + \frac{1}{3} + \frac{0}{1} + \frac{0}{2} + \frac{1}{1} + \frac{0}{2} + \frac{0}{2} + \frac{0}{2} = \frac{4}{3}$$

where, for example, $\pi(0, 0; 1) = 0/1$ since the individual vector $(0, 0)$ occurs no times out of a total of $|vir(d_1)| = 1$ vectors in $vir(d_1)$; likewise, $\pi(0, 0; 2) = 1/3$ since $(0, 0)$ is one of three individual vectors in $vir(d_2)$ and so on, and where the summation does not include the $u = 5$ term since $|vir(d_5)| = 0$. Similarly, we can show that

$$O(0, 1) = 0; \;\; O(0, 2) = 0; \;\; O(0, 3) = 0;$$
$$O(1, 0) = 1/3; \;\; O(1, 1) = 4/3; \;\; O(1, 2) = 5/2; \;\; O(1, 3) = 5/2.$$ □

4.2 Interval-Valued Variables

For the symbolic random variable $\mathbf{Y} = \{Y_1, \ldots, Y_p\}$, recall that an interval-valued variate Y_j takes values on a subset $\mathcal{B}_j$ of the real line $\Re$, $\mathcal{B}_j \subseteq \mathcal{R}$, with $\mathcal{B}_j = \{[\alpha, \beta], -\infty < \alpha, \beta < \infty\}$. We want to determine the joint histogram for two such variables. As before, rules may or may not be present. Let us suppose that the specific variables of interest $Y_{j_1} \equiv Z_1$ and $Y_{j_2} \equiv Z_2$ take values for the observation w_u on the rectangle $\mathbf{Z}(u) = Z_1(u) \times Z_2(u) = ([a_{u1}, b_{u1}], [a_{u2}, b_{u2}])$ for each $w_u \in E$. As before, we make the intuitive choice of assuming individual vectors $x \in vir(d_u)$ are each uniformly distributed over the respective intervals $Z_1(u)$ and $Z_2(u)$.

Definition 4.3: For interval-valued random variables (ZS_1, Z_2), the **empirical joint density function** is

$$f(\xi_1, \xi_2) = \frac{1}{m} \sum_{u \in E} \frac{I_u(\xi_1, \xi_2)}{||Z(u)||} \tag{4.3}$$

where $I_u(\xi_1, \xi_2)$ is the indicator function that (ξ_1, ξ_2) is or is not in the rectangle $Z(u)$ and where $||Z(u)||$ is the area of this rectangle. □

Definition 4.4: Analogously with Equation (3.20), we can find the **joint histogram** for (Z_1, Z_2) by graphically plotting $\{R_{g_1 g_2}, p_{g_1 g_2}\}$ over the rectangles $R_{g_1 g_2} = \{[\xi_{1,g_1-1}, \xi_{1g_1}) \times [\xi_{2,g_2-1}, \xi_{2g_2})\}$, $g_1 = 1, \ldots, r_1$, $g_2 = 1, \ldots, r_2$, where

$$f_{g_1 g_2} = \sum_{u \in E} \frac{||Z(u) \cap R_{g_1 g_2}||}{||Z(u)||}, \tag{4.4}$$

i.e., $f_{g_1 g_2}$ is the number of observations that fall in the rectangle $R_{g_1 g_2}$ and is not necessarily an integer value (except in the special case of classical data). The relative frequency that an arbitrary individual description vector lies in the rectangle $R_{g_1 g_2}$ is therefore

$$p_{g_1 g_2} = f_{g_1 g_2}/m. \tag{4.5}$$

□

Notice that each term in the summation of Equation (4.4) is in effect that fraction of the observed rectangle $Z(u)$ which overlaps with the histogram rectangle $R_{g_1 g_2}$ for given (g_1, g_2).

Example 4.2. We illustrate these concepts with the data of Table 3.5. In particular, let us consider the systolic and diastolic blood pressure levels, Y_2 and Y_3, respectively. Suppose further that there is a logical rule specifying that the diastolic blood pressure must be less than the systolic blood pressure, namely

$$v_0 : Y_3 \leq Y_2, \tag{4.6}$$

i.e., for each observation w_u, $b_{u3} \leq a_{u2}$. All the observations satisfy this rule, except the observation for $u = 11$ where the rule v is violated since

$Y_2(w_{11}) > Y_3(w_{11})$.Therefore, $|vir(d_{11})| = 0$, and so our subsequent calculations would not include this observation, making $m \equiv 14$.

Suppose we want to find the joint histogram for these two variables. Thus, $Z_1 \equiv Y_2$ and $Z_2 \equiv Y_3$. Let us suppose we want to construct the histogram on the rectangles $R_{g_2g_3} = [\zeta_{2,g_2-1}, \zeta_{2,g_2}) \times [\zeta_{3,g_3-1}, \zeta_{3,g_3})$, $g_2 = 1, \ldots, 6$, $g_3 = 1, \ldots, 4$, where the sides of these rectangles are the intervals $[90, 110) \times [45, 65), \ldots, [90, 110) \times [95, 125), \ldots, [190, 210) \times [95, 125)$; see Table 4.1.

Table 4.1 Systolic/diastolic blood pressures – rule v_0.

g_2	g_3 / $I_2 \backslash I_3$	Joint frequencies (Y_2, Y_3): $f_{g_2g_3}$ 1 [45, 65)	2 [65, 80)	3 [80, 95)	4 [95, 125]	Frequency Y_2	Marginal distribution Y_2
1	[90, 110)	1.500	1.167	0.333	0.000	3.000	0.214
2	[110, 130)	0.000	0.811	1.378	0.571	2.760	0.197
3	[130, 150)	0.000	0.357	2.248	1.338	3.943	0.282
4	[150, 170)	0.000	0.158	0.701	0.701	1.560	0.111
5	[170, 190)	0.000	0.062	0.673	0.792	2.027	0.145
6	[190, 210]	0.000	0.000	0.105	0.105	0.710	0.051
Frequency Y_3		1.500	2.555	5.438	4.507	14	
Marginal distribution Y_3		0.107	0.183	0.388	0.322		1.000

Then, from Equation (4.6), we can calculate the observed frequency $f_{g_2g_3}$ that an arbitrary description vector lies in the rectangle $R_{g_2g_3}$. Therefore, for $g_2 = 5$ and $g_3 = 3$, for example, the rectangle R_{53} is $R_{53} = \{[170, 190) \times [80, 95)\}$. From Equation (4.6), it follows that the frequency

$$
\begin{aligned}
f_{53} = & 0+0+\left(\frac{180-170}{180-140}\right)\left(\frac{95-90}{100-90}\right)_{u=3} +0+0+0+0 \\
& +\left(\frac{190-170}{190-110}\right)\left(\frac{95-80}{110-70}\right)_{u=8} +\left(\frac{180-170}{180-138}\right)\left(\frac{95-90}{110-90}\right)_{u=9} \\
& +0+0+0+\left(\frac{190-170}{194-175}\right)\left(\frac{95-90}{100-90}\right)_{u=14} +0 \\
= & 0.673;
\end{aligned}
$$

hence the relative frequency is

$$p_{53} = 0.673/14 = 0.048.$$

Similarly, for $g_2 = 3$ and $g_3 = 3$, the rectangle $R_{33} = \{[130, 150) \times [80, 95)\}$ has an observed frequency of

$$\begin{aligned}
f_{33} =& 0+0+\left(\frac{150-140}{180-140}\right)\left(\frac{95-90}{100-90}\right)_{u=3}+\left(\frac{142-130}{142-110}\right)\left(\frac{95-80}{108-80}\right)_{u=4}+0\\
&+\left(\frac{142-134}{142-134}\right)\left(\frac{95-80}{110-80}\right)_{u=6}+\left(\frac{150-130}{160-130}\right)\left(\frac{90-80}{90-76}\right)_{u=7}\\
&+\left(\frac{150-130}{190-110}\right)\left(\frac{95-80}{110-70}\right)_{u=8}+\left(\frac{150-138}{180-138}\right)\left(\frac{95-90}{110-90}\right)_{u=9}\\
&+\left(\frac{150-130}{150-110}\right)\left(\frac{95-80}{100-78}\right)_{u=10}+0+\left(\frac{138-130}{138-125}\right)\left(\frac{95-80}{99-78}\right)_{u=13}\\
&+(0)\left(\frac{95-90}{100-90}\right)_{u=14}+(0)\left(\frac{82-80}{82-70}\right)_{u=15}\\
=& 2.248;
\end{aligned}$$

hence

$$\begin{aligned}
p_{33} &= P\{x \in [130, 150) \times [80, 95)\}\\
&= 2.248/14 = 0.161.
\end{aligned}$$

Table 4.1 provides all the observed joint frequencies $f_{g_2 g_3}$. This table also gives the marginal observed frequencies as well as the marginal probabilities which correspond as they should to the observed frequencies f_{g_j} and the observed probabilities p_{g_j}, $j = 2, 3$, of the univariate case were these obtained directly from Equation (3.20). A plot of the joint histogram is given in Figure 4.1.

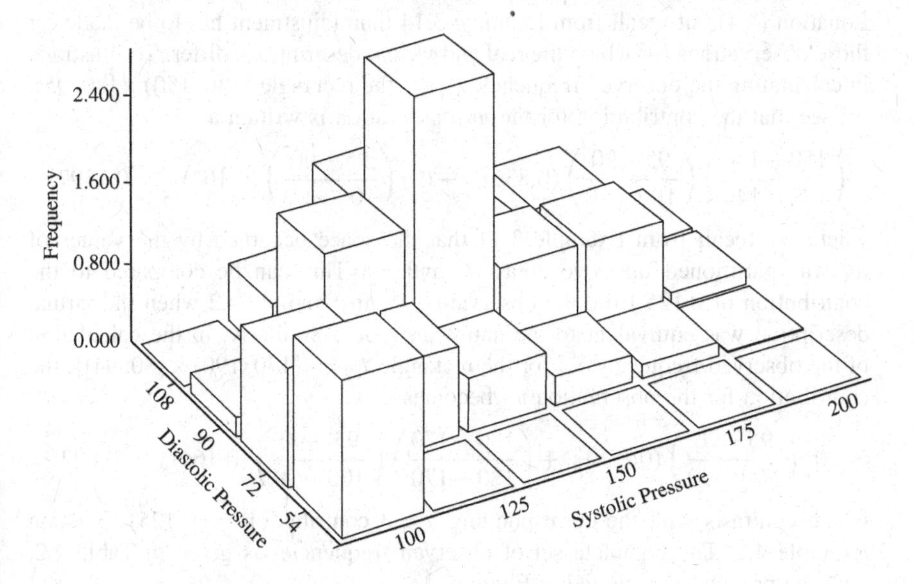

Figure 4.1 Joint histogram: blood data – rule v_0. □

Example 4.3. Suppose now that the rule v_4 (of Example 3.15) applies. That is,

$$v_4 = \begin{cases} v_0: & Y_3 \le Y_2, \\ v_3: & A = \{Y_1 \le 70\} \Rightarrow B = \{Y_2 \le 170\}. \end{cases} \tag{4.7}$$

As before, the observations w_u with $u = 11, 12$ are each logically invalid under v_4 and so have a virtual description $|vir(d_u)| = 0$, while the observations w_3 and w_{14} are partially valid under u_4; see Figures 3.4(d) and 4.2 for a graphical view of the virtual description of these observations.

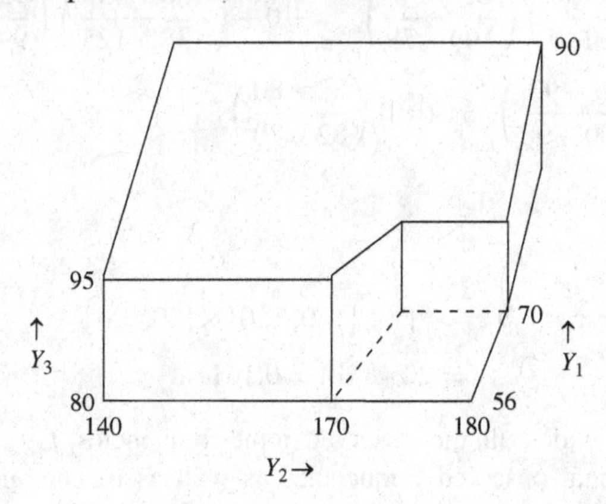

Figure 4.2 Virtual description of $u = 3$ under rule v_4.

To calculate the observed frequencies $f_{g_1g_2}$ on the rectangle $R_{g_1g_2}$, we use Equation (4.4) but recall from Example 3.14 that adjustment has to be made for those observations for which the real and virtual descriptions differ. To illustrate, in calculating the observed frequency f_{33} for the rectangle $[130, 150) \times [80, 95)$, we see that the contribution for the w_3 observation is written as

$$\left(\frac{150-140}{170-140}\right)\left(\frac{95-90}{100-90}\right)(0.836)_{R_1} + (0)\left(\frac{95-90}{100-90}\right)(0.164)_{R_2} = 0.1397,$$

where we recall from Example 3.14 that the space occupied by the values of w_3 was partitioned into two areas R_1 and R_2. This can be compared to the contribution of 0.125 from the observation w_3 in Example 4.2 when the virtual description was equivalent to the entire w_3 space. Similarly, in the calculation of the observed frequency f_{53} for the rectangle $I_{53} = \{[170, 190) \times [80, 94]\}$, the contribution for the observation w_3 becomes

$$(0)\left(\frac{95-90}{100-90}\right)(0.836)_{R_1} + \left(\frac{180-170}{180-170}\right)\left(\frac{95-90}{100-90}\right)(0.164)_{R_2} = 0.082$$

which contrasts with the corresponding $u = 3$ contribution $(= 0.125)$ to f_{53} in Example 4.2. The complete set of observed frequencies is given in Table 4.2, with the probabilities plotted in Figure 4.3.

Table 4.2 Systolic/diastolic blood pressures – rule v_4.

g_2		Joint frequencies (Y_2, Y_3): $f_{g_2 g_3}$				Frequency Y_2	Marginal distribution Y_2
		1	2	3	4		
g_1	$I_1 \backslash I_2$	[45, 65)	[65, 80)	[80, 95)	[95, 125]		
1	[90, 110)	1.500	1.167	0.333	0.000	3.000	0.231
2	[110, 130)	0.000	0.811	1.378	0.571	2.760	0.212
3	[130, 150)	0.000	0.357	**2.262**	**1.352**	3.971	0.306
4	[150, 170)	0.000	0.158	**0.730**	**0.730**	1.617	0.124
5	[170, 190)	0.000	0.062	**0.630**	**0.749**	1.442	0.111
6	[190, 210]	0.000	0.000	0.105	0.105	0.210	0.016
Frequency Y_3		1.500	2.555	5.938	4.007	13	
Marginal distribution Y_3		0.107	0.183	0.424	0.286		1.000

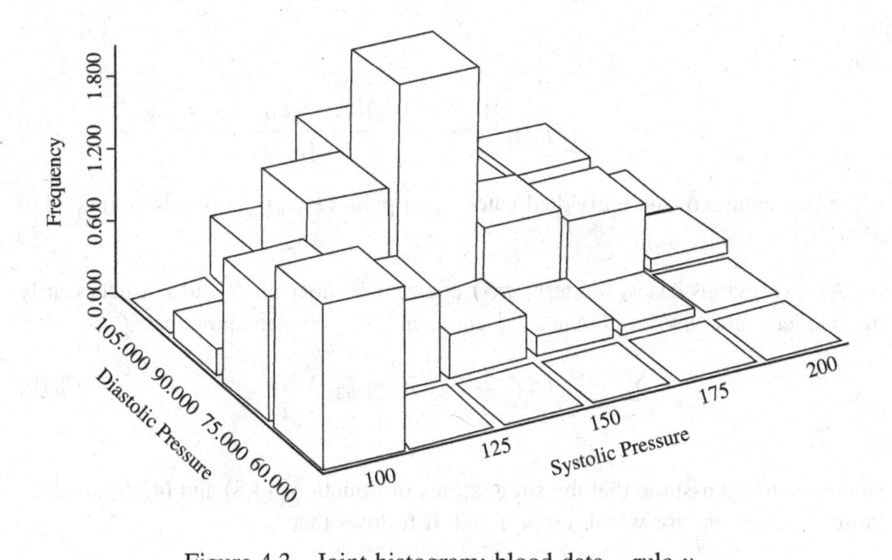

Figure 4.3 Joint histogram: blood data – rule v_4.

4.3 Modal Multi-Valued Variables

Modal multi-valued variables consist of a 'list' of categorical or qualitative variables with associated probabilities. The dataset of Table 4.3 is an example, where the variable Y_1 = Gender takes the realization $\xi = \{\text{male } p_1;\ \text{female } p_2\}$ with $p_1 + p_2 = 1$, i.e., an observation has Gender = male with probability p_1 and Gender = female with probability p_2. Definitions for the observed frequencies and empirical joint distributions given in Section 4.1 for non-modal multi-valued categorical variables can be extended to the present case.

For concreteness, let the modal multi-valued variable Y_j take values, for observations $w_u, u = 1, \ldots, m,\ j = 1, \ldots, p,$

$$Y_j(w_u) = \{[\xi_{uj1}, p_{uj1}; \ldots ; \xi_{ujs_j}, p_{ujs_j}]\}$$

with $\sum_{k=1}^{s_j} p_{ujk} = 1$. That is, the variable Y_j takes s_j possible values $\{\xi_{jk},\ k = 1, \ldots, s_j\}$ with respective probabilities $\{p_{ujk},\ k = 1, \ldots, s_j\}$ for observation $w_u, u = 1, \ldots, m$. Suppose our interest is in finding the joint histogram for the specific variables $Y_{j_1} \equiv Z_1$ and $Y_{j_2} \equiv Z_2$. Sometimes for notational simplicity, we shall write '$u \in E$' to mean the observations w_u from the set $E = \{w_1, \ldots, w_m\}$.

Definition 4.5: For the modal multi-valued variables $Z_j = \{\xi_{jk_j},\ k_j = 1, \ldots, s_j\},\ j = 1, 2$, the **observed frequency** that $(Z_1 = \xi_{1k_1},\ Z_2 = \xi_{2k_2})$ is

$$O(Z_1 = \xi_{1k_1}, Z_2 = \xi_{2k_2}) = \sum_{u \in E} \pi_{Z_1,Z_2}(\xi_{1k_1}, \xi_{2k_2}; u) \tag{4.8}$$

where

$$\pi_{Z_1,Z_2}(\xi_{1k_1}, \xi_{2k_2}; u) = p_{u1k_1} p_{u2k_2} \frac{|\{x \in vir(d_u) | x_{Z_1} = \xi_{1k_1}, x_{Z_2} = \xi_{1k_2}\}|}{|vir(d_u)|} \tag{4.9}$$

is the percentage of thè individual categorical pairs (ξ_{1k_1}, ξ_{2k_2}) that is contained in $vir(d_u)$. □

As in previous cases, the term $\pi(\cdot)$ is a real number on $\Re$ and not necessarily the integer value always obtained for classical data. We can show that

$$\sum_{k_1,k_2} \sum_{\substack{\xi_{1k_1} \in Z_1 \\ \xi_{2k_2} \in Z_2}} O(Z_1 = \xi_{1k_1}, Z_2 = \xi_{2k_2}) = m, \tag{4.10}$$

where again we assume that the summations in Equations (4.8) and (4.10) are over those observations for which $vir(d_u) \neq 0$. It follows that

Definition 4.6: The **empirical joint histogram** of Z_1 and Z_2 is the set of pairs

$$\{[\xi_{1k_1}, \xi_{2k_2}; O_{Z_1 Z_2}(\xi_{1k_1}, \xi_{2k_2})],\ k_j = 1, \ldots,\ s_j,\ j = 1, 2\}.$$ □

Weighted joint histograms can be obtained analogously. In this case, the $\pi(\cdot)$ term in Equation (4.9) becomes

$$\pi^*_{Z_1,Z_2}(\xi_{1k_1}, \xi_{2k_2}; u) = w^*_u \pi_{Z_1,Z_2}(\xi_{1k_1}, \xi_{2k_2}; u) \tag{4.11}$$

where w^*_u is the weight associated with the uth observation w_u.

Example 4.4. The dataset of Table 4.3 was obtained by aggregating a large dataset (containing several economic and demographic variables among others). In this case, the symbolic concept of interest was the occupation (secretaries, psychologists, domestic workers, office cleaners, teachers, librarians, social workers, nurses, and retail assistants). The random variables of interest were the gender (Y_1) distributions and the salary (Y_2) distributions. Gender was recorded as male ($\equiv 1$, say) or female ($\equiv 2$, say). The salaries were specified as $1, \ldots, 7$, representing seven ordered salary ranges, respectively. Also shown are the number of individual records aggregated to give these observations. Thus, for example, $n_1 = 476$ (of the overall 10000 total) individuals were secretaries and of these 5.5% ($p_{111} = 0.055$) were male and 94.5% ($p_{112} = 0.945$) were female.

Table 4.3 Occupations: Salary $\times$ Gender modal data.

w_u	Occupation	$Y_1 =$ Gender		$Y_2 =$ Salary							n_u
		Male	Female	1	2	3	4	5	6	7	
		p_{11u}	p_{12u}	p_{21u}	p_{22u}	p_{23u}	p_{24u}	p_{25u}	p_{26u}	p_{27u}	
w_1	Secretaries	0.055	0.945	0.050	0.175	0.189	0.193	0.174	0.116	0.103	476
w_2	Psychologists	0.138	0.862	0.049	0.187	0.218	0.218	0.132	0.101	0.095	1424
w_3	Domestics	0.075	0.925	0.060	0.185	0.209	0.217	0.124	0.116	0.089	733
w_4	Cleaners	0.109	0.891	0.065	0.164	0.220	0.209	0.129	0.120	0.093	1567
w_5	Teachers	0.228	0.772	0.050	0.182	0.230	0.215	0.123	0.107	0.093	1864
w_6	Librarians	0.221	0.779	0.075	0.153	0.213	0.218	0.157	0.083	0.101	671
w_7	Social Workers	0.149	0.851	0.055	0.177	0.216	0.207	0.144	0.096	0.105	1085
w_8	Nurses	0.034	0.966	0.034	2.04	0.217	0.194	0.135	0.105	0.111	668
w_9	Retail	0.071	0.929	0.060	0.163	0.204	0.232	0.132	0.109	0.100	1512

Let us construct the joint histogram for $(Z_1, Z_2) \equiv (Y_1, Y_2) =$ (Gender, Salary). Let us suppose there are no rules, so that in this case each virtual description space is the actual observation as given. Consider $(Z_1 = 1,\ Z_2 = 1)$. Then, from Equations (4.8) and (4.9), we have

$$\begin{aligned} O(Z_1 = 1, Z_2 = 1) &= (0.055)(0.050)_{u=1} + (0.138)(0.049)_{u=2} + \ldots \\ &\quad + (0.071)(0.060)_{u=9} \\ &= 0.063; \end{aligned}$$

similarly,

$$\begin{aligned} O(Z_1 = 2, Z_2 = 3) &= (0.945)(0.189)_{u=1} + (0.862)(0.218)_{u=2} + \ldots \\ &\quad + (0.929)(0.204)_{u=9} \\ &= 1.916. \end{aligned}$$

Likewise, the observed frequencies for all $(Z_1 = \xi_{1k_1},\ Z_2 = \xi_{2k_2})$ values can be calculated. The complete joint histogram is displayed in Table 4.4a.

Table 4.4a Joint histogram – occupations.

Y_1 = Gender	Y_2 = Salary							Frequency Y_1	Relative frequency Y_1
	1	2	3	4	5	6	7		
Male	0.063	0.187	0.234	0.231	0.149	0.110	0.106	1.080	0.120
Female	0.435	1.403	1.682	1.672	1.101	0.843	0.784	7.920	0.890
Frequency Y_2	0.498	1.590	1.916	1.903	1.250	0.953	0.890	9.000	
Relative frequency Y_2	0.055	0.177	0.213	0.211	0.139	0.106	0.099		1.0

Table 4.4b Weighted joint relative frequency – occupations.

Y_1 = Gender	Y_2 = Salary							Relative frequency Y_1
	1	2	3	4	5	6	7	
Male	0.007	0.023	0.029	0.028	0.107	0.014	0.013	0.131
Female	0.048	0.153	0.187	0.186	0.117	0.093	0.085	0.869
Relative frequency Y_2	0.055	0.176	0.216	0.214	0.134	0.107	0.098	1.000

Suppose now that we construct a weighted joint frequency function using weights proportional to the number of individuals (n_u) in the concept class $u = 1, \ldots, m$. Then, from Equations (4.9) and (4.11), we have, for example, Relative frequency $(Z_1 = 1,\ Z_2 = 4)$

$$= \left(\frac{476}{10000}\right)(0.055)\,(0.193)_{u=1} + \left(\frac{1424}{10000}\right)(0.135)\,(0.218)_{u=2} + \ldots$$
$$+ \left(\frac{1512}{10000}\right)(0.071)\,(0.232)_{u=10}$$
$$= 0.028,$$

where the weight for secretaries is $w_1^* = 476/10000$. The complete set of weighted joint relative frequencies is provided in Table 4.4b. □

4.4 Modal Interval-Valued Variables

Let us develop corresponding results for modal interval-valued variables where the data are recorded as histograms. Suppose interest centers on the two specific variables Z_1 and Z_2. Suppose for each observation w_u that each variable $Z_j(u)$ takes values

$$Z_j(w_u) = \{[a_{ujk}, b_{ujk}), p_{ujk}, k = 1, \ldots, s_{uj}\};$$

that is, the subintervals $\xi_{ujk} = [a_{ujk}, b_{ujk})$ have relative frequency p_{ujk}, $k = 1, \ldots, s_{uj}$, $j = 1, 2$, and $u = 1, \ldots, m$, with $\sum_{k=1}^{s_{uj}} p_{ujk} = 1$. When $s_{uj} = 1$ and $p_{ujk} = 1$ for all j, u, k values, the data are interval-valued realizations.

Then, by extending Section 4.2, we can define

Definition 4.7: Analogously with Equation (4.3), the **empirical joint density function** for (Z_1, Z_2) at the value (ξ_1, ξ_2) is

$$f(\xi_1, \xi_2) = \frac{1}{m} \sum_{u \in E} \left\{ \sum_{k_1=1}^{s_{u1}} \sum_{k_2=1}^{s_{u2}} \frac{p_{u1k_1} p_{u2k_2} I_{k_1,k_2}(\xi_1, \xi_2)}{||Z_{k_1 k_2}(u)||} \right\} \tag{4.12}$$

where $||Z_{k_1 k_2}(u)||$ is the area of the rectangle $Z_{k_1 k_2}(u) = [a_{u1k_1}, b_{u1k_1}) \times [a_{u2k_2}, b_{u2k_2})$, and $I_{k_1 k_2}(\xi_1, \xi_2)$ is the indicator variable that the point (ξ_1, ξ_2) is (is not) in the rectangle $Z_{k_1 k_2}(u)$. □

Definition 4.8: Analogously with Equation (4.4), the **joint histogram** for Z_1 and Z_2 is found by plotting $\{R_{g_1 g_2}, p_{g_1 g_2}\}$ over the rectangles $R_{g_1 g_2} = [\zeta_{1,g_1-1}, \zeta_{1g_1}) \times [\zeta_{2,g_2-1}, \zeta_{2g_2})$, $g_1 = 1, \ldots, r_1$, $g_2 = 1, \ldots, r_2$, with

$$f_{g_1 g_2} = \sum_{u \in E} \sum_{k_1 \in Z(g_1)} \sum_{k_2 \in Z(g_2)} \frac{||Z(k_1, k_2; u) \cap R_{g_1 g_2}||}{||Z(k_1, k_2; u)||} p_{u1k_1} p_{u2k_2} \tag{4.13}$$

and hence

$$p_{g_1 g_2} = f_{g_1 g_2}/m,$$

where $Z(g_j)$ represents all the intervals $Z(k_j; u) \equiv [a_{ujk_j}, b_{ujk_j})$, $j = 1, 2$, which overlap with the rectangle $R_{g_1 g_2}$ for each given u value. □

In effect, each term in the summations of Equation (4.13) is that proportion of the observed subrectangle $Z(k_1, k_2; u)$ which overlaps with the histogram rectangle $R_{g_1 g_2}$ for each (g_1, g_2).

Example 4.5. The data in Table 4.5 give histogram-valued observations for the variables Y_1 = Cholesterol and those in Table 4.6 are for Y_2 = Hemoglobin counts for $m = 14$ observations. Each observation here came about by considering the concept of gender × age group, where gender had values from $\mathcal{Y}_1 = \{\text{female, male}\}$, and age group took values $\mathcal{Y}_2 = \{\text{20s}, \ldots, \text{70s, 80s and over}\}$. These histogram values resulted from aggregating a portion of a very large dataset containing (classical) values for individuals living in a certain region of the USA. In this particular case, the variable age was recorded as a (classical) point by year. Hence, age group is itself a concept with its observation values (e.g., 20s)

Table 4.5 Cholesterol values for gender age groups.

w_u	Concept		Frequency histogram
	Gender	Age	
w_1	Female	20s	{[80, 100), 0.025; [100, 120), 0.075; [120, 135), 0.175; [135, 150), 0.250; [150, 165), 0.200; [165, 180), 0.162; [180, 200), 0.088; [200, 240], 0.025}
w_2	Female	30s	{[80, 100), 0.013; [100, 120), 0.088; [120, 135), 0.154; [135, 150), 0.253; [150, 165), 0.210; [165, 180), 0.177; [180, 195), 0.066; [195, 210), 0.026; [210, 240], 0.013}
w_3	Female	40s	{[95, 110), 0.012; [110, 125), 0.029; [125, 140), 0.113; [140, 155), 0.206; [155, 170), 0.235; [170, 185), 0.186; [185, 200), 0.148; [200, 215), 0.043; [215, 230), 0.020; [230, 245], 0.008}
w_4	Female	50s	{[105, 120), 0.009; [120, 135), 0.026; [135, 150), 0.046; [150, 165), 0.105; [165, 180), 0.199; [180, 195), 0.248; [195, 210), 0.199; [210, 225), 0.100; [225, 240), 0.045; [240, 260], 0.023}
w_5	Female	50s	{[115, 140), 0.012; [140, 160), 0.069; [160, 180), 0.206; [180, 200), 0.300; [200, 220), 0.255; [220, 240), 0.146; [240, 260], 0.012}
w_6	Female	70s	{[120, 140), 0.017; [140, 160), 0.083; [160, 180), 0.206; [180, 200), 0.294; [200, 220), 0.250; [220, 240), 0.117; [240, 260], 0.033}
w_7	Female	80+	{[120, 140), 0.036; [140, 160), 0.065; [160, 180), 0.284; [180, 200), 0.325; [200, 220), 0.213; [220, 240), 0.065; [240, 260], 0.012}
w_8	Male	20s	{[110, 135), 0.143; [135, 155), 0.143; [155, 165), 0.286; [165, 175), 0.214; [175, 185), 0.143; [185, 195], 0.071}
w_9	Male	30s	{[90, 100), 0.022; [100, 120), 0.044; [120, 140), 0.044; [140, 160), 0.333; [160, 180), 0.289; [180, 200), 0.179; [200, 220], 0.089}
w_{10}	Male	40s	{[120, 135), 0.018; [135, 150), 0.109; [150, 165), 0.327; [165, 180), 0.255; [180, 195), 0.182; [195, 210), 0.073; [210, 225], 0.036}
w_{11}	Male	50s	{[105, 125), 0.019; [125, 145), 0.020; [145, 165), 0.118; [165, 185), 0.216; [185, 205), 0.294; [205, 225), 0.137; [225, 245], 0.176; [245, 265], 0.020}

w_{12}	Male	60s	{[130, 150), 0.041; [150, 170), 0.042; [170, 190), 0.167; [190, 210), 0.375; [210, 230), 0.250; [230, 250), 0.083; [250, 270], 0.042}
w_{13}	Male	70s	{[165, 180), 0.105; [180, 195), 0.316; [195, 210), 0.158; [210, 225), 0.158; [225, 240), 0.210; [240, 255], 0.053}
w_{14}	Male	80+	{[155, 170), 0.067; [170, 185), 0.133; [185, 200), 0.200; [200, 215), 0.267; [215, 230), 0.200; [230, 245), 0.067; [245, 260], 0.066}

Table 4.6 Hemoglobin values for gender and age groups.

w_u	Concept		Frequency histogram
	Gender	Age	
w_1	Female	20s	{[12.0, 12.9), 0.050; [12.9, 13.2), 0.112; [13.2, 13.5), 0.212; [13.5, 13.8), 0.201; [13.8, 14.1), 0.188; [14.1, 14.4), 0.137; [14.4, 14.7), 0.075; [14.7, 15.0], 0.025}
w_2	Female	30s	{[10.5, 11.0), 0.007; [11.0, 11.3), 0.039; [11.3, 11.6), 0.082; [11.6, 11.9), 0.174; [11.9, 12.2), 0.216; [12.2, 12.5), 0.266; [12.5, 12.8), 0.157; [12.8, 14.0], 0.059}
w_3	Female	40s	{[10.5, 11.0), 0.009; [11.0, 11.5), 0.084; [11.5, 11.8), 0.148; [11.8, 12.1), 0.217; [12.1, 12.4), 0.252; [12.4, 12.7), 0.180; [12.7, 13.0), 0.087; [13.0, 14.0], 0.023}
w_4	Female	50s	{[10.5, 11.2), 0.046; [11.2, 11.6), 0.134; [11.6, 12.0), 0.222; [12.0, 12.4), 0.259; [12.4, 12.8), 0.219; [12.8, 13.2), 0.105; [13.2, 13.6), 0.012; [13.6, 14.0], 0.003}
w_5	Female	60s	{[10.8, 11.2), 0.028; [11.2, 11.5), 0.081; [11.5, 11.8), 0.133; [11.8, 12.1), 0.231; [12.1, 12.4), 0.219; [12.4, 12.7), 0.182; [12.7, 13.0), 0.061; [13.0, 13.3), 0.057; [13.3, 13.6], 0.008}
w_6	Female	70s	{[10.8, 11.1), 0.022; [11.1, 11.4), 0.050; [11.4, 11.7), 0.078; [11.7, 12.0), 0.183; [12.0, 12.3), 0.228; [12.3, 12.6), 0.233; [12.6, 12.9), 0.117; [12.9, 13.2), 0.067; [13.2, 13.6], 0.022}
w_7	Female	80+	{[10.8, 11.2), 0.029; [11.2, 11.5), 0.095; [11.5, 11.8), 0.148; [11.8, 12.1), 0.213; [12.1, 12.4), 0.207; [12.4, 12.7), 0.160; [12.7, 13.0), 0.077; [13.0, 13.3), 0.047; [13.3, 13.6], 0.024}
w_8	Male	20s	{[12.9, 13.1), 0.071; [13.1, 13.3), 0.143; [13.3, 13.5), 0.214; [13.5, 13.7), 0.217; [13.7, 13.9), 0.212; [13.9, 14.1], 0.143}
w_9	Male	30s	{[10.2, 10.7), 0.022; [10.7, 11.1), 0.045; [11.1, 11.5), 0.089; [11.5, 11.9), 0.222; [11.9, 12.3), 0.200; [12.3, 12.7), 0.267; [12.7, 13.1), 0.133; [13.1, 13.4], 0.022}

Table 4.6 Continued.

w_u	Concept		Frequency histogram
	Gender	Age	
w_{10}	Male	40s	{[10.8, 11.2), 0.018; [11.2, 11.6), 0.163; [11.6, 12.0), 0.273; [12.0, 12.4), 0.273; [12.4, 12.8), 0.182; [12.8, 13.2), 0.073; [13.2, 13.6], 0.018}
w_{11}	Male	50s	{[10.8, 11.2), 0.020; [11.2, 11.6), 0.118; [11.6, 12.0), 0.235; [12.0, 12.4), 0.275; [12.4, 12.8), 0.235; [12.8, 13.2), 0.059; [13.2, 13.6), 0.020; [13.6, 14.0], 0.038}
w_{12}	Male	60s	{[11.3, 11.6), 0.125; [11.6, 11.9), 0.166; [11.9, 12.2), 0.166; [12.2, 12.5), 0.167; [12.5, 128), 0.292; [12.8, 13.2), 0.042; [13.2, 13.5], 0.042}
w_{13}	Male	70s	{[11.4, 11.7), 0.053; [11.7, 12.0), 0.315; [12.0, 12.3), 0.316; [12.3, 12.6), 0.211; [12.6, 12.9], 0.105}
w_{14}	Male	80+	{[10.8, 11.2), 0.133; [11.2, 11.6), 0.067; [11.6, 12.0), 0.134; [12.0, 12.4), 0.333; [12.4, 12.8), 0.200; [12.8, 13.2], 0.133}

being the result of a prior level of aggregation (e.g., over all those with ages 20, . . . , 29).

To obtain the joint frequency distribution for (Y_1, Y_2), we first observe that cholesterol values span the interval (80, 270) while hemoglobin values span (10.5, 15). Therefore, let us build the joint frequency histogram on the rectangles $R_{g_1g_2}$, $g_1 = 1, \ldots, 6$, and $g_2 = 1, \ldots, 5$, given as $[75, 110) \times [10, 11), \ldots, [75, 110) \times [14, 15], \ldots, [250, 285] \times [14, 15]$ shown in Table 4.10.

To illustrate the application of Equation (4.13), let us first take the histogram rectangle $R_{11} = [75, 110) \times [10, 11)$ and consider the contribution $f_{11}(3)$, say, of the third observation $w_u = 3$. The double summation over k_1 and k_2 produces the $s_{31} \times s_{32} = 10 \times 8 = 80$ cross-product terms, from Table 4.5 and Table 4.6,

$$\begin{aligned} f_{11}(w_3) = & \left(\frac{110-95}{110-95}\right)\left(\frac{11.0-10.5}{11.0-10.5}\right)(0.012)(0.009) \\ & + \left(\frac{110-95}{110-95}\right)(0)(0.012)(0.084) + \ldots \\ & + (0)(0)(0.008)(0.023). \end{aligned}$$

These can be rearranged to give

$$\begin{aligned} f_{11}(w_3) = & \left[\left(\frac{110-95}{110-95}\right)(0.012) + \ldots + (0)(0.008)\right]_{u_1=3} \\ & \times \left[\left(\frac{11.0-10.5}{11.0-10.5}\right)(0.009) + \ldots + (0)(0.023)\right]_{u_2=3} \end{aligned}$$

where $u_j = 3$ represents the contributions from the Z_j subintervals for the observation $w_3 (u = 3)$.

Proceeding in this manner for all observations w_u, $u = 1, \ldots, m$, we obtain the total observed frequency for the rectangle $R_{11} = [75, 110) \times [10, 11)$ as

$$
\begin{aligned}
f_{11} = & \left\{ \left[\left(\frac{100-80}{100-80} \right) (0.025) + \left(\frac{110-100}{120-100} \right) (0.75) + 0 + \ldots + 0 \right]_{u_1=1} \right. \\
& \left. \times [0]_{u_2=1} \right\}_{u=1} + \left\{ \left[\left(\frac{100-80}{100-80} \right) (0.013) + \left(\frac{110-100}{120-100} \right) (0.088) \right. \right. \\
& \left. \left. + 0 + \ldots + 0 \right]_{u_1=2} \times \left[\left(\frac{11.0-10.5}{11.0-10.5} \right) (0.007) + 0 + \ldots + 0 \right]_{u_2=2} \right\}_{u=2} \\
& + \left\{ \left[\left(\frac{110-95}{110-95} \right) (0.012) + 0 + \ldots + 0 \right]_{u_1=3} \right. \\
& \left. \times \left[\left(\frac{11.0-10.5}{11.0-10.5} \right) (0.009) + 0 + \ldots + 0 \right]_{u_2=3} \right\}_{u=3} \\
& + \left\{ \left[\left(\frac{110-105}{120-105} \right) (0.009) + 0 + \ldots + 0 \right]_{u_1=4} \right. \\
& \left. \times \left[\left(\frac{11.0-10.5}{11.2-10.5} \right) (0.046) + 0 + \ldots + 0 \right]_{u_2=4} \right\}_{u=4} \\
& + 0 + 0 + 0 + 0 + \\
& \left\{ \left[\left(\frac{100-90}{100-90} \right) (0.22) + \left(\frac{110-100}{120-100} \right) (0.044) + 0 + \ldots + 0 \right]_{u_1=9} \right. \\
& \left. \times \left[\left(\frac{10.7-10.2}{10.7-10.2} \right) (0.22) + \left(\frac{11.0-10.7}{11.1-10.7} \right) (0.045) + 0 + \ldots + 0 \right]_{u_2=9} \right\}_{u=9} \\
& + 0 + \left\{ \left[\left(\frac{110-105}{125-105} \right) (0.19) + 0 + \ldots + 0 \right]_{u_1=11} \right. \\
& \left. \times \left[\left(\frac{11.0-10.8}{11.2-10.8} \right) (0.020) + 0 + \ldots + 0 \right]_{u_2=11} \right\}_{u=11} \\
& + 0 + 0 + 0 \\
= & \ 0 + 0.000399 + 0.000108 + 0.000099 + 0 + \ldots + 0 \\
& + 0.002553 + 0 + 0.000014 + 0 + 0 + 0 \\
= & \ 0.003163.
\end{aligned}
$$

Hence,

$$p_{11} = 0.003163/14 = 2.26 \times 10^{-4}.$$

All the $f_{g_1 g_2}$ values are shown in Table 4.7. Also shown are the marginal frequencies f_i, $i = 1, 2$, and the probabilities, by using Equations (3.34) and (3.37) directly. The joint distribution function is plotted in Figure 4.4.

Table 4.7 Histogram frequencies $\times 10^1$ (Cholesterol, Hemoglobin).

	Y_1 = Cholesterol	Y_2 = Hemoglobin					Frequency	Distribution
	$g_1 \backslash g_2$	1	2	3	4	5	Y_1	Y_1
		[10, 11)	[11, 12)	[12, 13)	[13, 14)	[14, 15]		
1	[75, 110)	0.031	0.446	0.723	0.444	0.187	1.733	0.013
2	[110, 145)	0.177	4.001	6.337	4.773	1.290	16.578	0.118
3	[145, 180)	0.794	14.992	22.411	10.685	1.794	50.676	0.362
4	[180, 215)	0.940	17.143	27.059	4.597	0.394	50.131	0.358
5	[215, 250)	0.352	6.640	11.000	1.156	0.047	19.195	0.137
6	[250, 285]	0.039	0.524	0.096	0.000	0.000	1.587	0.011
	Frequency Y_2	2.333	43.746	68.458	21.751	3.715	140	
	Distribution Y_2	0.233	0.312	0.489	0.155	0.027		1.0

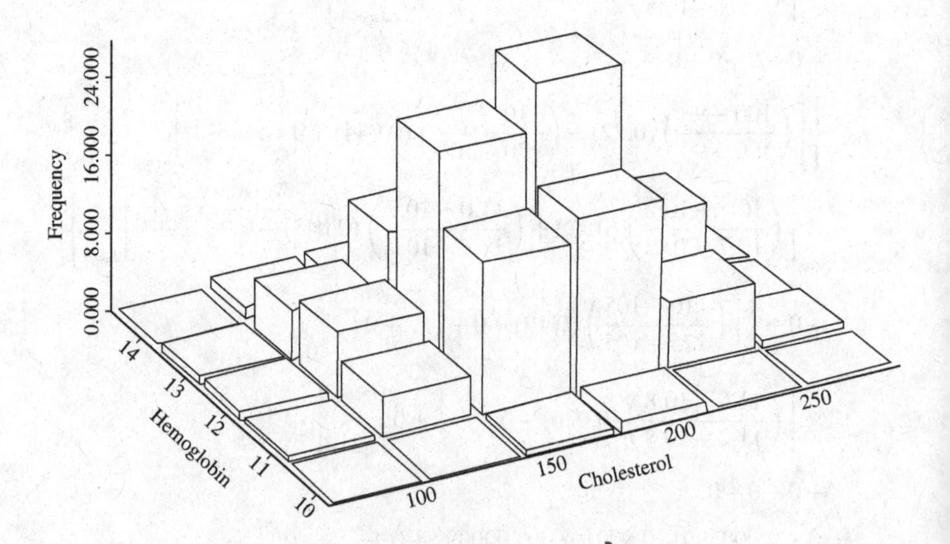

Figure 4.4 Joint histogram of histograms. □

4.5 Baseball Interval-Valued Dataset

As indicated earlier in Chapters 2 and 3, symbolic data by the very formation of the concepts underlying such data may necessitate the inclusion of special rules that maintain the integrity of the resulting symbolic values. Specific examples have been considered thus far. However, more complicated situations can arise from seemingly innocuous datasets, which situations must be considered in any subsequent analyses. This feature is apparent in the dataset of Table 4.8 relating to baseball team statistics. An obvious but simple rule that the number of hits cannot exceed the number of at-bats leads to non-simple virtual description spaces. This requires special attention, especially when determining joint histograms (but which also impact on the univariate functions). We consider this dataset in detail, and then conclude with some general principles of how to analyze interval-valued data in the presence of rules. More complete details can be found in Billard and Diday (2006).

Table 4.8 At-Bats and Hits by team.

w_u team	Y_1 #At-Bats	Y_2 #Hits	Pattern*	u team	Y_1 # At-Bats	Y_2 #Hits	Pattern[a]
w_1	(289, 538)	(75, 162)	B	w_{11}	(212, 492)	(57, 151)	B
w_2	(88, 422)	(49, 149)	I	w_{12}	(177, 245)	(189, 238)	G
w_3	(189, 223)	(201, 254)	F	w_{13}	(342, 614)	(121, 206)	B
w_4	(184, 476)	(46, 148)	B	w_{14}	(120, 439)	(35, 102)	B
w_5	(283, 447)	(86, 115)	B	w_{15}	(80, 468)	(55, 115)	I
w_6	(24, 26)	(133, 141)	A	w_{16}	(75, 110)	(75, 110)	C
w_7	(168, 445)	(37, 135)	B	w_{17}	(116, 557)	(95, 163)	I
w_8	(123, 148)	(137, 148)	E	w_{18}	(197, 507)	(52, 53)	B
w_9	(256, 510)	(78, 124)	B	w_{19}	(167, 203)	(48, 232)	H
w_{10}	(101, 126)	(101, 132)	D				

[a] Pattern of virtual observation under rule $v\,(\alpha = 1)$.

4.5.1 The data: actual and virtual datasets

The dataset consists of interval-valued values for the random variables Y_1 = Number of At-Bats and Y_2 = Number of Hits collected by $m = 19$ teams over a certain period of time. An observation takes values ξ_u in a $p = 2$ dimensional hypercube, or rectangle, bounded by $(a_{u1}, b_{u1}) \times (a_{u2}, b_{u2}) = R(u)$, say, $u = 1, \ldots, m$. (We note that this $R(u)$ represents the real, actual, observation, and is not the histogram rectangle of previous sections.)

Suppose there is a logical rule

$$v : Y_2 \leq \alpha Y_1. \tag{4.14}$$

When $\alpha = 1$, the number of hits cannot exceed the number of at-bats. When $\alpha = 0.350$ (say), then the batting average is 0.350 or less. Without such a rule, there may be individual descriptions x in $R(u)$ which imply that the number of hits is more than the number of at-bats. For example, under this rule, part of the actual rectangle $R(2)$ is valid (e.g., $x = (88, 49)$), and part of $R(2)$ is invalid (e.g., $x = (88, 149)$).

There are nine different possible patterns A, ..., I for the virtual space $V(u)$ which result from applying the rule v of Equation (4.14) to the actual rectangles $R(u)$. These are displayed in Figure 4.5. Notice that there are four different types of patterns. Patterns A and B are entirely outside or inside the valid region. In this case, $vir(\mathrm{A}) = \phi$, and $vir(\mathrm{B}) = V(\mathrm{B}) = R(\mathrm{B})$. Patterns C, D, E, F all produce triangular virtual spaces. Patterns G and H produce four-sided hyperrectangles that can be partitioned into a triangle and rectangular subregions, while pattern I is a five-sided hyperrectangle which decomposes into a triangular space plus three different subrectangular spaces.

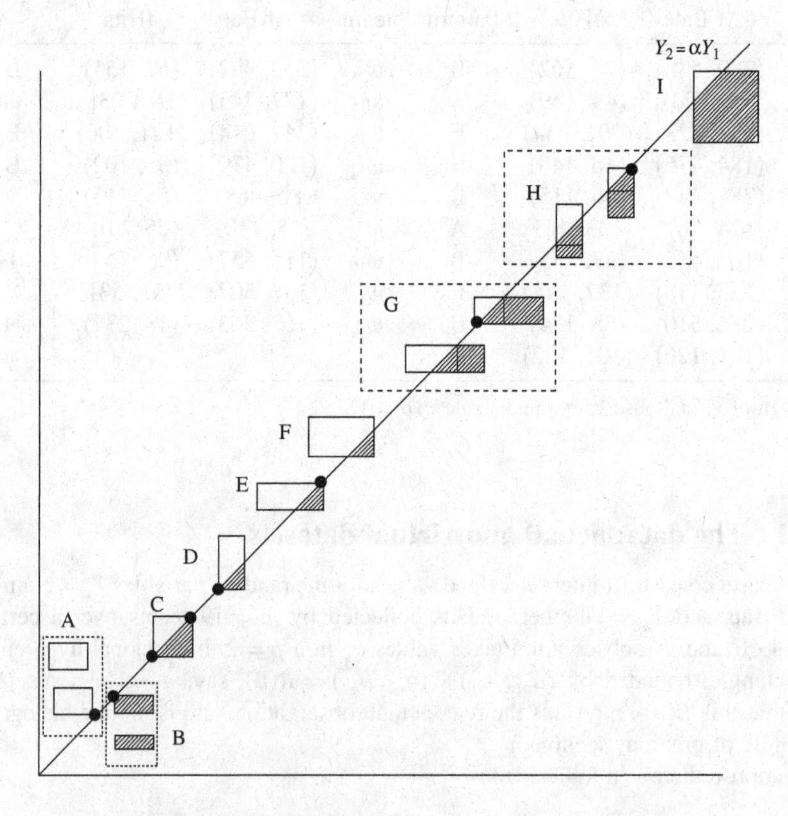

Figure 4.5 Patterns for virtual V – shaded regions.

The conditions on the actual $\xi = ([a_1, b_1], [a_2, b_2])$ values under this rule that produces these respective patterns are summarized in Table 4.9, where the subscript u is suppressed. For example, pattern F emerges whenever

$$\alpha a_1 < a_2 < \alpha b_1 \quad \text{and} \quad \alpha b_1 < b_2.$$

Table 4.9 Virtual patterns and conditions.

Pattern	Pattern V[a]	Conditions	Pattern	Pattern V[a]	Conditions
A	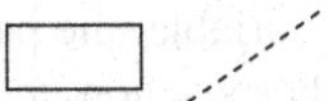	$\alpha b_1 \leq a_2$	F	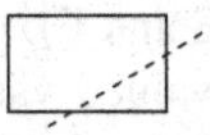	$\begin{cases}\alpha a_1 < a_2 < \alpha b_1 \\ \alpha b_1 < b_2\end{cases}$
B	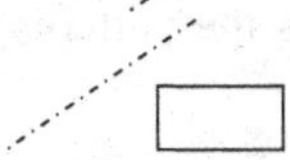	$\alpha a_1 \leq b_2$	G	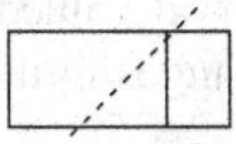	$\begin{cases}\alpha a_1 \leq a_2 \\ \alpha b_1 > b_2\end{cases}$
C	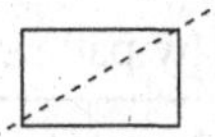	$\begin{cases}\alpha a_1 = a_2 \\ \alpha b_1 = b_2\end{cases}$	H	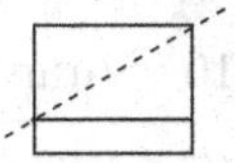	$\alpha a_1 > a_2$ $\alpha b_1 \leq b_2$
D	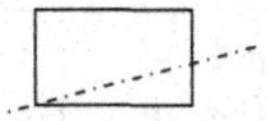	$\begin{cases}\alpha a_1 = a_2 \\ \alpha b_1 < b_2\end{cases}$	I	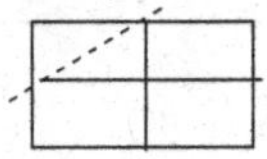	$\begin{cases}\alpha a_1 > a_2 \\ \alpha b_1 > b_2 > \alpha a_1\end{cases}$
E		$\begin{cases}\alpha a_1 < a_2 \\ \alpha b_1 = b_2\end{cases}$			

[a] The dashed line represents the line $Y_2 = \alpha Y_1$.

In order to calculate relevant descriptive statistics, the coordinates of the virtual space $V(u)$ must be found. These vary by pattern type. Consider pattern G. It is clear that the virtual Y_2 values are unchanged from those for the actual Y_2 values. However, those for Y_1 do change. There are two subspaces R_1 and R_2. The coordinates for Y_1 in the triangle R_1 are $Y_1 = [a_2/\alpha, b_2/\alpha]$. Its area is

$$|R_1| = (b_2/\alpha - \alpha_2/\alpha)(b_2 - a_2)/2 = (b_2 - a_2)^2/(2\alpha).$$

The rectangular region R_2 has coordinates

$$R_2 = ([b_2/\alpha, b_1], [a_2, b_2]),$$

and has area

$$|R_2| = (\alpha b_1 - b_2)(b_2 - a_2)/\alpha.$$

Therefore, the Y_1 variable is now a histogram-valued variable

$$Y_1 = \{[a_2/\alpha,\ b_2/\alpha),\ p_1;\ [b_2/\alpha,\ b_1],\ p_2\}$$

where the observed relative frequencies are

$$p_1 = |R_1|/|V|, \; p_2 = |R_2|/|V|$$

with

$$|V| = |R_1| + |R_2|.$$

The virtual description spaces for all patterns are provided in Table 4.10. We notice that relative to the random variable Y_1, patterns G and I produce a histogram-valued variable in Y_1, whereas for the other six non-empty patterns, Y_1 remains as an interval-valued variable even though its end point values may have changed (as they do in patterns E and F). Relative to the Y_2 random variable, the patterns H and I give a histogram-valued variable, whereas for the other six non-empty patterns, Y_2 stays as an interval-valued variable but this time it is the patterns D and F for which there are changes in the end point values.

Table 4.10 Virtual observation space values – by pattern.

Pattern	Virtual ξ_1'	Virtual ξ_2'	Apparent virtual ξ'
A	ϕ	ϕ	ϕ
B	(a_1, b_1)	(a_2, b_2)	$\{(a_1, b_1),\ (a_2, b_2)\}$
C	(a_1, b_1)	(a_2, b_2)	$\{(a_1, b_1),\ (a_2, b_2)\}$
D	(a_1, b_1)	$(a_2, \alpha b_1)$	$\{(a_1, b_1),\ (a_2, \alpha b_1)\}$
E	$(a_2/\alpha, b_1)$	(a_2, b_2)	$\{(a_2/\alpha, b_1),\ (a_2, b_2)\}$
F	$(a_2/\alpha, b_1)$	$(a_2, \alpha b_1)$	$\{(a_2/\alpha, b_1),\ (a_2, \alpha b_1)\}$
G	$\{(a_2/\alpha, b_2/\alpha)p_1,$ $(b_2/\alpha, b_1)p_2\}$	(a_2, b_2)	$\{[(a_2/\alpha, b_2/\alpha), (a_2, b_2)]p_1,$ $[(b_2/\alpha, b_1), (a_2, b_2)]p_2\}$
	$p_i = \lvert R_i\rvert/\lvert V\rvert, \lvert V\rvert = \lvert R_1\rvert + \lvert R_2\rvert, \lvert R_1\rvert = (b_2 - a_2)^2/(2\alpha),$ $\lvert R_2\rvert = (\alpha b_1 - b_2)(b_2 - a_2)/\alpha$		
H	(a_1, b_1)	$\{(a_2, \alpha a_1)p_1,$ $(\alpha a_1, \alpha b_1)p_2\}$	$\{[(a_1, b_1), (a_2, \alpha a_1)]p_1,$ $[(a_1, b_1), (\alpha a_1, \alpha b_1)]p_2\}$
	$p_1 = \lvert R_i\rvert/\lvert V\rvert, \lvert V\rvert = \lvert R_1\rvert + \lvert R_2\rvert, \lvert R_1\rvert = (b_1 - a_1)(\alpha a_1 - a_2),$ $\lvert R_2\rvert = \alpha(b_1 - a_1)^2/2$		
I	$\{(a_1, b_2/\alpha)p_1,$ $(b_2/\alpha, b_1)p_2\}$	$\{(a_2, \alpha a_1)p_1^*,$ $(\alpha a_1, b_2)p_2^*\}$	$\{[(a_1, b_2/\alpha), (\alpha a_1, b_2)]p_1^{**},$ $[(a_1, b_2/\alpha), (a_2, \alpha a_1)]p_2^{**},$ $[(b_2/\alpha, b_1), (a_2, \alpha a_1)]p_3^{**},$ $[(b_2/\alpha, b_1), (\alpha a_1, b_2)]p_4^{**}\}$
	$p_1 = (\lvert R_1\rvert + \lvert R_2\rvert)/\lvert V\rvert$ $p_2 = (\lvert R_2\rvert + \lvert R_4\rvert)/\lvert V\rvert$	$p_1^* = (\lvert R_2\rvert + \lvert R_3\rvert)/\lvert V\rvert$ $p_2^* = (\lvert R_1\rvert + \lvert R_4\rvert)/\lvert V\rvert$	$p_i^{**} = \lvert R_i\rvert/\lvert V\rvert, i = 1, \ldots, 4$
	$\lvert R_1\rvert = (b_2 - \alpha a_1)^2/(2\alpha), \lvert R_2\rvert = (b_2 - \alpha a_1)(\alpha a_1 - a_2)/\alpha, \lvert R_3\rvert = (\alpha b_1 - b_2)(\alpha a_1 - a_2)/\alpha,$ $\lvert R_4\rvert = (\alpha b_1 - b_2)(b_2 - \alpha a_1)/\alpha, \lvert V\rvert = \lvert R_1\rvert + \lvert R_2\rvert + \lvert R_3\rvert + \lvert R_4\rvert$		

The 'apparent' virtual description of $\boldsymbol{\xi} = (\text{virtual } \xi_1, \ \text{virtual } \xi_2)$ is also displayed in Table 4.10. As written, these take the virtual ξ_j, $j = 1, 2$, values, as calculated for each Y_j separately. As such, even though a given virtual ξ_j may now be histogram valued, the 'apparent' $\boldsymbol{\xi}$ may be misconstrued to be the original rectangle R. The real virtual $\boldsymbol{\xi}$ is the patterned region (A, . . . , or I). An effect of this is that when calculating joint histogram frequencies, the calculations corresponding to the valid triangular region need to be adjusted for these 'apparent' rectangles.

4.5.2 Joint histograms

We take the rule v of Equation (4.14) with $\alpha = 1$. Under this rule, team w_1 has pattern B, i.e., all its values in $R(1)$ are valid. Team w_2 has pattern I, while teams w_6 and w_{16} are deleted entirely since $vir(w_6) = vir(w_{16}) = 0$. The pattern associated with each team is given in Table 4.8. Then, from the virtual ξ_1, virtual ξ_2, and virtual ξ formulas of Table 4.10, we obtain the virtual values for these data as shown in Tables 4.11a and 4.11b.

Table 4.11a Virtual ξ_1' and ξ_2' under rule $v : Y_1 \geq Y_2$.

Team w_u	(a) Pattern	(b) $\xi_1' : Y_1 =$ # Hits	(c) $\xi_2' : Y_2 =$ # At-Bats
w_1	B	(289, 538)	(75, 162)
w_2	I	{(88, 149), 0.134; (149, 422), 0.866}	{(49, 88), 0.413; (88, 149), 0.587}
w_3	F	(201, 223)	(201, 223)
w_4	B	(184, 476)	(46, 148)
w_5	B	(283, 447)	(86, 115)
w_6	A	ϕ	ϕ
w_7	B	(168, 445)	(37, 135)
w_8	E	(137, 148)	(137, 148)
w_9	B	(256, 510)	(78, 124)
w_{10}	D	(101, 126)	(101, 126)
w_{11}	B	(212, 492)	(57, 151)
w_{12}	G	{(189, 238), 0.778; (238, 245), 0.222}	(189, 238)
w_{13}	B	(342, 614)	(121, 206)
w_{14}	B	(120, 439)	(35, 102)
w_{15}	I	{(80, 115), 0.066; (115, 468), 0.934}	{(55, 80), 0.428; (80, 115), 0.572}
w_{16}	C	(75, 110)	(75, 110)
w_{17}	I	{(116, 163), 0.072; (163, 557), 0.928}	{(95, 116), 0.321; (116, 163), 0.679}
w_{18}	B	(197, 507)	(52, 53)
w_{19}	H	(167, 203)	{(48, 167), 0.869; (167, 232), 0.131}

Table 4.11b Virtual $\xi' = (\xi'_1, \xi'_2)$ under rule $v: Y_1 \geq Y_2$.

Team w_u	(d) $\xi' = (\xi'_1, \xi'_2): Y = (Y_1, Y_2)$
w_1	{(289, 538), (75, 162)}
w_2	{[(88, 149), (88, 149)], 0.059; [(88, 149), (49, 88)], 0.075; [(149, 422), (49, 88)], 0.338; [(149, 422), (88, 149)], 0.528}
w_3	{(201, 223), (201, 223)}
w_4	{(184, 476), (46, 148)}
w_5	{(283, 447), (86, 115)}
w_6	ϕ
w_7	{(168, 445), (37, 135)}
w_8	{(137, 148), (137, 148)}
w_9	{(256, 510), (78, 124)}
w_{10}	{(101, 126), (101, 126)}
w_{11}	{(212, 492), (57, 151)}
w_{12}	{[(189, 238), (189, 238)], 0.778; [(238, 245), (189, 238)], 0.222}
w_{13}	{(342, 614), (121, 206)}
w_{14}	{(120, 439), (35, 102)}
w_{15}	{[(80, 115), (80, 115)], 0.027; [(80, 115), (55, 80)], 0.039; [(115, 468), (55, 80)], 0.389; [(115, 468), (80, 115)], 0.545}
w_{16}	{(75, 110), (75, 110)}
w_{17}	{[(116, 163), (116, 163)], 0.039; [(116, 163), (95, 116)], 0.034; [(163, 557), (95, 116)], 0.286; [(163, 557), (116, 163)], 0.641}
w_{18}	{(197, 507), (52, 53)}
w_{19}	{[(167, 203), (48, 167)], 0.869; [(167, 203), (167, 232)], 0.131}

Let us build the histogram for Y_1 = Number of At-Bats on the $r_1 = 5$ intervals $[0, 50), [50, 200), \ldots, [500, 650]$, and suppose the histogram for Y_2 = Number of Hits is constructed on the $r_2 = 5$ intervals $[0, 75), [75, 125), \ldots, [225, 275]$; see Table 4.12. We first construct the joint histogram for (Y_1, Y_2) without the rule v, for comparative purposes. Since the virtual space is the same as the actual space for all observations, all observations remain as the interval-valued variables of Table 4.8. Therefore, the methodology of Section 4.2 is used. The resulting relative frequencies are shown in Table 4.12 and the joint histogram is displayed in Figure 4.6.

However, under the rule v of Equation (4.14), we observe from Table 4.11 that the observations are now histogram-valued variables (albeit, in some cases, the particular case of an interval-valued variable pertains). Therefore, to construct the joint histogram for (Y_1, Y_2) we now need to use the methodology of Section 4.4. The resulting relative frequencies are given in Table 4.13 and are plotted in Figure 4.7.

Comparing Figure 4.6 to Figure 4.7 reveals the effect of this rule. For example, for the joint histogram rectangle $R_{g_1 g_2} = R_{13} = [0, 50) \times [125, 175)$, we see a change

Table 4.12 Joint histogram for (Y_1, Y_2) – no rules.

$g_2 \backslash g_1$	1	2	3	4	5	Frequency
$Y_2 \backslash Y_1$	[0, 50)	[50, 200)	[200, 350)	[350, 500)	[500, 650]	Y_2
1 [0, 75)	0.000	0.544	1.473	1.161	0.022	3.201
2 [75, 125)	0.000	2.668	2.556	2.783	0.204	8.211
3 [125, 175)	1.000	1.686	0.749	1.095	0.384	4.914
4 [175, 225)	0.000	0.644	0.826	0.201	0.153	1.824
5 [225, 275]	0.000	0.302	0.549	0.000	0.000	0.850
Frequency Y_1	1.000	5.844	6.153	5.240	0.763	19

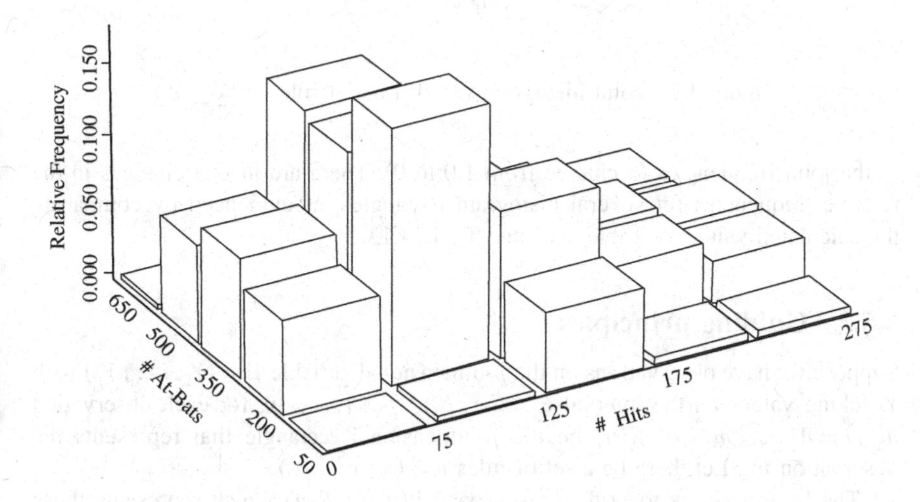

Figure 4.6 Joint histogram (Y_1, Y_2) – no rules.

Table 4.13 Joint histogram for (Y_1, Y_2) – rule $v : Y_2 \leq Y_1$.

$g_2 \backslash g_1$	1	2	3	4	5	Frequency
$Y_2 \backslash Y_1$	[0, 50)	[50, 200)	[200, 350)	[350, 500)	[500, 650]	Y_2
1 [0, 75)	0.000	0.599	1.487	1.167	0.023	3.275
2 [75, 125)	0.000	2.888	2.589	2.801	0.206	8.484
3 [125, 175)	0.000	1.501	0.767	1.105	0.386	3.761
4 [175, 225)	0.000	0.221	1.626	0.201	0.153	2.201
5 [225, 275]	0.000	0.059	0.220	0.000	0.000	0.279
Frequency Y_1	0.000	5.268	6.690	5.274	0.768	18

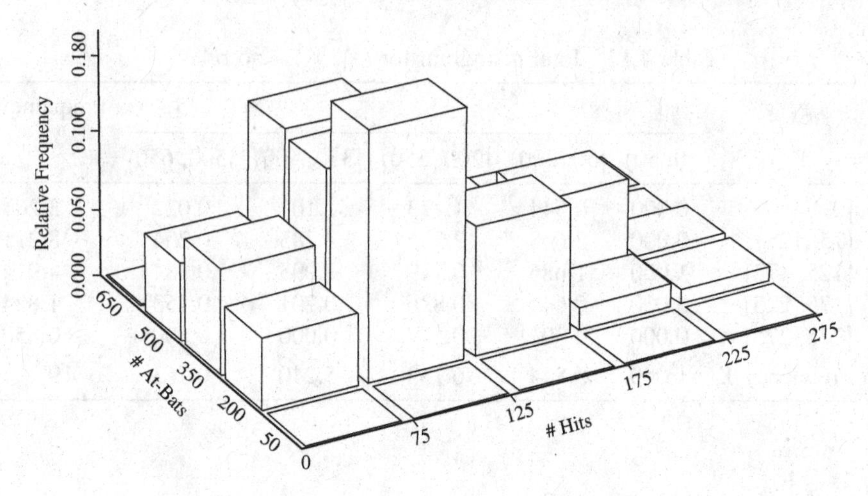

Figure 4.7 Joint histogram (Y_1, Y_2) under rule $v : Y_2 \geq Y_1$.

in the joint frequencies (a change from 1.0 to 0). There are in fact changes in the relative frequencies for several histogram rectangles, as evidenced by comparing the calculated values in Table 4.12 and Table 4.13.

4.5.3 Guiding principles

Suppose we have observations on the p-dimensional variable $Y = (Y_1, \ldots, Y_p)$ with Y_j taking values on the interval $\xi_j = (a_j, b_j)$, $j = 1, \ldots, p$, for each observation w_u, $u = 1, \ldots, m$. Let $R(u)$ be the p-dimensional rectangle that represents the observation w_u. Let there be a set of rules $v = (v_1, v_2, \ldots)$.

The basic issue is to find that subspace $V(u)$ of $R(u)$ which represents those values of $R(u)$ for which the rule v holds, i.e., those $x = (x_1, \ldots, x_p)$ such that $v_i(x) = 1$, for all rules v_i; see Equation (3.5). For some u, this $V(u) = \phi$, the empty set; for other u, this $V(u) = R(u)$, the original observation; and for others, this $V(u)$ is a non-empty p-dimensional rectangle $V(u) \equiv R^*(u)$ contained in $R(u)$, i.e., $V(u) \equiv R^*(u) \subseteq R(u)$. For these observations, the adjustment to the relevant calculation for the descriptive statistic of interest is routine.

Frequently, the virtual observation $V(u)$ is a p-dimensional non-rectangular hypercube. However, it is usually the case that this virtual space can be partitioned into components each of which is itself a rectangle (or a shape such as a triangle which is clearly a half-rectangle). For example, pattern I observed for the baseball data under rule v of Equation (4.14) (see Table 4.8) can be partitioned into components R_j, $j = 1, \ldots, 4$. Each component, R_j, is a proportion, p_j, of the whole virtual space V, with $V = \bigcup_j R_j$, and $\sum p_j = 1$, for a given u. Each R_j component is then added to the dataset as though it were an 'observation' but is an observation with probability weight p_j. This necessitates a probability weight of

$p = 1$ for those observations u for which $V(u)$ is itself a p-dimensional rectangle. When all virtual components $\{R_j, j = 1, 2, \ldots\}$ are rectangles, then the direct use of the methodologies presented in Section 4.2 and Section 4.4 described above applies.

When (as in the baseball example) an R_j is a triangle, adjustment has to be made to ensure that the calculated area of the 'triangle' is indeed that, and not the area of the corresponding rectangle. Components R_j that are not rectangles are different. In some instances, this non-rectangular shape in and of itself is not a problem, though calculating the probability might (but should not in general) be tricky. This present discussion assumes $V(u)$ can be partitioned into rectangular components (with appropriate adjustment for triangular pieces).

4.6 Measures of Dependence

4.6.1 Moment dependence

The prime dependence measure between any two variables Y_{j_1} and Y_{j_2} is through the covariance function or functions of the covariance itself (such as the product-moment correlation function, regression coefficient, and so on). An alternative measure using Spearman's rho and copulas is presented in Section 4.6.2.

In this section, we consider the usual product-moment-based covariance function typically used for classical data but adapted to symbolic data. The various types of symbolic data are considered in turn. In each case, we assume interest is centered on the two symbolic variables $Y_{j_1} \equiv Z_1$ and $Y_{j_2} \equiv Z_2$.

Definition 4.9: For quantitative multi-valued variables Z_1 and Z_2, the **empirical covariance function** $Cov(Z_1, Z_2)$ is given by

$$Cov(Z_1, Z_2) = \frac{1}{m}[\sum_{\xi_1, \mathcal{Z}_1} \sum_{\xi_2 \in \mathcal{Z}_2} (\xi_1 \times \xi_2) O_{Z_1, Z_2}(\xi_1, \xi_2)] - \bar{Z}_1 \bar{Z}_2 \tag{4.15}$$

where the joint observed frequency $O_{Z_1, Z_2}(\xi_1, \xi_2)$ is found from Equation (4.1), and $\bar{Z}_i$ is the symbolic sample mean for Z_i given by Equation (3.11). □

Example 4.6. We use the cancer dataset (of Table 3.2) for which the joint observed frequencies $O_{Z_1, Z_2}(\xi_1, \xi_2)$ were calculated in Example 4.1 and the sample means were calculated in Example 3.8. Then, by substituting into Equation (4.15), we have

$$\begin{aligned} Cov(Y_1, Y_2) = &\frac{1}{8}[(0\times 0)(4/3) + (0\times 1)(0) + (0\times 2)(0) + (0\times 3)(0) \\ &+ (1\times 0)(1/3) + (1\times 1)(4/3) + (1\times 2)(5/2) + (1\times 3)(5/2)] \\ &- (0.833)(1.729) \\ =& 0.288. \end{aligned}$$

□

Definition 4.10: For interval-valued variables Z_1 and Z_2, the **empirical covariance function** $Cov(Z_1, Z_2)$ is given by

$$Cov(Z_1, Z_2) = \frac{1}{3m}\sum_{u\in E} G_1 G_2 [Q_1 Q_2]^{1/2} \tag{4.16}$$

where, for $j = 1, 2$,

$$Q_j = (a_{uj} - \bar{Z}_j)^2 + (a_{uj} - \bar{Z}_j)(b_{uj} - \bar{Z}_j) + (b_{uj} - \bar{Z}_j)^2,$$

$$G_j = \begin{cases} -1, & \text{if } \bar{Z}_{uj} \le \bar{Z}_j, \\ 1, & \text{if } \bar{Z}_{uj} > \bar{Z}_j, \end{cases}$$

and where $\bar{Z}_1$ and $\bar{Z}_2$ are as defined in Equation (3.22) and where $\bar{Z}_{uj} = (a_{uj} + b_{uj})/2$ is the midpoint (mean) value of the variable Z_j for observation w_u. □

This is analogous with the empirical variance function of Equation (3.23). Clearly, when $Z_1 = Z_2$, the function in Equation (4.16) reduces to the variance of Z_1 as a special case. Also, when the data are classically valued with $a_1 = b_1$ and $a_2 = b_2$, the function of Equation (4.16) reduces to the familiar formula for classical data.

Definition 4.11: For interval-valued variables Z_1 and Z_2, the empirical **correlation coefficient** $R(Z_1, Z_2)$ is given by

$$r(Z_1, Z_2) = Cov(Z_1, Z_2)/S_{Z_1} S_{Z_2} \tag{4.17}$$

where the covariance $Cov(Z_1, Z_2)$ is given in Equation (4.16) and the standard deviations for Z_1 and Z_2 are obtained from Equation (3.23). □

Example 4.7. Let us calculate the covariance function for Y_2 = Stipe Length and Y_3 = Stipe Width in the mushroom data of Table 3.3; see Example 3.10. Then, from Equation (4.16), we have

$$
\begin{aligned}
&Cov(Y_2, Y_3)\\
&= \frac{1}{3\times 23}\{(-1)(-1)[(\{4.0-7.391\}^2+\{4.0-7.391\}\{9.0-7.391\}+\{9.0-7.391\}^2)\\
&\quad \times(\{0.50-1.824\}^2+\{0.50-1.824\}\{2.5-1.824\}+\{2.5-1.824\}^2)]^{1/2}_{u=1}\\
&\quad +\ldots\\
&\quad +(1)(1)[(\{4.0-7.391\}^2+\{4.0-7.391\}\{14.0-7.391\}+\{14.0-7.391\}^2)\\
&\quad \times(\{1.0-1.824\}^2+\{1.0-1.824\}\{3.5-1.824\}+\{3.5-1.8245\}^2)]^{1/2}_{u=23}\}\\
&=2.134,
\end{aligned}
$$

where $\bar{Y}_2 = 7.391$ and $\bar{Y}_3 = 1.824$ were given in Table 3.4. Also, from Equation (4.17), we obtain

$$r(Y_2, Y_3) = 2.134/[(3.557)(1.041)] = 0.576,$$

where $S_{Y_2} = 3.557$ and $S_{Y_3} = 1.041$ are as given in Table 3.4. □

Were we to calculate a covariance and correlation coefficient based on the interval midpoints (a classical point value), the resulting covariance function would not reflect the internal cross-variations between Z_1 and Z_2. We observed this phenomenon in Section 3.3 on the symbolic variances. The symbolic covariance function of Equation (4.16) (and likewise the symbolic correlation function of Equation (4.17)) reflects both the internal variation and that across observations.

Example 4.8. The covariance function based on the midpoint values for Y_2 = Stipe Length and Y_3 = Stipe Width for the mushrooms of Table 3.4 is

$$Cov(Y_2, Y_3) = 1.877.$$

The standard deviations based on the midpoints are

$$S_{Y_2} = 3.089, \; S_{Y_3} = 0.994.$$

All three of these statistics are smaller than their symbolic counterparts given in Example 4.7. This reflects the loss of information that results when using only the midpoint values instead of the entire interval. The correlation coefficient based on the midpoints is $r(Y_2, Y_3) = 0.611$. □

It is implicit in the definition of Equations (4.16) and (4.17) that the calculations are based on those individual descriptions for which associated rules v apply, i.e., on the resulting virtual description spaces. Observations for which the virtual space is non-rectangular require suitable adjustments.

Example 4.9. Let us take the blood pressure data of Table 3.5 and calculate the covariance function of the Pulse Rate Y_1 and the Systolic Pressure Y_2 under the aegis of the rule v_4; see Equation (4.7) and Example 3.15. Under this rule, the observations w_{11} and w_{12} have $vir(d_u) = 0$ and so do not enter into the calculations. The observations w_3 and w_{14} have virtual descriptions that differ from the real descriptions (see Figure 4.8), and so adjustments need to be made in the calculations.

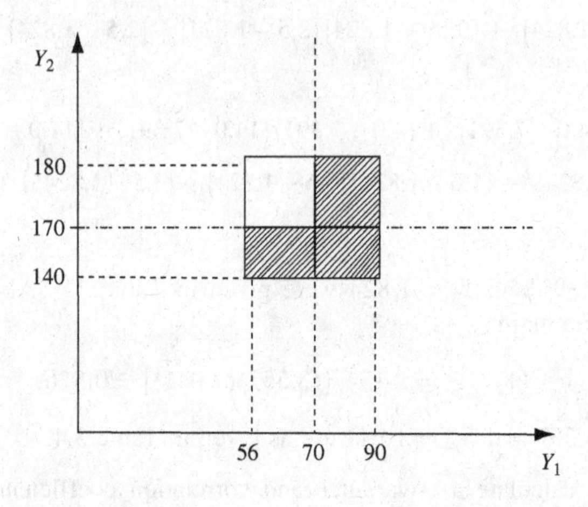

Figure 4.8 Blood pressure data, rule $v_4 : u = 3$.

We first calculate $\bar{Y}_1 = 76.050$ and $\bar{Y}_2 = 133.637$, from Equation (3.22) with adjustment for virtual spaces. Thus, for example,

$$\begin{aligned}\bar{Y}_1 = &\frac{1}{2 \times 13}\{(68+44)_{u=1} + (72+60)_{u=2} + [(56+70)(0.344)_{R'_1} \\ &+ (90+70)(0.656)_{R'_2}]_{u=3} + \ldots + (100+86)_{u=10} + (81+73)_{u=13} \\ &+ (75+70)_{u=14} + (52+42)_{u=15}\} \\ = &76.050.\end{aligned}$$

Therefore, the covariance function from Equation (4.16) becomes

$$\begin{aligned}&Cov(Y_1, Y_2) \\ &= \frac{1}{3 \times 13}\{(-1)(-1)[(\{68-76.050\}^2 + \{68-76.050\}\{44-76.050\} \\ &\quad + \{44-76.050\}^2) \times (\{110-133.637\}^2 + \{110-133.637\}\{90-133.637\} \\ &\quad + \{90-133.637\}^2)]^{1/2}_{u=1} + \ldots + \{(-1)(-1)[(\{90-76.050\}^2 \\ &\quad + \{90-76.050\}\{56-76.050\} + \{56-76.040\}^2) \\ &\quad \times (\{170-133.637\}^2 + \{170-133.637\}\{140-133.637\} \\ &\quad + \{140-133.637\}^2)]^{1/2}(0.836)_{R_1} \\ &\quad + (-1)(-1)[\{90-76.050\}^2 + \{90-76.050\}\{70-76.050\} + \{70-76.050\}^2)\end{aligned}$$

$$\begin{aligned}
&\times (\{180-133.637\}^2+\{180-133.637\}\{170-133.637\}\\
&+\{170-133.637\}^2)]^{1/2}(0.164)_{R_2}\}_{u=3}+\ldots+(1)(-1)[\{100-76.050\}^2\\
&+\{100-76.050\}\{86-76.050\}+\{86-76.050\}^2)\times(\{150-133.637\}^2\\
&+\{150-133.637\}\{110-133.637\}+\{110-133.637\}^2)]_{u=10}^{1/2}\\
&+(1)(-1)[(\{81-76.050\}^2+\{81-76.050\}\{73-76.050\}+\{73-76.050\}^2)\\
&\times(\{138-133.637\}^2+\{138-133.637\}\{125-133.637\}\\
&+\{125-133.637\}^2)]_{u=13}^{1/2}+(-1)(1)[(\{75-76.050\}^2\\
&+\{75-76.050\}\{70-76.050\}+\{70-76.050\}^2)\\
&\times(\{194-133.637\}^2+\{194-133.637\}\{175-133.637\}\\
&+\{175-133.637\}^2)]_{u=14}^{1/2}+(-1)(-1)[(\{52-76.050\}^2\\
&+\{52-76.050\}\{42-76.050\}+\{42-76.050\}^2)\times(\{115-133.637\}^2\\
&+\{115-133.637\}\{105-133.637\}+\{105-133.637\}^2)]_{u=15}^{1/2}\}\\
=&\,215.297.
\end{aligned}$$

Notice that the subareas R_1 and R_2 for observation w_3 used in the cross-product terms in calculating the covariance differ from the subareas R_1' and R_2' used in calculating $\bar{Y}_1$. The moment correlation coefficient is

$$r(Y_1, Y_2) = 0.517.$$

Likewise, we can obtain the covariance and correlation between Y_2 = Systolic Pressure and Y_3 = Diastolic Pressure as

$$Cov(Y_2, Y_3) = 273.496, \quad r(Y_2, Y_3) = 0.727;$$

and between Y_1 = Pulse Rate and Y_3 = Diastolic Pressure as

$$Cov(Y_1, Y_3) = 168.965, \quad r(Y_1, Y_3) = 0.780. \qquad \square$$

Definition 4.12: For quantitative modal multi-valued variables Z_1 and Z_2, the **empirical covariance function** $Cov(Z_1, Z_2)$ is given by

$$Cov(Z_1, Z_2) = \frac{1}{m}[\sum_{\xi_{1k_1}\in\mathcal{Z}_1}\sum_{\xi_1 k_2\in\mathcal{Z}_2}(\xi_{1k_1}\times\xi_{2k_2})O(Z_1=\xi_{1k_1}, Z_2=\xi_{2k_2})]-\bar{Z}_1\bar{Z}_2 \quad (4.18)$$

where the observed frequency of $\mathbf{Z} = (\xi_{1k_1}, \xi_{2k_2})$ is given by Equation (4.8), and where the empirical sample means $\bar{Z}_i$, $i = 1, 2$, are found from Equation (3.11). $\square$

Example 4.10. Consider the unweighted age and gender data of Table 4.6 and let us recode the gender values to male = 1 and female = 2. From Equation (4.16) and Table 4.7, it follows that

$$\begin{aligned}
Cov(Y_1, Y_2) &= \frac{1}{9}[(1\times 1)(0.063) + \ldots + (1\times 7)(0.106)\\
&\quad + (2\times 1)(0.435) + \ldots + (2\times 7)(0.890)] - (1.900)(3.916)\\
&= (66.262)/9 - 7.4404 = -0.078
\end{aligned}$$

where $\bar{Y}_1 = 1.900$ and $\bar{Y}_2 = 3.916$ were calculated from Equation (3.11). □

Definition 4.13: For modal interval-valued (i.e., histogram-valued) variables Z_1 and Z_2, the **empirical covariance function** is given by

$$Cov(Z_1, Z_2) = \frac{1}{3m}\sum_{u\in E}\sum_{k_1=1}^{s_{u1}}\sum_{k_2=1}^{s_{u2}} p_{u1k_1}p_{u2k_2}G_1G_2[Q_1Q_2]^{1/2} \tag{4.19}$$

where, for $j = 1, 2$,

$$Q_j = (a_{ujk_j} - \bar{Z}_j)^2 + (a_{ujk_j} - \bar{Z}_j)(b_{ujk_j} - \bar{Z}_j) + (b_{ujk_j} - \bar{Z}_j)^2,$$

$$G_j = \begin{cases} -1, & \text{if } \bar{Z}_{uj} \le \bar{Z}_j,\\ 1, & \text{if } \bar{Z}_{uj} > \bar{Z}_j,\end{cases}$$

and where $\bar{Z}_1$ and $\bar{Z}_2$ are as given in Equation (3.39) and where

$$\bar{Z}_{uj} = \frac{1}{2}\sum_{k_j=1}^{s_{uj}} p_{ujk_j}(a_{ujk_j} + b_{ujk_j})$$

is the mean value of the variable Z_j for the observation w_u. □

Example 4.11. Consider the data of Tables 4.6 and 4.14. Table 4.14 gives values for Y_1 = Hematocrit, and Table 4.6 gave values for Y_2 = Hemoglobin (used in Example 4.5) for gender × age groups. We calculate the covariance function between hematocrit (Y_1) and hemoglobin (Y_2) by substituting into Equation (4.19) as

$$\begin{aligned}
Cov(Y_1, Y_2) &= \frac{1}{3\times 14}(\{(1)(1)\{(0.025)(0.050)[(\{35 - 37.157\}^2\\
&\quad + \{35 - 37.157\}\{37.5 - 37.157\} + \{37.5 - 37.157\}^2)\\
&\quad \times (\{12 - 12.363\}^2 + \{12 - 12.363\}\{12.9 - 12.363\}\\
&\quad + \{12.9 - 12.363\}^2)]^{1/2} + \ldots + (0.038)(0.025)[(\{45.5 - 37.157\}^2\\
&\quad + \{45.5 - 37.157\}\{47.0 - 37.157\} + \{47.0 - 37.157\}^2)\\
&\quad \times (\{14.7 - 12.363\}^2 + \{14.7 - 12.363\}\{15.0 - 12.363\}
\end{aligned}$$

$$
\begin{aligned}
&+\{15.0-12.363\}^2)]^{1/2}\}_{u=1}+\ldots\\
&+(1)(-1)\{(0.133)(0.133)[(\{33.5-37.157\}^2\\
&+\{33.5-37.157\}\{35.5-37.157\}+\{35.5-37.157\}^2)(\{10.8-12.363\}^2\\
&+\{10.8-12.363\}\{11.2-12.363\}+\{11.2-12.363\}^2)]^{1/2}+\ldots\\
&+(1)(-1)\{(0.200)(0.133)[(\{41.5-37.157\}^2\\
&+\{41.5-37.157\}\{43.0-37.157\}+\{43.0-37.157\}^2)\\
&\times(\{12.8-12.363\}^2+\{12.8-12.363\}\{13.2-12.363\}\\
&+\{13.2-12.363\}^2)]^{1/2}\}_{u=14})=1.347,
\end{aligned}
$$

where $\bar{Y}_1 = 37.157$ and $\bar{Y}_2 = 12.363$ were obtained by use of Equation (3.39).

For these data, from Equation (3.40) we obtain

$$S_{Y_1} = 2.743, \quad S_{Y_2} = 0.739;$$

hence from Equation (4.17), we can calculate the correlation coefficient as

$$r = 1.347/(2.743 \times 0.739) = 0.664. \qquad \square$$

Table 4.14 Hematocrit values for gender and age groups.

w_u	Concept		Frequency histogram
	Gender	Age	
w_1	Female	20s	{[35.0, 37.5), 0.025; [37.5, 39.0), 0.075; [39.0, 40.5), 0.188; [40.5, 42.0), 0.387; [42.0, 45.5), 0.287; [45.5, 47.0), 0.038}
w_2	Female	30s	{[31.0, 33.0), 0.046; [33.0, 35.0), 0.171; [35.0, 36.5), 0.295; [36.5, 38.0), 0.243; [38.0, 39.5), 0.170; [39.5, 41.0), 0.072; [41.0, 44.0], 0.003}
w_3	Female	40s	{[31.0, 33.0), 0.049; [33.0, 35.0), 0.203; [35.0, 36.5), 0.223; [36.5, 38.0), 0.241; [38.0, 39.5), 0.209; [39.5, 41.0), 0.069; [41.0, 43.5], 0.006}
w_4	Female	50s	{[31.0, 32.0), 0.011; [32.0, 33.5), 0.066; [33.5, 35.0), 0.194; [35.0, 36.5), 0.231; [36.5, 38.0), 0.248; [38.0, 39.5), 0.168; [39.5, 41.0), 0.068; [41.0, 42.5], 0.014}
w_5	Female	60s	{[31.0, 33.0), 0.037; [33.0, 34.5), 0.178; [34.5, 36.0), 0.215; [36.0, 37.5), 0.247; [37.5, 38.0), 0.101; [38.0, 39.5), 0.182; [39.5, 41.0), 0.028; [41.0, 42.5], 0.012}
w_6	Female	70s	{[31.0, 32.5), 0.011; [32.5, 34.0), 0.089; [34.0, 35.5), 0.200; [35.5, 37.0); 0.272; [37.0, 38.5), 0.228; [38.5, 40.0), 0.122; [40.0, 41.5), 0.072; [41.5, 43.5], 0.006}

Table 4.14 Continued.

w_u	Concept		Frequency histogram
	Gender	Age	
w_7	Female	80+	{[31.0, 32.5), 0.018; [32.5, 34.0), 0.107; [34.0, 35.5), 0.195; [35.5, 37.0), 0.302; [37.0, 38.5), 0.189; [38.5, 40.0), 0.136; [40.0, 41.5), 0.041; [41.5, 43.0], 0.012}
w_8	Male	20s	{[37.5, 39.0), 0.214; [39.0, 40.5), 0.286; [40.5, 42.0), 0.286; [42.0, 43.5], 0.214}
w_9	Male	30s	{[30.0, 32.0), 0.022; [32.0, 33.5), 0.111; [33.5, 35.0), 0.178; [35.0, 36.5), 0.311; [36.5, 38.0), 0.222; [38.0, 39.5), 0.111; [39.5, 41.0], 0.045}
w_{10}	Male	40s	{[30.0, 32.0), 0.018; [32.0, 33.5), 0.109; [33.5, 35.0), 0.073; [35.0, 36.5), 0.327; [36.5, 38.0), 0.218; [38.0, 39.5), 0.164; [39.5, 41.0], 0.091}
w_{11}	Male	50s	{[33.5, 35.0), 0.215; [35.0, 36.5), 0.294; [36.5, 38.0), 0.255; [38.0, 39.5), 0.118; [39.5, 41.0), 0.098; [41.0, 42.0], 0.020}
w_{12}	Male	60s	{[32.0, 33.5), 0.125; [33.5, 35.0), 0.208; [35.0, 36.5), 0.375; [36.5, 38.0), 0.125; [38.0, 39.5), 0.125; [39.5, 41.0], 0.042}
w_{13}	Male	70s	{[32.0, 33.5), 0.158; [33.5, 35.0), 0.263; [35.0, 36.5), 0.263; [36.5, 38.0), 0.053; [38.0, 39.5], 0.263}
w_{14}	Male	80+	{[33.5, 35.5), 0.133; [35.5, 37.5), 0.267; [37.5, 39.5), 0.267; [39.5, 41.5), 0.133; [41.5, 43.0], 0.200}

4.6.2 Spearman's rho and copulas

Another common measure of dependence between two random variables Y_1 and Y_2 is Spearman's rho. This measure is extended by Billard (2004) to interval-valued variables by exploiting the concept of copulas first introduced by Sklar (1959).

Definition 4.14: Let $H(y_1, y_2)$ be the joint distribution function of (Y_1, Y_2), and let $F_j(y_j)$ be the marginal distribution function of Y_j, $j = 1, 2$. Then, there exists a function $C(y_1, y_2)$ called a **copula** which satisfies

$$H(y_1, y_2) = C(F_1(y_1), F_2(y_2)). \tag{4.20}$$

□

Notice in particular that the functions $H(\cdot)$ and $F(\cdot)$ are cumulative distribution functions rather than probability density functions. Copulas play an important role in incorporating dependence into the relationship between joint distribution functions and marginal distribution functions since the same marginal distributions can arise from different joint distributions. Properties of copulas and their construction are treated in Nelsen (1999). Deheuvels (1979) provided a method for calculating the empirical copula function. The adaption to interval-valued data proceeds as follows.

For conciseness, assume the two variables of interest are $Y_{j_1} \equiv Z_1$ and $Y_{j_2} \equiv Z_2$. Let

$$X_{uj} = (a_{uj} + b_{uj})/2 \tag{4.21}$$

be the midpoints of the interval $[a_{uj}, b_{uj}]$ for variable Z_j, for each observation w_u, $u = 1, \ldots, m$. Let W_{jk_j} be the (X_{uj}) value of the k_jth rank order of the statistics X_{uj}, $k_j = 1, \ldots, m$, $j = 1, 2$. Then, let us define the rectangles

$$R(w_{1k_1}, w_{2k_2}) = (q_1, w_{1k_1}) \times (q_2, w_{2k_2}) \tag{4.22}$$

for $k_1, k_2 = 1, \ldots, m$, where

$$q_1 \le \min_u(a_{u1}), \quad q_2 \le \min_u\{a_{u2}\},$$

and for $k_j = m + 1$, the rectangle is

$$R(w_{1,m+1}, w_{2,m+1}) = (q_1, t_1) \times (q_2, t_2) \tag{4.23}$$

where

$$t_1 \ge \max_u\{b_{u1}\}, \quad t_2 \ge \max_u\{b_{u2}\}.$$

We can then define the following.

Definition 4.15: The **empirical copula** for interval-valued data is given by

$$C'(w_{1k_1}, w_{2k_2}) = \frac{1}{m} \sum_{u \in E} \frac{||Z(u) \cap R(w_{1k_1}, w_{2k_2})||}{||Z(u)||} \tag{4.24}$$

for $k_1, k_2 = 1, \ldots, m + 1$, where $Z(u)$ is the rectangle $(a_{u1}, b_{u1}) \times (a_{u2}, b_{u2})$ associated with the observation $\boldsymbol{\xi}_u$ and where $||A||$ is the area of the rectangle A. □

Example 4.12. To illustrate let us take the mushroom data of Table 4.15, and let us consider the variables $Z_1 \equiv Y_2 =$ Stipe Length and $Z_2 \equiv Y_3 =$ Stipe Thickness. The ranked interval midpoints are shown in Table 4.15. For example, the fifth ranked Z_1 value and the fourth ranked Z_2 value are, respectively, from Equation (4.21),

$$W_{1,5} \equiv X_{8,1} = (6.00 + 3.00)/2 = 4.50$$

$$W_{2,4} \equiv X_{20,2} = (0.80 + 0.40)/2 = 0.60.$$

Applying Equation (4.23) to the data in Table 3.3 and Table 4.15, we have

$$C'(w_{1,5}, w_{2,4}) = C'(4.50, 0.60)$$

$$= \frac{1}{23}\left\{\left(\frac{4.50 - 4.00}{9.00 - 4.00}\right)\left(\frac{0.60 - 0.50}{2.50 - 0.50}\right)_{u=1} + \left(\frac{4.50 - 4.00}{14.00 - 4.00}\right)(0)_{u=2}\right.$$

$$+0+\ldots+0+\left(\frac{4.50-3.00}{7.00-3.00}\right)\left(\frac{0.60-0.40}{1.00-0.40}\right)_{u=7}+0$$
$$+\left(\frac{4.50-3.00}{5.00-3.00}\right)\left(\frac{0.60-0.40}{0.70-0.40}\right)_{u=9}$$
$$+0+\left(\frac{4.50-3.00}{6.00-3.00}\right)\left(\frac{0.35-0.25}{0.35-0.25}\right)_{u=11}$$
$$+0+\ldots+0+\left(\frac{4.50-2.50}{4.50-2.50}\right)\left(\frac{0.60-0.40}{0.70-0.40}\right)_{u=16}$$
$$+0+0+0+\left(\frac{4.50-3.00}{7.00-3.00}\right)\left(\frac{0.60-0.40}{0.80-0.40}\right)_{u=20}+0+0+0\Bigg\}$$
$$=1.98417/23=0.0863.$$

Table 4.15 Ranked midpoints.

Stipe Length			Stipe Thickness		
k_1	u	W_{1k_1}	k_2	u	W_{2k_2}
1	10	2.50	1	11	0.30
2	5	3.50	2	16	0.55
3	16	3.50	3	9	0.55
4	9	4.00	4	20	0.60
5	8	4.50	5	7	0.70
6	11	4.50	6	10	1.25
7	7	5.00	7	8	1.50
8	20	5.00	8	1	1.50
9	4	5.50	9	13	1.50
10	1	6.50	10	3	1.50
11	6	7.00	11	22	1.50
12	13	7.00	12	19	1.75
13	3	8.00	13	21	1.75
14	2	9.00	14	5	2.00
15	19	9.00	15	12	2.00
16	21	9.00	16	14	2.00
17	23	9.00	17	2	2.25
18	12	9.50	18	23	2.25
19	18	10.50	19	17	2.75
20	14	11.0	20	6	3.00
21	22	11.00	21	18	3.00
22	17	11.50	22	4	3.75
23	15	14.00	23	15	4.00

The copula frequency is therefore $mC'(w_{1,5}, w_{2,4}) = 1.98417$. A portion of the set of empirical copula frequency values is provided in Table 4.16 and the complete set is plotted in Figure 4.9. □

Clearly the marginal distributions can likewise be determined. In this case, they correspond to the $C(w_{1i_1}, w_{2,m+1})$ and $C(w_{1,m+1}, w_{2i_2})$ column/row respectively.

Table 4.16 Copula frequencies: mushroom data.

$k_1 \backslash k_2$		0	1	2	⋯	22	23	24	Relative
	$W_{1k_1} \backslash W_{2k_2}$	0	0.30	0.55	⋯	3.75	4.00	4.50	frequency
0	0	0	0	0	⋯	0	0	**0**	**0**
1	2.50	0	0	0	⋯	0.667	0.667	**0.667**	**0.029**
2	3.50	0	0.083	0.620	⋯	2.833	2.833	**2.833**	**0.123**
⋮	⋮	⋮	⋮	⋮		⋮	⋮	⋮	⋮
22	11.50	0	0.500	2.650	⋯	18.728	19.051	**19.510**	**0.848**
23	14.00	0	0.500	2.650	⋯	21.018	21.372	**21.955**	**0.955**
24	19.00	**0**	**0.500**	**2.650**	⋯	**21.750**	**22.167**	**23.000**	**1.000**
Relative frequency		**0**	**0.022**	**0.115**		**0.946**	**0.964**	**1.000**	

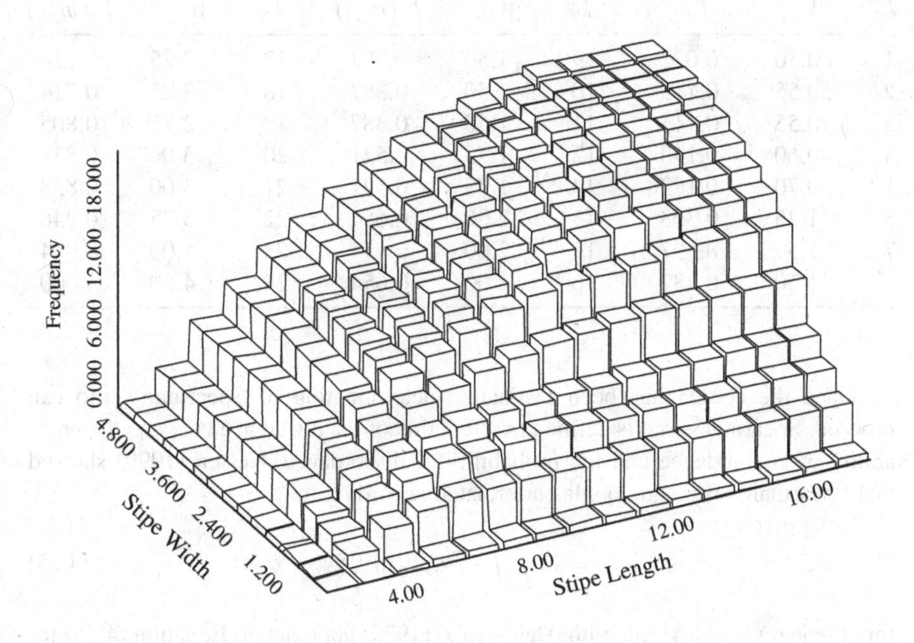

Figure 4.9 Empirical copula function: mushroom data.

Example 4.13. In Example 4.12, the column $C(\cdot, 24)$ shown in bold in Table 4.16 is the marginal distribution of Y_2 = Stipe Length and the row $C(24, \cdot)$ shown in bold in Table 4.16 is the marginal distribution of Y_3 = Stipe Thickness. The complete cumulative marginal distributions of Y_2 and Y_3 are given in Table 4.17. □

Table 4.17 Marginal distributions.

Y_2: Stipe Length								
k_1	W_{1k_1}	$F_1(w_{1k_1})$	k_1	W_{1k_1}	$F_1(w_{1k_1})$	k_1	W_{1k_1}	$F(w_{1k_1})$
1	2.5	0.029	9	5.5	0.380	17	9.0	0.684
2	3.5	0.123	10	6.5	0.485	18	9.5	0.722
3	3.5	0.123	11	7.0	0.537	19	10.5	0.791
4	4.0	0.178	12	7.0	0.537	20	11.0	0.821
5	4.5	0.257	13	8.0	0.604	21	11.0	0.821
6	4.5	0.257	14	9.0	0.684	22	11.5	0.848
7	5.0	0.326	15	9.0	0.684	23	14.0	0.955
8	5.0	0.326	16	9.0	0.684	24	19.0	1.000
Y_3: Stipe Width								
k_2	W_{2k_2}	$F_2(w_{2k_2})$	k_2	W_{2k_2}	$F_2(w_{2k_2})$	k_2	W_{2k_2}	$F_2(w_{2k_2})$
1	0.30	0.022	9	1.50	0.387	17	2.25	0.714
2	0.55	0.115	10	1.50	0.387	18	2.25	0.714
3	0.55	0.115	11	1.50	0.387	19	2.75	0.805
4	0.60	0.140	12	1.75	0.521	20	3.00	0.838
5	0.70	0.189	13	1.75	0.521	21	3.00	0.838
6	1.25	0.308	14	2.00	0.654	22	3.75	0.946
7	1.50	0.397	15	2.00	0.654	23	4.00	0.964
8	1.50	0.387	16	2.00	0.654	24	4.50	1.000

Once the copula has been evaluated, determination of Spearman's rho can proceed. Spearman's rho is defined as a function of the difference between the probability of concordance and the probability of discordance. Nelsen (1999) showed that Spearman's rho and copulas are related according to

$$\rho = 12 \int \int C(y_1, y_2) dy_1 dy_2 \qquad (4.25)$$

for classical variables. Adapting Deheuvels' (1979) approach to Equation (4.25) for symbolic data, we have the following definition.

Definition 4.16: The **empirical Spearman's rho** for two interval-valued variables Y_1 and Y_2 is given by

$$r = \frac{12}{m^2-1}\{[\sum_{k_1=1}^{m+1}\sum_{k_2=1}^{m+1} C'(w_{1k_1}, w_{2k_2})] - [\sum_{k_1=1}^{m+1} F_1(w_{1k_1})][\sum_{k_2=1}^{m+1} F_2(w_{wk_2})]\}. \quad (4.26)$$

where $\{F_j(w_{jk_j}), i_j = 1, \ldots, m+1\}$, $j = 1, 2$, is the marginal distribution function of Y_j. □

Example 4.14. Let us calculate the empirical Spearman's rho function for Y_2 and Y_3 in the mushroom data. The empirical copula function $C'(w_1, w_2)$ was calculated in Example 4.12, and the respective marginal distribution functions $F_i(w_i)$, $i = 1, 2$, were observed in Example 4.13. Applying those results to Equation (4.26), we have

$$\begin{aligned} r = & \left(\frac{12}{23^2-1}\right)\{[(0+0.250+\ldots+0.667)+(0.083+\ldots+2.833) \\ & +\ldots+(0.500+\ldots+23.00)]/23 \\ & -[0.029+\ldots+0.955+1.000][0.022+\ldots+0.964+1.00]\} \\ = & 0.578. \end{aligned}$$

□

Exercises

E4.1. Take the health insurance data of Tables 2.2a.

(i) Build the joint histogram for the random variables (Y_5, Y_6) where Y_5 = Marital Status and Y_6 = Number of Parents living.

(ii) Calculate the covariance function of Y_5 and Y_6.

E4.2. Redo Exercise E4.1 by using the data from Table 2.6. Compare the results.

E4.3. Build the joint histogram for (Y_2, Y_3) with Y_2 = Systolic Blood Pressure and Y_3 = Diastolic Blood Pressure for the blood data of Table 3.5.

E4.4. Redo Exercise E4.3 by adding the rule (v_3 of Example 3.14)

$$v : A = \{Y_1 \leq 70\} \Rightarrow B = \{Y_2 \leq 170\}.$$

E4.5. Calculate the covariance function between Y_2 and Y_3 for each of the two situations described in Exercises E4.3 and E4.4, respectively.

E4.6. Consider the cholesterol data of Table 4.5 and the hematocrit data of Table 4.14. Take the observations $w_1, \ldots, w_7$ for the women only. Calculate (i) the joint histogram for cholesterol (Y_1) and hematocrit (Y_2), and (ii) the covariance function of (Y_1, Y_2).

E4.7. Refer to Exercise E4.6. Repeat this using the observations $w_8, \ldots, w_{14}$ for the men only. Compare the results for men and women.

E4.8. Repeat Exercise E4.6 now using the hemoglobin data of Table 4.6 and the hematocrit data of Table 4.14, for women only.

E4.9. Repeat Exercise E4.7 using Table 4.6 and Table 4.14 for the observations for men only, and compare your results to those for women in Exercise E4.8.

E4.10. Take the first 10 observations $w_1, \ldots, w_{10}$ of the Iris data of Table 5.9.

(i) Build the joint histogram on the two random variables Y_1 = Sepal Length and Y_2 = Sepal Width.

(ii) Calculate the covariance function of (Y_1, Y_2).

E4.11. Repeat Exercise E4.10 using the second set of 10 observations $w_{11}, \ldots, w_{20}$.

E4.12. Repeat Exercise E4.10 using the third set of observations $w_{21}, \ldots, w_{30}$. Compare the results from Exercises E4.10–E4.12.

E4.13. Calculate an empirical copula function for the two variables Y_2 = Systolic Blood Pressure and Y_3 = Diastolic Blood Pressure using the first 10 observations only from Table 3.5.

E4.14. Consider the first 10 teams in the baseball data of Table 4.8. Calculate the empirical copula function for Y_1 = Number of At-Bats and Y_2 = Number of Hits under the rule that $Y_1 \geq Y_2$.

References

Billard, L. (2004). Dependencies in Bivariate Interval-Valued Symbolic Data. In: *Classification, Clustering, and Data Mining Applications* (eds. D. Banks, L. House, F. R. McMorris, P. Arabie, and W. Gaul). Springer-Verlag, Berlin, 319–324.

Billard, L. and Diday, E. (2006). Descriptive Statistics for Interval-Valued Observations in the Presence of Rules. *Computational Statistics* 21, 187–210.

Deheuvels, P. (1979). La Fonction de Dependence Empirique et ses Proprietes. Un Test Non Parametrique d'Independence. *Academic Royale Belgique Bulletin de la Classe des Science* 65, 274–292.

Nelsen, R. B. (1999). *An Introduction to Copulas.* Springer-Verlag, New York.

Sklar, A. (1959). Fonctions de rèpartition â *n* dimensions et leurs marges. *Publication Institute Statistique Université Paris* 8, 229–231.

5

Principal Component Analysis

A principal component analysis is designed to reduce p-dimensional observations into s-dimensional components (where usually $s \ll p$). More specifically, a principal component is a linear combination of the original variables, and the goal is to find those s principal components which together explain most of the underlying variance–covariance structure of the p variables. Two methods currently exist for performing principal component analysis on symbolic data, the vertices method covered in Section 5.1 and the centers methods treated in Section 5.2. These are for interval-valued data only. No methodology yet exists for other types of symbolic data.

5.1 Vertices Method

Cazes *et al.* (1997), Chouakria (1998), and Chouakria *et al.* (1998) developed a method of conducting principal component analysis on symbolic data for which each symbolic variable Y_j, $j = 1, \ldots, p$, takes interval values $\xi_{uj} = [a_{uj}, b_{uj}]$, say, for each observation $u = 1, \ldots, m$, and where each observation may represent the aggregation of n_u individuals. Or equivalently, each observation is inherently interval valued. When $a_{uj} < b_{uj}$, we refer to the interval $[a_{uj}, b_{uj}]$ as a non-trivial interval. For each observation $\boldsymbol{\xi}_u = (\xi_{u1}, \ldots, \xi_{up})$, let q_u be the number of non-trivial intervals. Each symbolic data point, $\boldsymbol{\xi}_u = ([a_{u1}, b_{u1}], \ldots, [a_{up}, b_{up}])$, is represented by a hyperrectangle H_u with $m_u = 2^{q_u}$ vertices. When $a_{uj} < b_{uj}$ for all $j = 1, \ldots, p$, then $m_u = 2^p$.

Example 5.1. The first two observations (w_1, w_2) in Table 5.1 are examples of observations with no trivial intervals; these form $p = 3$ dimensional hyperrectangles in $\Re^3$ (see Figure 5.1). The observation w_3, however, with $a_{33} = b_{33}$, is a

Symbolic Data Analysis: Conceptual Statistics and Data Mining L. Billard and E. Diday

two-dimensional hyperrectangle, i.e., it is a plane in $\Re^3$ as seen in Figure 5.1. In this case, $q_3 = 2$, and so there are $m_u = 2^2 = 4$ vertices. Likewise, the observation w_4 is a line in $\Re^3$ with $q_4 = 1$ and so $m_4 = 2^1 = 2$ vertices, and that of w_5 is a point in $\Re^3$ with $q_5 = 0$ and so there is $m_5 = 2^0 = 1$ vertex in $\Re^3$. □

Table 5.1 Trivial and non-trivial intervals

w_u	$Y_1 = [a_1, b_1]$	$Y_2 = [a_2, b_2]$	$Y_3 = [a_3, b_3]$
w_1	[1, 3]	[1, 4]	[2, 3]
w_2	[2, 4]	[3, 4]	[4, 6]
w_3	[1, 3]	[2, 4]	[3, 3]
w_4	[3, 5]	[3, 3]	[2, 2]
w_5	[3, 3]	[2, 2]	[5, 5]

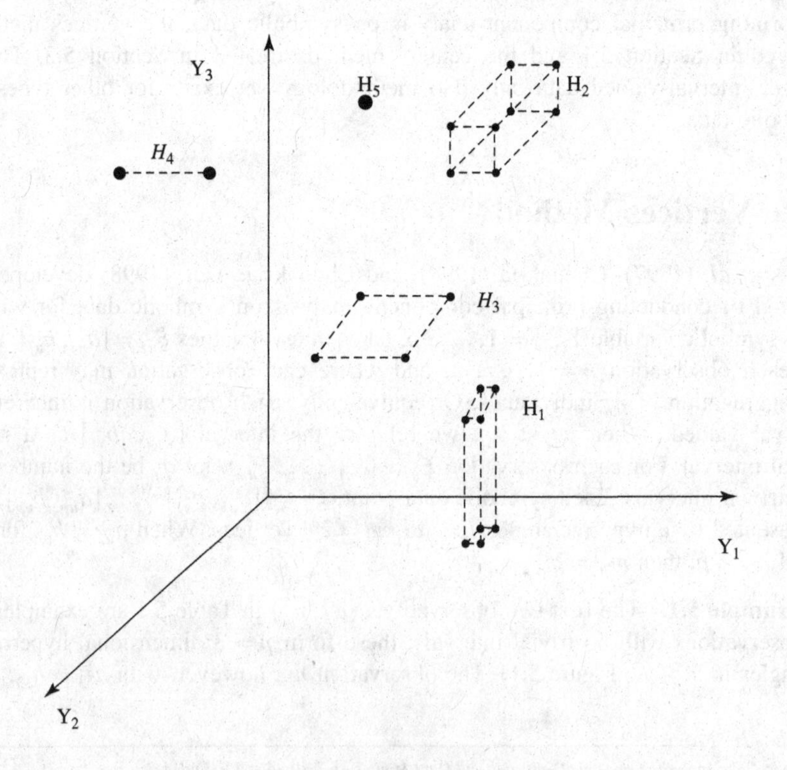

Figure 5.1 Types of hypercubes: clouds of vertices.

Each hyperrectangle H_u can be represented by a $2^{q_u} \times p$ matrix $\boldsymbol{M}_u$ with each row containing the coordinate values of a vertex V_{k_u}, $k_u = 1, \ldots, 2^{q_u}$, of the hyperrectangle. There are

$$n = \sum_{u=1}^{m} m_u = \sum_{u=1}^{m} 2^{q_u} \tag{5.1}$$

vertices in total. Then, an $(n \times p)$ data matrix $\boldsymbol{M}$ is constructed from the $\{\boldsymbol{M}_u, u = 1, \ldots, m\}$ according to

$$\boldsymbol{M} = \begin{pmatrix} \boldsymbol{M}_1 \\ \vdots \\ \boldsymbol{M}_m \end{pmatrix} = \left(\begin{matrix} \begin{bmatrix} a_{11} & \ldots & a_{1q_1} \\ \vdots & & \vdots \\ b_{11} & \ldots & b_{1q_1} \end{bmatrix} \\ \vdots \\ \begin{bmatrix} a_{m1} & \ldots & a_{mq_m} \\ \vdots & & \vdots \\ b_{m1} & \ldots & b_{mq_m} \end{bmatrix} \end{matrix} \right). \tag{5.2}$$

For example, if $p = 2$, the observation $\xi_u = ([a_{u1}, b_{u1}], [a_{u2}, b_{2u}])$ is transformed to the $2^p \times p = 2^2 \times 2$ matrix

$$\boldsymbol{M}_u = \begin{bmatrix} a_{u1} & a_{u2} \\ a_{u1} & b_{u2} \\ b_{u1} & a_{u2} \\ b_{u1} & b_{u2} \end{bmatrix},$$

and likewise for $\boldsymbol{M}$.

Example 5.2. Consider the artificial data of Table 5.1. We saw in Example 5.1 that the first two $(u = 1, 2)$ observations formed a hyperrectangle in $p = 3$ dimensional space. Therefore, for $u = 1$, $u = 2$, the $m_u = 2^3 = 8$ vertices are represented by the rows in $\boldsymbol{M}_1$ and $\boldsymbol{M}_2$ given as

$$\boldsymbol{M}_1 = \begin{pmatrix} 1 & 1 & 2 \\ 1 & 1 & 3 \\ 1 & 4 & 2 \\ 1 & 4 & 3 \\ 3 & 1 & 2 \\ 3 & 1 & 3 \\ 3 & 4 & 2 \\ 3 & 4 & 3 \end{pmatrix}, \quad \boldsymbol{M}_2 = \begin{pmatrix} 2 & 3 & 4 \\ 2 & 3 & 6 \\ 2 & 4 & 4 \\ 2 & 4 & 6 \\ 4 & 3 & 4 \\ 4 & 3 & 6 \\ 4 & 4 & 4 \\ 4 & 4 & 6 \end{pmatrix}.$$

In this case $k_u = 2^3 = 8$, and $\boldsymbol{M}_1$ and $\boldsymbol{M}_2$ are both 8×3 matrices. However, for $u = 3$, we have a $2^2 \times 3$ matrix since the hyperrectangle H_3 associated with $\boldsymbol{\xi}_3$ is a rectangle in $\mathfrak{R}^3$, i.e., $m_3 = 2^2 = 4$, and therefore

$$\boldsymbol{M}_3 = \begin{pmatrix} 1 & 2 & 3 \\ 1 & 4 & 3 \\ 3 & 2 & 3 \\ 3 & 4 & 3 \end{pmatrix};$$

for $u = 4$, the matrix $\boldsymbol{M}_4$ is $2^1 \times 3$ since $\boldsymbol{\xi}_4$ represented by H_4 is a line in $\mathfrak{R}^3$, i.e., $m_4 = 2^1 = 2$, and therefore

$$\boldsymbol{M}_4 = \begin{pmatrix} 3 & 3 & 2 \\ 5 & 3 & 2 \end{pmatrix};$$

and for $m = 5$, $\boldsymbol{M}_5$ is a 1×3 matrix since $\boldsymbol{\xi}_5$ is a single point in $\mathfrak{R}^3$, i.e., $m_5 = 2^0 = 1$, and therefore

$$\boldsymbol{M}_5 = (3 \quad 2 \quad 5).$$

Then, the complete set of vertices is represented by the rows of $\boldsymbol{M}$, where, from Equation (5.2),

$$\boldsymbol{M} = \begin{pmatrix} \boldsymbol{M}_1 \\ \boldsymbol{M}_2 \\ \boldsymbol{M}_3 \\ \boldsymbol{M}_4 \\ \boldsymbol{M}_5 \end{pmatrix}$$

and where $\boldsymbol{M}$ is an $n \times p = 23 \times 3$ matrix since here

$$n = \sum_{u=1}^{n} m_u = \sum_{u=1}^{m} 2^{q_u} = 8+8+4+2+1 = 23. \qquad \square$$

Example 5.3. Consider the two observations

$$\boldsymbol{\xi}(w_1) = ([20, 25], [60, 80]),$$
$$\boldsymbol{\xi}(w_2) = ([21, 28], [65, 86]).$$

Then,

$$\boldsymbol{M}_1 = \begin{pmatrix} 20 & 60 \\ 20 & 80 \\ 25 & 60 \\ 25 & 80 \end{pmatrix}, \quad \boldsymbol{M}_2 = \begin{pmatrix} 21 & 65 \\ 21 & 86 \\ 28 & 65 \\ 28 & 86 \end{pmatrix},$$

and hence

$$M' = \begin{pmatrix} M_1 \\ M_2 \end{pmatrix}'$$

$$= \begin{pmatrix} 20 & 20 & 25 & 25 & 21 & 21 & 28 & 28 \\ 60 & 80 & 60 & 80 & 65 & 86 & 65 & 86 \end{pmatrix}.$$

□

The matrix M is treated as though it represents classical p-variate data for n individuals. Chouakria (1998) has shown that the basic theory for a classical analysis carries through; hence, a classical principal component analysis can be applied.

Briefly, a classical principal component analysis finds linear combinations of the classical variables $X_1, \ldots, X_p$. Thus, the νth principal component $PC\nu$, $\nu = 1, \ldots, s$, is

$$PC\nu = e_{\nu 1}X_1 + \ldots + e_{\nu p}X_p \tag{5.3}$$

with $\sum_{j=1}^{p} l_{\nu j}^2 = 1$, $\boldsymbol{l}_\nu = (l_{\nu 1}, \ldots, l_{\nu p})$ are the eigenvalues $\nu = 1, \ldots, p$, and where $\boldsymbol{e}_\nu = (\boldsymbol{e}_{\nu 1}, \ldots, \boldsymbol{e}_{\nu p})$ is the νth eigenvector associated with the underlying covariance matrix. The resulting $PC\nu$, $\nu = 1, \ldots, s$, are uncorrelated. The νth principal component is that value which maximizes the variance $Var(PC\nu)$ subject to $\boldsymbol{e}'_\nu \boldsymbol{e}_\nu = 1$. The total variance underlying the $(X_1, \ldots, X_p)$ values can be shown to be $\lambda = \sum_{\nu=1}^{p} \lambda_\nu$. Hence, the proportion of the total variance that is explained by the νth principal component is λ_ν / λ. Often by standardizing the covariance matrix, the resulting correlation matrix is used instead. A detailed discussion of the application of and theory of principal components can be found in most texts devoted to multivariate analyses; see, for example, Anderson (1984) for a theoretical treatise and Johnson and Wichern (1998) for an applied presentation.

Observations can be weighted by a factor p_u say, with $\sum_{u=1}^{m} p_u = 1$. In classical analyses these weights may typically be $p_u = 1/m$. However, for our symbolic data, it may be desirable to have weights that reflect in some way the differences between, for example, $\xi_1 = [3, 7]$ and $\xi_2 = [1, 9]$, i.e., two observations that by spanning different interval lengths have different internal variations. This suggests the following options.

Definition 5.1: The **weight** associated with an observation $\boldsymbol{\xi}_u$, $u = 1, \ldots, m$, is defined by p_u. The **vertex weight** associated with the vertex k_u of $\boldsymbol{\xi}_u$, $k_u = 1, \ldots, m_u$, is defined by $p_{k_u}^u$, with

$$\sum_{k_u=1}^{m_u} p_{k_u}^u = p_u, \quad u = 1, \ldots, m, \tag{5.4}$$

and

$$\sum_{u=1}^{m} p_u = 1. \tag{5.5}$$

□

There are many possible choices for p_u and $p^u_{k_u}$. The simplest scenarios are when all observations are equally weighted, and all vertices in the cluster of vertices in H_u are equally weighted. In these cases,

$$p_u = 1/m, \quad u = 1, \ldots, m, \tag{5.6}$$

and

$$p^u_{k_u} = p_u(1/q_u), \quad k_u = 1, \ldots, m_u. \tag{5.7}$$

Other possible weights for p_u could take into account the 'volume' occupied in $\Re^p$ by an observation ξ_u, as suggested by the following definitions.

Definition 5.2: The **volume** associated with observations $\boldsymbol{\xi}_u$, $u = 1, \ldots, m$, is

$$V(\boldsymbol{\xi}_u) = \prod_{\substack{j=1 \\ b_{uj} \neq a_{uj}}}^{p} (b_{uj} - a_{uj}). \tag{5.8}$$

□

Definition 5.3: An observation $\boldsymbol{\xi}_u$, $u = 1, \ldots, m$, is said to be **proportionally weighted by volume** when

$$p_u = V(\boldsymbol{\xi}_u)/V \tag{5.9}$$

where $V = \sum_{u=1}^{m} V(\boldsymbol{\xi}_u)$. □

Definition 5.4: An observation $\boldsymbol{\xi}_u$, $u = 1, \ldots, m$, is said to be inversely proportionally weighted by volume when

$$p'_u = [1 - V(\boldsymbol{\xi}_u)/V] / \sum_{u=1}^{m} [1 - V(\boldsymbol{\xi}_u)/V]. \tag{5.10}$$

□

Clearly, in both definitions (5.3) and (5.4), $\sum_{u=1}^{m} p_u = 1$ as required. The use of the weight p_u in Equation (5.9) clearly gives larger weight to an observation, e.g., $\xi = [1, 9]$ which embraces more volume in $\Re^p$ reflecting the larger internal variation of such an observation over an observation with smaller volume, such as $\xi = [3, 7]$. The use of the p'_u in Equation (5.10) might be preferable when the interval length(s) reflect a measure of uncertainty δ, e.g., $\delta = 6 \pm \delta$. The larger the degree of uncertainty, the larger the value of δ. In this case, however, it would be desirable to give such an observation a lower weight than one for which the degree of uncertainty δ is small.

Example 5.4. For the data of Table 5.1, it follows that the volume of each data point is, from Equation (5.8),

$$V(\xi_1) = (3-1)(4-1)(3-1) = 12, \ V(\xi_2) = (5-2)(4-3)(6-4) = 6,$$

$$V(\xi_3) = (3-1)(4-2) = 4, \ V(\xi_4) = (5-3) = 2, \ V(\xi_5) = 0.$$

Then, $\sum V(\xi_u) = 24$. Hence the weights p_u are, from Equation (5.9),

$$p_1 = 12/24 = 0.5;$$

likewise,

$$p_2 = 0.25, \ p_3 = 0.167, \ p_4 = 0.089, \ p_5 = 0.$$

Clearly, $\sum p_u = 1$.

The inverse volume weights for the data of Table 5.1 follow from Equation (5.10) as

$$\begin{aligned} p_1' &= \frac{(1-12/24)}{[(1-12/24)+(1-6/24)+(1-2/24)+(1-0/24)]} \\ &= (1-12/24)/4 = 0.125; \end{aligned}$$

likewise,

$$p_2' = 0.188, \ p_3' = 0.208, \ p_4' = 0.229, \ p_5' = 0.250. \qquad \square$$

Weights reflecting the number of observations n_u that made up the symbolic observation w_u could be defined by

$$p_u'' = p_u(n_u / \sum_{u=1}^{m} n_u). \tag{5.11}$$

Example 5.5. Let us assume the symbolic observations of Table 5.1 were the result of aggregating $n_u = 3, 2, 4, 2, 1$ classical observations for the categories $u = 1, \ldots, 5$, respectively. Then, the weighted volume-based weights p_u'' are

$$\begin{aligned} p_1'' &= 3(0.5)/[(3)(0.5) + 2(0.25) + 4(0.167) + 2(0.089) + 1(0)] \\ &= 1.5/2.846 = 0.527; \end{aligned}$$

likewise,

$$p_2'' = 0.177, \ p_3'' = 0.234, \ p_4'' = 0.062, \ p_5'' = 0. \qquad \square$$

The weight p_u (and likewise for p_u' and p_u'') is associated with the observation $\boldsymbol{\xi}_u$. This weight is then distributed across the $m_u = 2^{p_u}$ vertices associated with $\boldsymbol{\xi}_u$. This leads us to the following.

Definition 5.5: Under the assumption that values across the intervals $[a_{uj}, b_{uj}]$, $j = 1, \ldots, p$, $u = 1, \ldots, m$, are uniformly distributed, the weights associated with the observation w_u are equally distributed across its vertices V_{k_u}, $k_u = 1, \ldots, 2^{p_u}$, i.e., the weight on V_{k_u} is

$$p^u_{k_u}(u) = p_u/2^{q_u}, \; k_u = 1, \ldots, 2^{q_u}, \; u = 1, \ldots, m. \tag{5.12}$$

□

Therefore, a classical (weighted or unweighted as desired) principal component analysis is conducted on the n observations in $\boldsymbol{M}$. Let $Y^*_1, \ldots, Y^*_s$, $s \leq p$, denote the first s principal components with associated eigenvalues $\lambda_1 \geq \ldots \geq \lambda_s \geq 0$ which result from this classical analysis of the n vertices. We then construct the interval vertices principal components $Y^V_1, \ldots, Y^V_s$ as follows.

Let L_u be the set of row indices in $\boldsymbol{M}$ identifying the vertices of the hyperrectangle H_u, i.e., L_u represents the rows of $\boldsymbol{M}_u$ describing the symbolic data observation $\boldsymbol{\xi}_u$. For each row $k_u = 1, \ldots, 2^{q_u}$ in L_u, let $y_{\nu u k_u}$ be the value of the principal component $Y^*_{\nu u}$, $\nu = 1, \ldots, s$, for that row k_u. Then, the νth interval vertices principal component $Y^V_{\nu u}$ for observation w_u is given by

$$Y^V_{\nu u} = y_{\nu u} = [y^a_{\nu u}, y^b_{\nu u}], \; \nu = 1, \ldots, s, \tag{5.13}$$

where

$$y^a_{\nu u} = \min_{k_u \in L_u} \{y_{\nu u k_u}\} \text{ and } y^b_{\nu u} = \max_{k_u \in L_u} \{y_{\nu u k_u}\}. \tag{5.14}$$

Then, plotting these values against each ν principal component axis $PC\nu$ gives us the data represented as hyperrectangles of principal components in s-dimensional space. For example, take the $p = 3$ dimensional hyperrectangle for a single observation $\boldsymbol{\xi}$, represented by H in Figure 5.2 (adapted from Chouakria, 1998). Figure 5.2 contains $s = 3$ axes corresponding to the first, second, and third principal components calculated from Equations (5.13)–(5.14) on all m observations, using all n vertices. The projection of points in the hyperrectangle H onto the first and second principal component plane, and also onto the second and third principal component plane, are displayed.

It is easily verified that the rectangle formed by the two interval-valued principal components constitutes a maximal envelope of the projection points from H. Therefore, it follows that every point in the hypercube H when projected onto the plane lies in this envelope. However, depending on the actual value of H, there can be some (exterior) points within the envelope that may not be projections of points in H.

Equations (5.13)–(5.14) can be verified as follows. Take any point $\tilde{\boldsymbol{x}}_i$ with $\tilde{x}_{ij} \in [a_{ij}, b_{ij}]$. Then, the νth principal component associated with this $\tilde{\boldsymbol{x}}_i$ is

$$\tilde{PC}\nu = \sum_{j=1}^{p} e_{\nu j}(\tilde{x}_{uj} - \bar{X}^*_j).$$

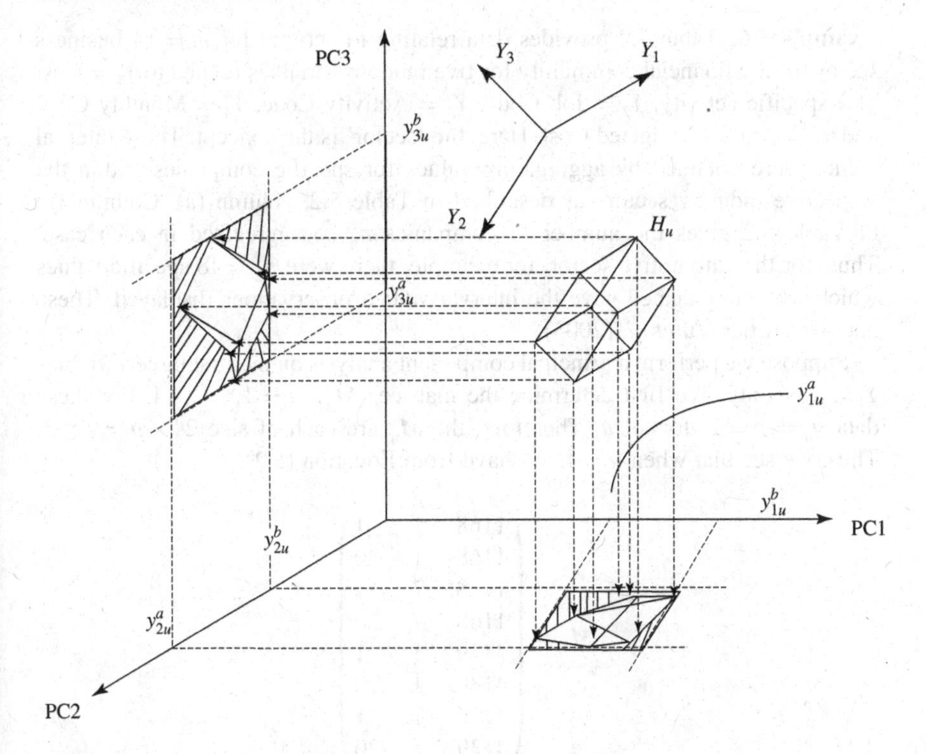

Figure 5.2 Projection hypercube H_u to principal component ($\nu = 1, 2, 3$) axes.

It follows that

$$\sum_{j=1}^{p} e_{\nu j}(\tilde{x}_{uj} - \bar{X}_j^*) \geq \sum_{j \in J^+}^{p} e_{\nu j}(a_{uj} - \bar{X}_j^*) + \sum_{j \in J^-}^{p} e_{\nu j}(b_{uj} - \bar{X}_j^*) \tag{5.15}$$

and

$$\sum_{j=1}^{p} e_{\nu j}(\tilde{x}_{uj} - \bar{X}_j^*) \leq \sum_{j \in J^-}^{p} e_{\nu j}(a_{uj} - \bar{X}_j^*) + \sum_{j \in J^+}^{p} e_{\nu j}(b_{uj} - \bar{X}_j^*) \tag{5.16}$$

where $J^+ = \{j | e_{\nu j} > 0\}$ and $J^- = \{j | e_{\nu j} < 0\}$. However, by definition, the right-hand side of Equation (5.15) is

$$\min_{k_u \in L_u} y_{\nu u k_u} = y^a_{\nu u}$$

and the right-hand side of Equation (5.16) is

$$\max_{k_u \in L_u} y_{\nu u k_u} = y^b_{\nu u}.$$

Hence, for all $j = 1, \ldots, p$,

$$P\tilde{C}\nu \in [y^a_{\nu u}, y^b_{\nu u}];$$

and so $Y^*_{\nu u}$ as in Equations (5.13)–(5.14) holds for all $\tilde{\boldsymbol{x}}_i$ with $x_{ij} \in [a_{ij}, b_{ij}]$.

Example 5.6. Table 5.2 provides data relating to activity for $m = 14$ business sectors of the financial community for five random variables related to $Y_1 =$ Cost of a specific activity, $Y_2 =$ Job Code, $Y_3 =$ Activity Code, $Y_4 =$ Monthly Cost, and $Y_5 =$ Annual Budgeted Cost. Here, the 'sector' is the concept. These interval values were obtained by aggregating values for specific companies within the respective industry sectors as described in Table 5.2, column (a). Column (g) of Table 5.2 gives the number of companies n_u so aggregated in each case. Thus, for the automotive sector, for example, there were $n_2 = 48$ specific values which when aggregated gave the interval-valued observations displayed. These data were after Vu *et al.* (2003).

Suppose we perform a principal component analysis on the first three variables Y_1, Y_2, Y_3 only. We first determine the matrices $\boldsymbol{M}_u$, $u = 1, \ldots, 14$. For these data $q_u = p = 2$, for all u. Therefore, the $\boldsymbol{M}_u$ are each of size $2^p \times p = 8 \times 3$. Thus, we see that when $u = 1$, we have from Equation (5.2),

$$\boldsymbol{M}_1 = \begin{pmatrix} 1168 & 1 & 1 \\ 1168 & 1 & 20 \\ 1168 & 7 & 1 \\ 1168 & 7 & 20 \\ 4139 & 1 & 1 \\ 4139 & 1 & 20 \\ 4139 & 7 & 1 \\ 4139 & 7 & 20 \end{pmatrix}.$$

Likewise, we find $\boldsymbol{M}_u$ for all u; and hence we find the matrix $\boldsymbol{M}$ which is of size $(m2^p) \times p = 112 \times 3$. Each row of $\boldsymbol{M}$ represents a classical vertex value in the set of $(m = 14)$ hypercubes.

A classical principal component analysis on these $n = 112$ classical values is then performed in the usual manner. We denote the νth principal component for the vertex k_u of the observation w_u by $PC\nu = y_{\nu u k_u}$, $\nu = 1, \ldots, s$, $u = 1, \ldots, m$, $k_u = 1, \ldots, 8$. Since we ultimately want to find the interval vertices principal components for each observation $\boldsymbol{\xi}_u$ through Equations (5.13) and (5.14), it is advantageous to sort the classical analysis by u minimally, and preferably by both u and the νth principal component value. The results when sorted by u and the first principal component ($\nu = 1$) are displayed for the FMCG (w_1) and durable goods (w_{13}) sectors in Table 5.3. Thus, the rows $i = 1, \ldots, 8$ correspond to the ($k_1 = 1, \ldots, 8$) vertices of $\boldsymbol{\xi}_1$ ($u = 1$), the rows $i = 97, \ldots, 104$ correspond to the ($k_{13} = 1, \ldots, 8$) vertices of $\boldsymbol{\xi}_{13}$ ($u = 13$), and so on.

Then, to find the first vertices principal component interval Y_{11}^V, i.e., $\nu = 1$, we take Table 5.3 and Equation (5.14) which give

$$y_{11}^a = \min_{k_1 \in L_1} \{y_{11k_1}\} = -1.535$$

Table 5.2 Finance data.

	(a)	(b) Job Cost	(c) Job Code	(d) Activity Code	(e) Monthly Cost	(f) Annual Budget	(g)
w_u	Sector	$[a_1, b_1]$	$[a_2, b_2]$	$[a_3, b_3]$	$[a_4, b_4]$	$[a_5, b_5]$	n_u
w_1	FMCG	[1168, 4139]	[1, 7]	[1, 20]	[650, 495310]	[1000, 1212300]	605
w_2	Automotive	[2175, 4064]	[1, 6]	[1, 20]	[400, 377168]	[400, 381648]	48
w_3	Telecommunications	[2001, 4093]	[1, 7]	[1, 20]	[1290, 250836]	[1290, 500836]	63
w_4	Publishing	[2465, 3694]	[1, 3]	[4, 20]	[5000, 42817]	[5000, 79367]	12
w_5	Finance	[2483, 4047]	[1, 7]	[2, 20]	[6500, 248629]	[6500, 313501]	40
w_6	Consultants	[2484, 4089]	[1, 6]	[9, 20]	[4510, 55300]	[4510, 74400]	15
w_7	Consumer goods	[2532, 4073]	[1, 6]	[2, 20]	[2900, 77500]	[2901, 77500]	34
w_8	Energy	[2542, 3685]	[1, 6]	[2, 20]	[2930, 54350]	[2930, 54350]	17
w_9	Pharmaceutical	[2547, 3688]	[1, 7]	[1, 20]	[1350, 49305]	[12450, 1274700]	31
w_{10}	Tourism	[2604, 3690]	[1, 5]	[9, 20]	[1600, 31700]	[1600, 31700]	13
w_{11}	Textiles	[2697, 4012]	[1, 5]	[9, 9]	[12800, 54850]	[12800, 94000]	15
w_{12}	Services	[3481, 4058]	[1, 1]	[9, 9]	[8400, 31500]	[8400, 41700]	3
w_{13}	Durable goods	[2726, 4068]	[1, 5]	[10, 20]	[5800, 28300]	[3700, 55400]	3
w_{14}	Others	[3042, 4137]	[1, 1]	[1, 20]	[458, 19400]	[458, 19400]	34

Table 5.3 Classical principal components, $u = 1,\ 13$.

i	k_u	u	$Y1$	$Y2$	$Y3$	$PC1$	$PC2$	$PC3$
1	1	1	4139	1	1	−1.535	−1.064	0.253
2	2	1	4139	1	20	−1.177	1.275	0.131
3	3	1	4139	7	1	0.137	−1.22	1.943
4	4	1	4139	7	20	0.496	1.107	1.820
5	5	1	1168	1	1	0.959	−1.579	−2.266
6	6	1	1168	1	20	1.318	0.756	−2.389
7	7	1	1168	7	1	2.632	−1.747	−0.577
8	8	1	1168	7	20	2.990	0.592	−0.699
⋮	⋮	⋮	⋮	⋮	⋮	⋮	⋮	⋮
97	1	13	4068	1	10	−1.306	0.031	0.135
98	2	13	4068	1	20	−1.117	1.262	0.070
99	3	13	4068	5	10	−0.191	−0.080	1.262
100	4	13	2726	1	10	−0.179	−0.201	−1.003
101	5	13	4068	5	20	−0.002	1.151	1.197
102	6	13	2726	1	20	0.010	1.030	−1.068
103	7	13	2726	5	10	0.936	−0.313	0.123
104	8	13	2726	5	20	1.125	0.918	0.059
⋮	⋮	⋮	⋮	⋮	⋮	⋮	⋮	⋮

and

$$y_{11}^b = \max_{k_1 \in L_1}\{y_{11k_1}\} = 2.990$$

where $L_1 = \{1, \ldots, 8\}$ and hence, from Equation (5.13), the first vertices principal component interval Y_{11}^V for the w_1 observation, i.e., for the FMCG sector, is

$$Y_{11}^V = y_{11} = [-1.535, 2.990].$$

Similarly, for the durable goods sector, i.e., for the w_{13} observation $\boldsymbol{\xi}_{13}$, we apply Equation (5.14) to Table 5.3 to obtain

$$y_{1,13}^a = \min_{k_{13} \in L_{13}}\{y_{11k_{13}}\} = -1.306$$

$$y_{1,13}^b = \max_{k_{13} \in L_{13}}\{y_{11k_{13}}\} = 1.125$$

where $L_{13} = \{97, \ldots, 104\}$ and hence, from Equation (5.13), the first vertices principal component interval $Y_{1,13}^V$ for the durable goods sector is

$$Y_{1,13}^V = y_{1,13} = [-1.306, 1.125].$$

The complete set of the first, second, and third vertices principal component intervals based on the Y_1, Y_2, Y_3 variables for all observations is provided in Table 5.4.

Table 5.4 Vertices principal components – finance data (based on Y_1, Y_2, Y_3).

w_u	Sector	$PC1 = Y_{1u}^V$	$PC2 = Y_{2u}^V$	$PC3 = Y_{3u}^V$
w_1	FMCG	[−1.535, 2.990]	[−1.747, 1.275]	[−2.389, 1.943]
w_2	Automotive	[−1.473, 1.866]	[−1.544, 1.262]	[−1.535, 1.598]
w_3	Telecommunications	[−1.497, 2.291]	[−1.602, 1.267]	[−1.683, 1.904]
w_4	Publishing	[−1.105, 0.786]	[−1.041, 1.197]	[−1.289, 0.420]
w_5	Finance	[−1.439, 1.886]	[−1.396, 1.259]	[−1.274, 1.859]
w_6	Consultants	[−1.343, 1.606]	[−0.506, 1.266]	[−1.273, 1.568]
w_7	Consumer goods	[−1.461, 1.566]	[−1.359, 1.263]	[−1.232, 1.599]
w_8	Energy	[−1.135, 1.558]	[−1.358, 1.196]	[−1.224, 1.270]
w_9	Pharmaceutical	[−1.157, 1.832]	[−1.508, 1.196]	[−1.220, 1.561]
w_{10}	Tourism	[−1.007, 1.227]	[−0.457, 1.197]	[−1.171, 0.947]
w_{11}	Textiles	[−1.278, 0.941]	[−0.441, −0.101]	[−1.021, 1.221]
w_{12}	Services	[−1.316, −0.832]	[−0.193, −0.093]	[−0.356, 0.133]
w_{13}	Durable goods	[−1.306, 1.125]	[−0.313, 1.262]	[−1.068, 1.262]
w_{14}	Others	[−1.534, −0.256]	[−1.254, 1.274]	[−0.800, 0.252]

A plot of the first and second vertices principal component intervals, i.e., of the $Y_{1,u}^V$ and $Y_{2,u}^V$, is shown in Figure 5.3. Four clusters emerge. The first contains the observation w_1 corresponding to the FMCG sector. The second contains the automotive, telecommunications, finance, consumer goods and energy sectors (i.e., w_2, w_3, w_5, w_7, w_8). In fact, the $PC1$ and $PC2$ principal components for the consumer goods (w_7) and energy (w_8) sectors are almost identical, as are those for the automotive (w_2) and finance (w_5)' sectors. The remaining observations (w_4, w_6, w_9, w_{10}, w_{11}, w_{12}, w_{13}) form a third cluster. However, it may be advantageous to identify a fourth cluster for the textile and services sectors corresponding to the w_{11}, w_{12} observations, with the third cluster consisting of the remaining five sectors. These clusters are nested clusters in that the first cluster spans the other clusters through the w_1 observation for the FMCG sector. The data values for this sector were the result of over 600 observations and have large internal variation. Therefore, it is not surprising that the related principal component takes this form. In contrast, the observations in the fourth cluster, the textile and services sectors, have much smaller internal variation across the three variables (Y_1, Y_2, Y_3) collectively, and this is reflected in their comparably clustered principal components.

Figure 5.4 shows the corresponding plot for the classical analysis of the 112 vertices values. Three distinct clusters are apparent. First, recall that here each symbolic observation is represented by its eight vertex values. Thus, the middle cluster of Figure 5.4 essentially contains the vertex values of the w_{11} and w_{12}

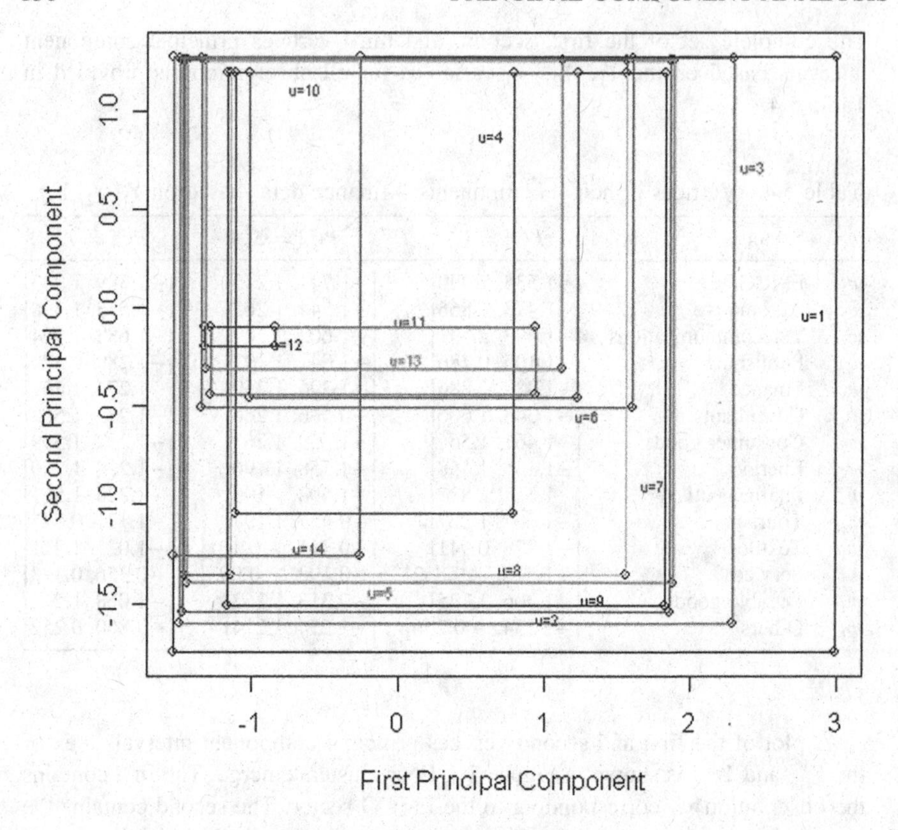

Figure 5.3 Vertices principal components: finance data.

symbolic values of the inner fourth cluster along with the lower vertex values for the w_u for $u = 6,\ 10,\ 13$ observations. The top outer cluster of Figure 5.4 contains the respective upper vertex values of all, except the w_{11} and w_{12} observations of the first three clusters of Figure 5.3, while the bottom outer cluster of Figure 5.4 contains the lower vertex values of all but the w_u for $u = 6,\ 10,\ 11,\ 12,\ 13$ observations.

We can also find the percentages of the overall variation, λ_ν/λ, accounted for by the νth principal component. For this analysis of these data, the first vertex principal component accounts for 36.3% of the variation while the first two vertices principal components together account for 69.6% of the total variation. □

As discussed earlier, observations can be weighted by $p_u \geq 0$. Then in the transformation of the symbolic observation to the matrix $\boldsymbol{M}$ of classical observations in Example 5.6, each vertex of the hyperrectangle H_u in effect has the same weight $(p_u 2^{-p})$. Weights can be arbitrarily assigned, or can take many forms. Let us consider an analysis where weights are proportional to the number of individual

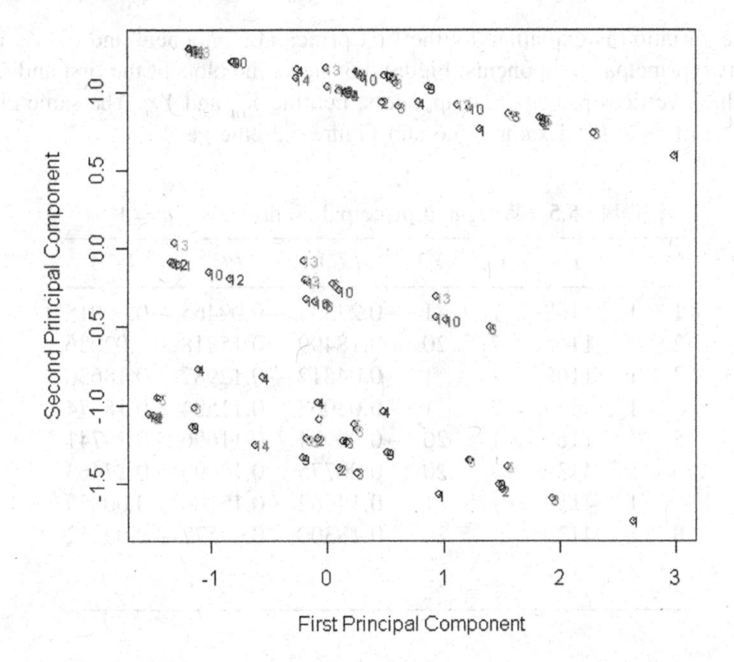

Figure 5.4 Classical principal components: vertices of finance data.

observations n_u that were aggregated to produce the symbol observation $\boldsymbol{\xi}_u$, $u = 1, \ldots, m$.

Example 5.7. Suppose now that instead of the unweighted principal component analysis of Example 5.6, we perform a weighted analysis. Let us do this for the same data of Table 5.2, and let us assume that the weights are proportional to the number of specific companies n_u that generated the symbolic values; see Table 5.2, column (g). That is, we use the weights of Equation (5.11).

We perform the weighted classical analysis on the vertices of matrix $\boldsymbol{M}$, and again sort the output by u and by the first principal component values. The results for the FMCG sector (w_1), i.e., for $k = 1, \ldots, 8$, are shown in Table 5.5. From this table and Equation (5.14), we obtain for the first ($\nu = 1$) principal component

$$y_{11}^a = \min_{k_1 \in L_1} \{y_{11k_1}\} = -0.223$$

$$y_{11}^b = \max_{k_1 \in L_1} \{y_{11k_1}\} = 0.183;$$

hence the first weighted principal component interval for the FMCG sector (w_1) is

$$Y_{11}^V = y_{11} = (-0.223, 0.183).$$

The respective first, second, and third weighted vertices principal components for all w_u, $u = 1, \ldots, 14$, are displayed in Table 5.6. For this analysis, 34.5%

of the variation is explained by the first principal component and 67.8% by the first two principal components. Figure 5.5 shows the plots of the first and second weighted vertices principal components, i.e., the Y^V_{1u} and Y^V_{2u}. The same clusters as were observed in Example 5.6 and Figure 5.3 emerge. □

Table 5.5 Weighted principal components, $u = 1$.

k	u	$Y1$	$Y2$	$Y3$	$PC1$	$PC2$	$PC3$
1	1	1168	7	1	−0.22337	−0.09465	−0.00915
2	1	1168	7	20	−0.18499	0.15218	−0.02026
3	1	1168	1	1	−0.04812	−0.12987	−0.18630
4	1	4139	7	1	−0.03063	−0.11584	0.18574
5	1	1168	1	20	−0.00974	0.11696	−0.19741
6	1	4139	7	20	0.00775	0.13099	0.17463
7	1	4139	1	1	0.14462	−0.15106	0.00859
8	1	4139	1	20	0.18300	0.09577	−0.00252
⋮	⋮	⋮	⋮	⋮	⋮	⋮	⋮

Table 5.6 Weighted vertices principal components – finance data (based on Y_1, Y_2, Y_3).

w_u	Sector	PCI $= Y^V_{1u}$	PC2 $= Y^V_{2u}$	PC3 $= Y^V_{3u}$
w_1	FMCG	[−0.223, 0.183]	[−0.151, 0.152]	[−0.197, 0.186]
w_2	Automotive	[−0.129, 0.178]	[−0.151, 0.139]	[−0.131, 0.151]
w_3	Telecommunications	[−0.169, 0.180]	[−0.151, 0.146]	[−0.143, 0.183]
w_4	Publishing	[−0.016, 0.154]	[−0.109, 0.119]	[−0.112, 0.037]
w_5	Finance	[−0.136, 0.177]	[−0.137, 0.143]	[−0.111, 0.179]
w_6	Consultants	[−0.093, 0.180]	[−0.047, 0.137]	[−0.111, 0.148]
w_7	Consumer goods	[−0.104, 0.179]	[−0.138, 0.137]	[−0.108, 0.151]
w_8	Energy	[−0.103, 0.154]	[−0.135, 0.137]	[−0.107, 0.126]
w_9	Pharmaceutical	[−0.134, 0.154]	[−0.148, 0.142]	[−0.107, 0.156]
w_{10}	Tourism	[−0.056, 0.154]	[−0.044, 0.130]	[−0.103, 0.093]
w_{11}	Textiles	[−0.050, 0.153]	[−0.046, −0.013]	[−0.091, 0.114]
w_{12}	Services	[0.118, 0.156]	[−0.047, −0.042]	[−0.039, −0.001]
w_{13}	Durable goods	[−0.046, 0.178]	[−0.034, 0.129]	[−0.095, 0.117]
w_{14}	Others	[0.073, 0.183]	[−0.151, 0.104]	[−0.074, 0.008]

When the number of variables p is large, it can be difficult to visualize features in the data. The reduction to $s < p$ principal components is one avenue that assists this visualization. Further interpretations can be obtained by looking at associations between the observations and the principal components. Interpretations of the parameters underlying classical analyses can be extended to the symbolic analysis. Typically, quality measures of the principal components are represented by their contribution functions and variances.

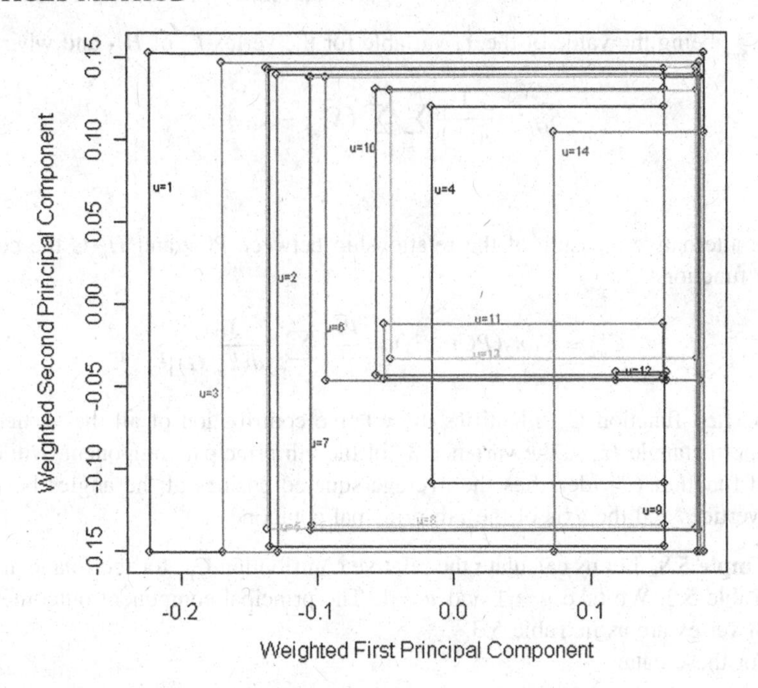

Figure 5.5 Weighted vertices principal components: finance data.

Definition 5.6: The **relative contribution** between a given principal component $PC\nu$ and an observation $\boldsymbol{\xi}_u$ represented here by its observed hypercube H_u can be measured by

$$C_{\nu u} = Con_1(PC\nu, H_u) = \frac{\sum_{k_u=1}^{n_u} p_{k_u}^u y_{\nu u k_u}^2}{\sum_{k_u=1}^{n_u} p_{k_u}^u [d(k_u, \boldsymbol{G})]^2} \tag{5.17}$$

where $y_{\nu i k_u}$ is the νth principal component for the vertex k_u of H_u (see Equation (5.14)), $w_{k_u}^u$ is the weight of that vertex (see Definition 5.1), and where $d(k_u, \boldsymbol{G})$ is the Euclidean distance between the row k_u of $\boldsymbol{X}$ and $\boldsymbol{G}$ defined as the centroid of all n rows of $\boldsymbol{X}$ given by

$$d^2(k_u, \boldsymbol{G}) = \sum_{j=1}^{p} \left(\frac{X_{k_u j} - \bar{X}_{Gj}}{S_{Gj}} \right)^2 \tag{5.18}$$

where

$$\bar{X}_{Gj} = \frac{1}{n} \sum_{u=1}^{m} \sum_{k_u=1}^{m_u} X_{k_u j}$$

with $X_{k_u j}$ being the value of the Y_j variable for the vertex k_u of H_u, and where

$$S_{Gj}^2 = \frac{1}{n-1} \sum_{u=1}^{m} \sum_{k_u=1}^{m_u} (X_{k_u j} - \bar{X}_{Gj})^2.$$

□

An alternative measure of the relationship between PCν and H_u is the contribution function

$$C_{\nu u}^* = Con_2(PC\nu, H_u) = \frac{p_{k_u}^u}{p_u} \sum_{k_u=1}^{m_u} \frac{y_{\nu u k_u}^2}{[d(k_u, \mathbf{G})]^2}.$$

The first function $C_{\nu u}$ identifies the relative contribution of all the vertices of the hyperrectangle H_u to the variance λ_ν of the νth principal component, while the second function $C_{\nu u}^*$ identifies the average squared cosines of the angles between these vertices and the axis of the νth principal component.

Example 5.8. Let us calculate the relative contribution $C_{\nu u}$ for the financial data of Table 5.2. We take $u = 1$ and $\nu = 1$. The principal component outcomes for each vertex are as in Table 5.3.

For these data,

$$\bar{X}_G = (\bar{X}_{G1}, \bar{X}_{G2}, \bar{X}_{G3}) = (3231.57, 3.07, 11.39),$$

and

$$S_G = (S_{G1}, S_{G2}, S_{G3}) = (829.17, 2.52, 8.02).$$

Therefore, for the first vertex ($k_1 = 1$), the squared Euclidean distance is, from Equation (5.18),

$$d^2\,(k_1 = 1, \mathbf{G}) = \left(\frac{1168 - 3231.57}{829.17}\right)^2 + \left(\frac{1 - 3.07}{2.52}\right)^2 + \left(\frac{1 - 11.39}{8.02}\right)^2$$
$$= 8.547,$$

and for the last vertex $k_1 = 8$, this distance is

$$d^2\,(k_1 = 8, \mathbf{G}) = \left(\frac{4139 - 3231.57}{829.17}\right)^2 + \left(\frac{7 - 3.07}{2.52}\right)^2 + \left(\frac{20 - 11.39}{8.02}\right)^2$$
$$= 4.782.$$

Hence, by summing over all vertices in this hyperrectangle H_1, we obtain from Equation (5.17),

$$C_{11} = Con(PC1, H_1) = \frac{(0.959)^2 + \ldots + (0.496)^2}{(8.547) + \ldots + (4.782)} = 0.422.$$

Note that the weights $p^u_{k_u}$ cancel out here as they take the same value $(1/p_u)$ for all k_u.

The complete set of these contributions between each observation ξ_u (H_u) and the νth vertices principal component $u = 1, \ldots, 14$, $\nu = 1, 2, 3$, is displayed in Table 5.7.

Table 5.7 Contributions and inertia $(H_u, PC\nu)$.

w_u	Sector	Contributions			Inertia			Total
		PC1	PC2	PC3	PC1	PC2	PC3	I
w_1	FMCG	0.422	0.224	0.353	0.185	0.107	0.184	0.159
w_2	Automotive	0.316	0.379	0.305	0.078	0.102	0.089	0.089
w_3	Telecommunications	0.360	0.311	0.329	0.109	0.103	0.119	0.110
w_4	Publishing	0.204	0.514	0.282	0.026	0.071	0.042	0.046
w_5	Finance	0.326	0.339	0.335	0.079	0.090	0.097	0.088
w_6	Consultants	0.376	0.243	0.381	0.063	0.045	0.077	0.061
w_7	Consumer goods	0.297	0.398	0.305	0.061	0.090	0.075	0.075
w_8	Energy	0.285	0.450	0.265	0.052	0.089	0.057	0.066
w_9	Pharmaceutical	0.310	0.407	0.283	0.070	0.101	0.077	0.082
w_{10}	Tourism	0.320	0.359	0.321	0.035	0.043	0.043	0.040
w_{11}	Textiles	0.469	0.065	0.465	0.042	0.006	0.050	0.033
w_{12}	Services	0.927	0.018	0.055	0.079	0.002	0.006	0.031
w_{13}	Durable goods	0.337	0.324	0.340	0.042	0.044	0.051	0.046
w_{14}	Others	0.385	0.507	0.108	0.068	0.099	0.023	0.065

Observe that for the services sector ($u = 12$), this contribution with the first (second) principal component $PC1$ is high (low) for $\nu = 1$ ($\nu = 2$), with

$$C_{1,12} = Con(PC1, H_{12}) = 0.927, \; C_{2,12} = Con(PC2, H_{12}) = 0.018,$$

respectively, whereas for the other sectors ($u \neq 12$), the differences are much less dramatic. These contributions are confirmed by the plots of Figure 5.3, where the plot corresponding to $u = 12$ is a thin rectangle largely confined to the first principal component axis, while the plots for $u \neq 12$ are squarer reflecting the relative values for these functions when $\nu = 1, 2$. □

Definition 5.7: A **correlation measure** between the νth principal component $PC\nu$ and the random variable X_j is

$$C_{\nu j} = Cor(PC\nu, X_j) = e_{\nu j}\sqrt{\lambda_\nu / \sigma_j^2} \tag{5.19}$$

where $e_{\nu j}$ is the jth component of the νth eigenvector $\boldsymbol{e}_\nu$ associated with X_j (see Equation (5.3)), where

$$\lambda_\nu = Var(PC\nu) \tag{5.20}$$

is the νth eigenvalue, and where σ_j^2 is the variance of X_j. Note that when the variance–covariance matrix is standardized these σ_j^2 reduce to $\sigma_j^2 = 1$. □

Example 5.9. Take the analysis of the financial data of Table 5.2. We obtain the eigenvalue

$$\boldsymbol{\lambda} = (1.090, 0.998, 0.912)$$

and the eigenvectors

$$\boldsymbol{e}_1 = (-0.696, 0.702, 0.151),$$
$$\boldsymbol{e}_2 = (0.144, -0.070, 0.987),$$
$$\boldsymbol{e}_3 = (0.703, 0.709, -0.052).$$

Substituting the relevant λ_ν and $e_{\nu j}$ values into Equation (5.19) gives the correlations $C_{\nu j}$, $j = 1, \ldots, 3$, $\nu = 1, \ldots, 3$. For example, the correlation between the first random variable Y_1 (Job Cost) and the first vertex principal component *PC1* ($\nu = 1$) is, with $\sigma^2 = 1$,

$$C_{11} = Con(PC1, Y_1) = e_{11}/\lambda_1 = (-0.696)\sqrt{1.090} = -0.727.$$

The complete set is given in Table 5.8. From these we observe, for example, that Job Cost (Y_1) and Job Code (Y_2), but not Activity (Y_3), are strongly correlated to the first principal component, while Activity Y_3 is strongly related to the second principal component. □

Table 5.8 Correlations (*PC*, *Y*).

Variable	*PC*1	*PC*2	*PC*3
Y_1 Job Cost	−0.727	0.144	0.672
Y_2 Job Code	0.732	−0.070	0.677
Y_3 Activity	0.158	0.986	−0.050

Definition 5.8: The absolute contribution of a single observation through the vertices of H_u to the variance λ_ν is measured by the **inertia**

$$I_{\nu u} = \text{Inertia}(PC\nu, H_u) = \left[\sum_{k_u=1}^{m_u} p_{k_u}^u y_{\nu u k_u}^2\right] \Big/ \lambda_\nu \tag{5.21}$$

and the contribution of this observation to the total variance is

$$I_u = \text{Inertia}(H_u) = \left\{\sum_{k_u=1}^{m_u} p_{k_u}^u \left[d(k_u, \boldsymbol{G})\right]^2\right\} \Big/ I_T, \tag{5.22}$$

where $I_T = \sum_{\nu=1}^{p} \lambda_\nu$ is the total variance of all the vertices in $\Re^p$. It is easily verified that

$$\sum_{u=1}^{m} I_{\nu u} = \lambda_\nu, \quad \sum_{u=1}^{m} I_u = I_T. \tag{5.23}$$

□

Example 5.10. The proportions of the variance of the νth principal component (i.e., of λ_ν) that are attributed to each of the observations ξ_u (through its H_u) for the financial data of Table 5.2 are shown in Table 5.7. For example, for the FMCG sector (w_1) and $PC1$ ($\nu = 1$), Equation (5.21) becomes

$$\begin{aligned} I_{11} &= \text{Inertia}(PC1, H_1) \\ &= \left(\frac{1}{14 \times 3}\right)[(0.959)^2 + \cdots + (0.496)^2]/(1.090) = 0.185. \end{aligned}$$

These results confirm what was observed in the plots of Figure 5.3: that the FMCG sector is responsible for a larger share of the overall variance than are the other sectors. □

In a different direction, an alternative visual aid in interpreting the results is that make only those vertices which a contribution to the principal component $PC\nu$ exceeding some prespecified value α are to be used in Equations (5.13)–(5.14). That is, we set

$$Y^*_{\nu u}(\alpha) = [y^a_{\nu u}(\alpha), y^b_{\nu u}(\alpha)]$$

where

$$y^a_{\nu u}(\alpha) = \min_{k_u \in L_u}\{y_{\nu u k_u} | Con(k_u, \nu) \geq \alpha\}, \; y^b_{\nu u}(\alpha) = \max_{k_u \in L_u}\{y_{\nu u k_u} | Con(k_u, \nu) \geq \alpha\}, \tag{5.24}$$

where

$$Con(k_u, PC\nu) = \frac{y^2_{\nu u k_u}}{[d(k_u, G)]^2} \tag{5.25}$$

is the contribution of a single vertex k_u to the νth principal component. When for a given $\nu = \nu_1$ (say) all $Con(k_u, PC\nu_1) < \alpha$, then to keep track of the position of the hypercube H_u on the principal component plane (ν_1, ν_2), say, we project the center of the hypercube onto that axis. In this case there is no variability on that principal component ν_1, whereas if there is variability for the other principal component ν_2, there is a line segment on its (ν_2) plane.

An alternative to the criterion of (5.24) is to replace $Con(k_i, \; PC\nu)$ with

$$Con(k_i, PC\nu_1, PC\nu_2) = Con(k_i, PC\nu_1) + Con(k_i, PC\nu_2). \tag{5.26}$$

In this case, vertices that make contributions to either of the two principal components $PC\nu_1$ and $PC\nu_2$ are retained, rather than only those vertices that contribute to just one principal component.

Example 5.11. Table 5.9 displays interval-valued observations for 30 categories of irises. These categories were generated by aggregating consecutive groups of five from the 150 individual observations published in Fisher (1936). Such

Table 5.9 Iris data.

w_u	Name	Sepal Length Y_1	Sepal Width Y_2	Petal Length Y_3	Petal Width Y_4
w_1	*S*1	[4.6, 5.1]	[3.0, 3.6]	[1.3, 1.5]	[0.2, 0.2]
w_2	*S*2	[4.4, 5.4]	[2.9, 3.9]	[1.4, 1.7]	[0.1, 0.4
w_3	*S*3	[4.3, 5.8]	[3.0, 4.0]	[1.1, 1.6]	[0.1, 0.2]
w_4	*S*4	[5.1, 5.7]	[3.5, 4.4]	[1.3, 1.7]	[0.3, 0.4]
w_5	*S*5	[4.6, 5.4]	[3.3, 3.7]	[1.0, 1.9]	[0.2, 0.5]
w_6	*S*6	[4.7, 5.2]	[3.0, 3.5]	[1.4, 1.6]	[0.2, 0.4]
w_7	*S*7	[4.8, 5.5]	[3.1, 4.2]	[1.4, 1.6]	[0.1, 0.4]
w_8	*S*8	[4.4, 5.5]	[3.0, 3.5]	[1.3, 1.5]	[0.1, 0.2]
w_9	*S*9	[4.4, 5.1]	[2.3, 3.8]	[1.3, 1.9]	[0.2, 0.6]
w_{10}	*S*10	[4.6, 5.3]	[3.0, 3.8]	[1.4, 1.6]	[0.2, 0.3]
w_{11}	*Ve*1	[5.5, 7.0]	[2.3, 3.2]	[4.0, 4.9]	[1.3, 1.5]
w_{12}	*Ve*2	[4.9, 6.6]	[2.4, 3.3]	[3.3, 4.7]	[1.0, 1.6]
w_{13}	*Ve*3	[5.0, 6.1]	[2.0, 3.0]	[3.5, 4.7]	[1.0, 1.5]
w_{14}	*Ve*4	[5.6, 6.7]	[2.2, 3.1]	[3.9, 4.5]	[1.0, 1.5]
w_{15}	*Ve*5	[5.9, 6.4]	[2.5, 3.2]	[4.0, 4.9]	[1.2, 1.8]
w_{16}	*Ve*6	[5.7, 6.8]	[2.6, 3.0]	[3.5, 5.0]	[1.0, 1.7]
w_{17}	*Ve*7	[5.4, 6.0]	[2.4, 3.0]	[3.7, 5.1]	[1.0, 1.6]
w_{18}	*Ve*8	[5.5, 6.7]	[2.3, 3.4]	[4.0, 4.7]	[1.3, 1.6]
w_{19}	*Ve*9	[4.0, 6.1]	[2.3, 3.0]	[3.3, 4.6]	[1.0, 1.4]
w_{20}	*Ve*10	[4.1, 6.2]	[2.5, 3.0]	[3.0, 4.3]	[1.1, 1.3]
w_{21}	*Vi*1	[5.8, 7.1]	[2.7, 3.3]	[5.1, 6.0]	[1.8, 2.5]
w_{22}	*Vi*2	[4.9, 7.6]	[2.5, 3.6]	[4.5, 6.6]	[1.7, 2.5]
w_{23}	*Vi*3	[5.7, 6.8]	[2.5, 3.2]	[5.0, 5.5]	[1.9, 2.4]
w_{24}	*Vi*4	[6.0, 7.7]	[2.2, 3.8]	[5.0, 6.9]	[1.5, 2.3]
w_{25}	*Vi*5	[5.6, 7.7]	[2.7, 3.3]	[4.9, 6.7]	[1.8, 2.3]
w_{26}	*Vi*6	[6.1, 7.2]	[2.8, 3.2]	[4.8, 6.0]	[1.6, 2.1]
w_{27}	*Vi*7	[6.1, 7.9]	[2.6, 3.8]	[5.1, 6.4]	[1.4, 2.2]
w_{28}	*Vi*8	[6.0, 7.7]	[3.0, 3.4]	[4.8, 6.1]	[1.8, 2.4]
w_{29}	*Vi*9	[5.8, 6.9]	[2.7, 3.3]	[5.1, 5.9]	[1.9, 2.5]
w_{30}	*Vi*10	[5.0, 6.7]	[2.5, 3.4]	[5.0, 5.4]	[1.8, 2.3]

an aggregation might occur if the five observations in each category corresponded to the measurements of five different flowers at the same 'site', or five different flowers on the same plant, etc. There are three species: *setosa (S)*, *versicolor (Ve)*, and *virginica (Vi)*. The data therefore list 30 observations (*S*1, ..., *S*10, *Ve*1, ..., *Ve*10, *Vi*1, ..., *Vi*10), 10 for each of the three species. There are $p = 4$ random variables, $Y_1 =$ Sepal Length, $Y_2 =$ Sepal Width, $Y_3 =$ Petal Length, and $Y_4 =$ Petal Width.

We perform the vertices principal component analysis on these data. The resulting principal components obtained from Equation (5.13) are displayed in Table 5.10. Also shown are the relative contributions for each observation calculated from Equation (5.17). The plot of the principal components on the $\nu = 1$ and 2 axes is shown in Figure 5.6. These correspond to $\alpha = 0$ in Equation (5.25). While some form of clustering is apparent, greater clarity can be gleaned by taking $\alpha > 0$ in Equation (5.24).

Table 5.10 Vertices *PC* – iris data.

w_u		Principal components		Contribution	
		PC1	PC2	PC1	PC2
w_1	*S*1	[−2.471, −1.810]	[−0.800, 0.423]	0.923	0.062
w_2	*S*2	[−2.784, −1.384]	[−1.054, 1.087]	0.814	0.140
w_3	*S*3	[−2.990, −1.410]	[−0.932, 1.370]	0.785	0.133
w_4	*S*4	[−2.578, −1.559]	[0.256, 2.057]	0.673	0.300
w_5	*S*5	[−2.624, −1.466]	[−0.298, 0.766]	0.914	0.042
w_6	*S*6	[−2.330, −1.577]	[−0.760, 0.320]	0.920	0.058
w_7	*S*7	[−2.740, −1.475]	[−0.566, 1.635]	0.766	0.201
w_8	*S*8	[−2.627, −1.600]	[−0.891, 0.395]	0.885	0.055
w_9	*S*9	[−2.687, −0.995]	[−2.079, 0.844]	0.610	0.348
w_{10}	*S*10	[−2.549, −1.598]	[−0.796, 0.859]	0.875	0.098
w_{11}	*Ve*1	[−0.081, 1.648]	[−1.397, 0.766]	0.391	0.353
w_{12}	*Ve*2	[−0.902, 1.392]	[−1.517, 0.799]	0.210	0.410
w_{13}	*Ve*3	[−0.617, 1.277]	[−2.161, 0.088]	0.130	0.709
w_{14}	*Ve*4	[−0.227, 1.416]	[−1.581, 0.466]	0.277	0.487
w_{15}	*Ve*5	[0.055, 1.445]	[−0.923, 0.593]	0.530	0.309
w_{16}	*Ve*6	[−0.250, 1.558]	[−0.875, 0.382]	0.452	0.154
w_{17}	*Ve*7	[−0.342, 1.207]	[−1.318, 0.085]	0.256	0.497
w_{18}	*Ve*8	[−0.192, 1.500]	[−1.397, 1.007]	0.324	0.494
w_{19}	*Ve*9	[−0.683, 1.004]	[−1.653, 0.069]	0.131	0.632
w_{20}	*Ve*10	[−0.654, 0.773]	[−1.273, 0.076]	0.146	0.545
w_{21}	*Vi*1	[0.751, 2.579]	[−0.476, 1.169]	0.739	0.110
w_{22}	*Vi*2	[−0.158, 3.148]	[−1.187, 1.895]	0.480	0.200
w_{23}	*Vi*3	[0.795, 2.295]	[−0.847, 0.851]	0.733	0.116
w_{24}	*Vi*4	[0.323, 3.316]	[−1.313, 2.262]	0.543	0.311
w_{25}	*Vi*5	[0.580, 2.974]	[−0.557, 1.390]	0.684	0.111
w_{26}	*Vi*6	[0.718, 2.279]	[−0.238, 0.976]	0.782	0.096
w_{27}	*Vi*7	[0.334, 2.962]	[−0.597, 2.296]	0.521	0.311
w_{28}	*Vi*8	[0.702, 2.686]	[0.099, 1.548]	0.660	0.186
w_{29}	*Vi*9	[0.825, 2.441]	[−0.461, 1.093]	0.742	0.107
w_{30}	*Vi*10	[0.243, 2.136]	[−1.113, 1.141]	0.484	0.197

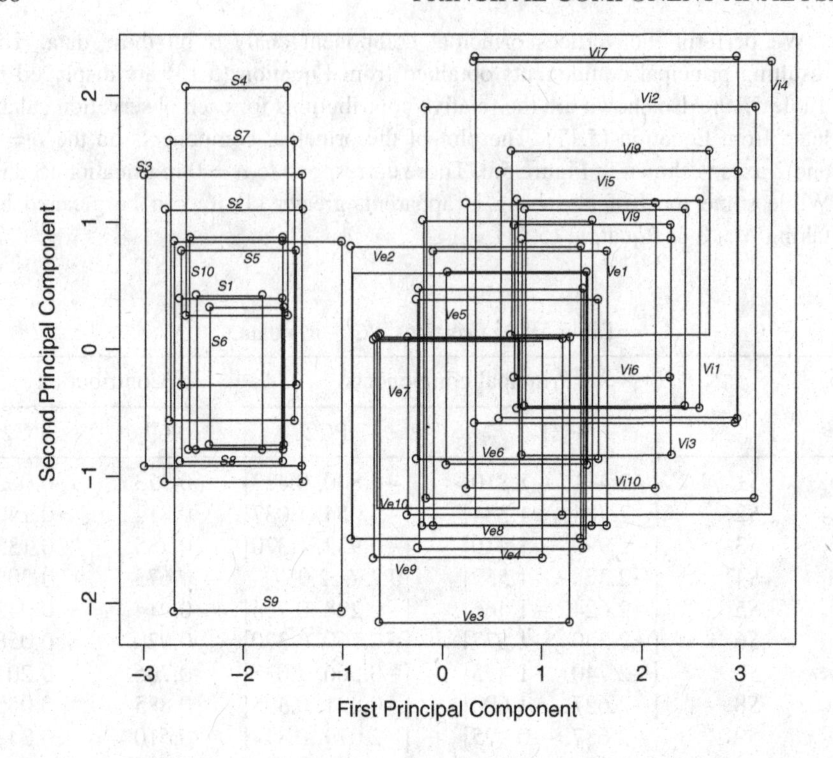

Figure 5.6 Iris data: principal components – $\alpha = 0$.

The first two principal components ($s = 1, 2$) on the vertices of the fourth category $S1$ (w_4) of the *setosa* iris are given in Table 5.11. The contributions between the vertices and the principal components are calculated from Equation (5.24). For example, consider the first vertex in H_4 (i.e., $k_4 = 1,\ u = 4$). Then,

$$Con(k_4 = 1,\ PC1) = (-2.079)^2/4.495 = 0.971$$

where the squared Euclidean distance for this vertex is

$$d^2(k_4 = 1, \boldsymbol{G}) = \left(\frac{5.1 - 5.807}{0.937}\right)^2 + \left(\frac{3.5 - 3.062}{0.540}\right)^2 + \left(\frac{1.3 - 3.730}{1.785}\right)^2 + \left(\frac{0.3 - 1.207}{0.774}\right)^2$$

$$= 4.454.$$

These contributions for all vertices in H_4 are given in Table 5.11 along with their first and second principal components.

Suppose $\alpha = 0.1$, and take $\nu = 1$. Then, from Equation (5.24), we see that no vertices are deleted from consideration since their contributions are all ≥ 0.1, and hence the first vertex principal component becomes

$$PC1(\alpha = 0.1) = [-2.578, -1.559].$$

Table 5.11 $Con(k_u, PCv)$, $u = 4$.

Vertex	Y_1	Y_2	Y_3	Y_4	PC1	PC2	$Con(k_4, PCv)$	
k_4							PC1	PC2
1	5.1	3.5	1.3	0.3	−2.079	0.256	0.971	0.015
2	5.1	3.5	1.3	0.4	−2.005	0.270	0.965	0.017
3	5.1	3.5	1.7	0.3	−1.948	0.275	0.975	0.019
4	5.1	3.5	1.7	0.4	−1.874	0.289	0.984	0.023
5	5.1	4.4	1.3	0.3	−2.578	1.807	0.669	**0.329**
6	5.1	4.4	1.3	0.4	−2.504	1.822	0.650	**0.344**
7	5.1	4.4	1.7	0.3	−2.448	1.827	0.639	**0.356**
8	5.1	4.4	1.7	0.4	−2.374	1.841	0.620	**0.373**
9	5.7	3.5	1.3	0.3	−1.764	0.471	0.798	0.057
10	5.7	3.5	1.3	0.4	−1.690	0.486	0.791	0.065
11	5.7	3.5	1.7	0.3	−1.633	0.491	0.799	0.072
12	5.7	3.5	1.7	0.4	−1.559	0.505	0.797	0.084
13	5.7	4.4	1.3	0.3	−2.263	2.023	0.546	**0.436**
14	5.7	4.4	1.3	0.4	−2.189	2.037	0.526	**0.456**
15	5.7	4.4	1.7	0.3	−2.133	2.042	0.516	**0.473**
16	5.7	4.4	1.7	0.4	−2.059	2.057	0.497	**0.496**

However, for $\nu = 2$, the vertices corresponding to $k_4 = 1, \ldots, 4, 9, \ldots, 12$, are deleted. Hence, the second vertex principal component is found by taking the minimum and maximum of the $PC2$ values that remain, i.e., we use Equation (5.24) to give

$$PC2(\alpha = 0.1) = [1.807, 2.057].$$

The principal component values that emerge are given in Table 5.12.

When $\alpha = 0.5$, for $PC1$, all vertices of H_4 are again retained since their contribution is ≥ 0.5. Thus, from Equation (5.24) we have

$$PC1(\alpha = 0.5) = [-2.578, -1.559].$$

However, we see that all vertices are such that

$$Con(k_i, PC2) \leq \alpha = 0.5$$

for this observation H_4. Therefore, to locate this observation on the $PC2$ axes of the $PC1 \times PC2$ plane, we calculate the average $PC2$ value from the $PC2$ values over all vertices. Thus we obtain

$$PC2 = 0.226.$$

The same procedure is followed for all hypercubes H_u, $u = 1, \ldots, 30$. The resulting principal component values are shown in Table 5.12. The single values

Table 5.12 Vertices PC, $\alpha = 0.1, 0.5$ – iris data.

w_u		$\alpha = 0.1$		$\alpha = 0.5$	
		PC1	PC2	PC1	PC2
w_1	*S*1	[−2.471, −1.810]	[−0.800, −0.791]	[−2.471, −1.810]	0.066
w_2	*S*2	[−2.784, −1.384]	[−1.054, 1.087]	[−2.784, −1.384]	0.149
w_3	*S*3	[−2.990, −1.410]	[−0.932, 1.370]	[−2.990, −1.410]	0.129
w_4	*S*4	[−2.578, −1.559]	[1.807, 2.057]	[−2.578, −1.559]	0.226
w_5	*S*5	[−2.624, −1.466]	[0.722, 0.766]	[−2.624, −1.466]	0.042
w_6	*S*6	[−2.330, −1.577]	[−0.760, −0.722]	[−2.330, −1.577]	0.061
w_7	*S*7	[−2.740, −1.475]	[1.331, 1.635]	[−2.740, −1.475]	0.161
w_8	*S*8	[−2.627, −1.600]	[−0.891, −0.863]	[−2.627, −1.600]	0.056
w_9	*S*9	[−2.687, −0.995]	[−2.079, 0.844]	[−2.687, −1.827]	[−2.079, −1.742]
w_{10}	*S*10	[−2.549, −1.598]	[−0.796, −0.859]	[−2.549, −1.598]	0.101
w_{11}	*Ve*1	[0.361, 1.648]	[−1.397, 0.766]	[1.149, 1.648]	[−1.397, −1.325]
w_{12}	*Ve*2	[−0.902, 1.392]	[−1.517, 0.799]	[−0.902, 1.392]	[−1.517, −1.364]
w_{13}	*Ve*3	[−0.671, 1.277]	[−2.161, −0.366]	[0.722, 0.722]	[−2.161, −0.636]
w_{14}	*Ve*4	[−0.227, 1.416]	[−1.581, 0.466]	[0.916, 1.416]	[−1.581, −1.481]
w_{15}	*Ve*5	[0.318, 1.445]	[−0.923, 0.593]	[0.762, 1.445]	[−0.923, 0.284]
w_{16}	*Ve*6	[−0.250, 1.558]	[−0.875, −0.185]	[−0.250, 1.558]	[−0.875, −0.775]
w_{17}	*Ve*7	[−0.342, 1.207]	[−1.318, −0.283]	[0.417, 1.207]	[−1.318, −1.017]
w_{18}	*Ve*8	[0.439, 1.500]	[−1.397, 1.007]	[1.271, 1.500]	[−1.397, 1.007]
w_{19}	*Ve*9	[−0.683, 1.004]	[−1.653, −0.327]	[−0.683, 0.615]	[−1.653, −1.138]
w_{20}	*Ve*10	[−0.654, 0.773]	[−1.273, −0.320]	[−0.654, 0.496]	[−1.273, −0.849]
w_{21}	*Vi*1	[0.751, 2.579]	[−0.476, 1.169]	[1.084, 2.579]	0.119
w_{22}	*Vi*2	[1.045, 3.148]	[−1.187, 1.895]	[1.870, 3.148]	[−1.187, 1.680]
w_{23}	*Vi*3	[0.795, 2.295]	[−0.847, 0.851]	[1.184, 2.295]	0.126
w_{24}	*Vi*4	[0.916, 3.316]	[−1.313, 2.262]	[1.804, 3.316]	[−1.313, 2.171]
w_{25}	*Vi*5	[0.580, 2.974]	[−0.557, 1.390]	[0.913, 2.974]	0.116
w_{26}	*Vi*6	[0.718, 2.279]	[0.451, 0.976]	[0.718, 2.279]	0.101
w_{27}	*Vi*7	[0.759, 2.962]	[−0.597, 2.296]	[1.000, 2.962]	[1.472, 2.233]
w_{28}	*Vi*8	[0.702, 2.686]	[0.711, 1.548]	[0.924, 2.686]	0.177
w_{29}	*Vi*9	[0.825, 2.441]	[−0.461, 1.093]	[1.158, 2.441]	0.115
w_{30}	*Vi*10	[0.613, 2.136]	[−1.113, 1.141]	[1.135, 2.136]	0.207

(such as $PC2 = 0.226$ for w_4) are those for which no vertex had $Con(k_i, PCv) > \alpha$. The resulting plots will be lines. Notice that the plot for $Ve3(w_{13})$ is also a line parallel to the $PC2$ axis. However, in this case there is one vertex for which $Con(k_i, PC1) > 0.5$, so that the first principal component takes the value [0.722, 0.722].

Clearly, when $\alpha = 0$, all vertices are included, so that for H_4

$$PC1(\alpha = 0) = [-2.578, -1.559], \ \ PC2(\alpha = 0) = [0.256, 2.057].$$

The plots for all observations for $\alpha = 0.1$ and $\alpha = 0.5$ are shown in Figure 5.7 and Figure 5.8, respectively, and can be compared along with that for $\alpha = 0$ in Figure 5.6.

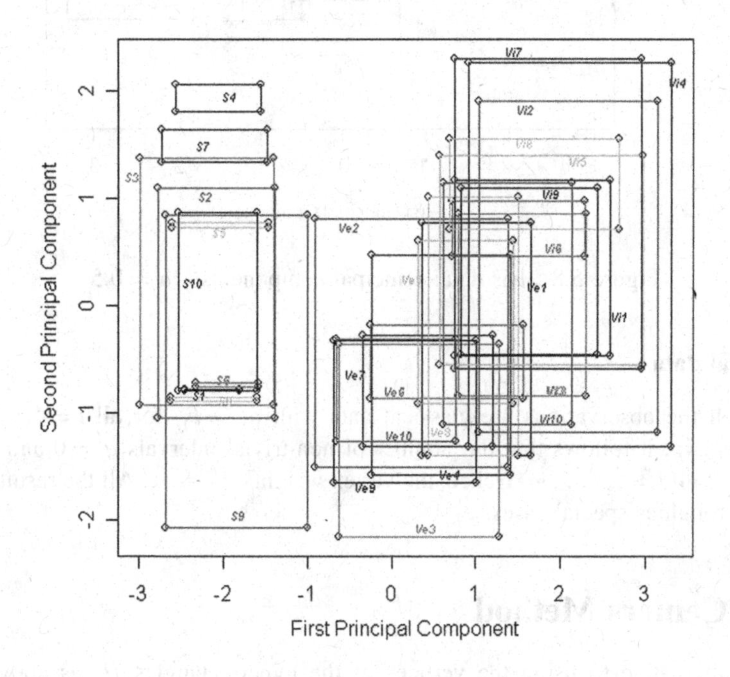

Figure 5.7 Iris data: principal components – $\alpha = 0.1$.

As α increases from 0 to 0.5, the two *versicolor* and *virginica* species separate out with no overlap between them except for the *Ve5* and *Ve8* categories which essentially link the two flora. The *setosa* species are all quite distinct from both the *versicolor* and *virginica*. The *S9* category becomes identified as a cluster of its own, separate from the other *setosa* categories. More importantly, instead of the harder-to-identify clusters when $\alpha = 0$, we have from $\alpha = 0.5$ more distinct clusters. That is, use of Equation (5.24) has enhanced our visual ability to identify clusters. □

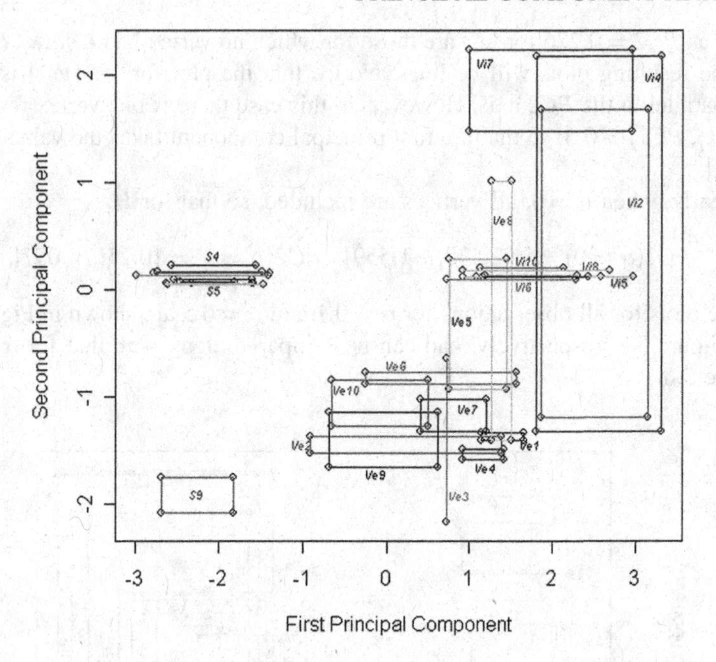

Figure 5.8 Iris data: principal components – $\alpha = 0.5$.

Classical data

When all the observations are classical data with $a_{ij} = b_{ij}$ for all $i = 1, \ldots, m$, $j = 1, \ldots, p$, it follows that the number of non-trivial intervals $q_i = 0$ and hence $n_i = 1$ for all $i = 1, \ldots, m$. Hence, the vertex weights $w^i_{k_i} \equiv w_i$. All the results here carry through as special cases.

5.2 Centers Method

As an alternative to using the vertices of the hyperrectangles H_u as above, the centers of the hyperrectangles could be used. In this case, each observation $\boldsymbol{\xi}_u = ([a_{1u}, b_{1u}], \ldots, [a_{pu}, b_{pu}])$ is transformed to

$$\boldsymbol{X}^c_u = (X^c_{u1}, \ldots, X^c_{up}), u = 1, \ldots, m,$$

where

$$X^c_{uj} = (a_{uj} + b_{uj})/2, \ j = 1, \ldots, p. \tag{5.27}$$

This implies that the symbolic data matrix has been transformed to a classical $m \times p$ matrix $\boldsymbol{X}^c$ with classical variables $X^c_1, \ldots, X^c_p$, say.

Then, the classical principal component analysis is applied to the data matrix $\boldsymbol{X}^c$. The νth centers principal component for observation $\boldsymbol{\xi}_u$ is, for $\nu = 1, \ldots, s$, $u = 1, \ldots, m$,

$$y^c_{\nu u} = \sum_{j=1}^{p} (X^c_{ju} - \bar{X}^c_j) w_{\nu j} \tag{5.28}$$

where the mean of the values for the variable X^c_j is

$$\bar{X}^c_j = \frac{1}{m} \sum_{u=1}^{m} X^c_{uj} \tag{5.29}$$

and where $\boldsymbol{w}_\nu = (w_{\nu 1}, \ldots, w_{\nu p})$ is the νth eigenvector of the variance–covariance matrix associated with $\boldsymbol{X}^c$.

Since each centered coordinate X^c_{uj} lies in the interval $[a_{uj}, b_{uj}]$ and since the principal components are linear functions of X^c_{ju}, we can obtain the interval center principal components as

$$Y^c_{\nu u} = [y^{ca}_{\nu u}, y^{cb}_{\nu u}] \tag{5.30}$$

where

$$y^{ca}_{\nu u} = \sum_{j=1}^{p} \min_{a_{uj} \le x_{uj} \le b_{uj}} (X^c_{uj} - \bar{X}^c_j) w_{\nu u} \tag{5.31}$$

and

$$y^{cb}_{\nu u} = \sum_{j=1}^{p} \max_{a_{uj} \le x_{uj} \le b_{uj}} (X^c_{uj} - \bar{X}^c_j) w_{\nu j}. \tag{5.32}$$

It follows that

$$y^{ca}_{\nu u} = \sum_{j=1|j-}^{p} (b_{uj} - \bar{X}^c_j) w_{\nu j} + \sum_{j=1|j+}^{p} (a_{uj} - \bar{X}^c_j) w_{\nu j} \tag{5.33}$$

and

$$y^{cb}_{\nu u} = \sum_{j=1|j-}^{p} (a_{uj} - \bar{X}^c_j) w_{\nu j} + \sum_{j=1|j+}^{p} (b_{uj} - \bar{X}^c_j) w_{\nu j} \tag{5.34}$$

where $j-$ are those values of j for which $w_{\nu j} < 0$ and $j+$ are those values of j for which $w_{\nu j} > 0$.

Example 5.12. Let us take the financial data of Table 5.2 and consider all five variables $Y_1, \ldots, Y_5$. The centroid values $X^c_{uj}, j = 1, \ldots, p = 5,\ u = 1, \ldots, m = 14$ are obtained from Equation (5.27) and displayed in Table 5.13. Also given are the mean centered values $\bar{X}^c_j$ along with the centered standard deviations S^c_j

for each variable X_j^c, $j = 1, \ldots, 5$. A classical principal component analysis on these centroid values gives the eigenvalues

$$\boldsymbol{\lambda} = (2.832, 1.205, 0.430, 0.327, 0.206);$$

and hence the eigenvector associated with the first principal component is

$$\boldsymbol{e}_1 = (-0.528, 0.481, -0.086, 0.491, 0.491).$$

Table 5.13 Centers method statistics – finance data.

w_u	X_1^c	X_2^c	X_3^c	X_4^c	X_5^c
w_1	2653.5	4.0	10.5	247980.0	606650.0
w_2	3119.5	3.5	10.5	188784.0	191024.0
w_3	3047.0	4.0	10.5	126063.0	251063.0
w_4	3079.5	2.0	12.0	23908.5	42183.5
w_5	3265.0	4.0	11.0	127564.5	160000.5
w_6	3286.5	3.5	14.5	29905.0	39455.0
w_7	3302.5	3.5	11.0	40200.0	40200.5
w_8	3113.5	3.5	11.0	28640.0	28640.0
w_9	3117.5	4.0	10.5	25327.5	643575.0
w_{10}	3147.0	3.0	14.5	16650.0	16650.0
w_{11}	3354.5	3.0	9.0	33825.0	53400.0
w_{12}	3769.5	1.0	9.0	19950.0	25050.0
w_{13}	3397.0	3.0	15.0	17050.0	29550.0
w_{14}	3589.5	1.0	10.5	9929.0	9929.0
$\bar{X}_j^c$	3231.57	3.07	11.39	66841.18	152669.32
S_j^c	264.16	1.04	1.93	75193.08	213311.91

We substitute this result into Equations (5.33) and (5.34) to obtain the interval centers principal component. For example, from Equation (5.33), we have the centers principal component, standardized values by S_j^c, for the FMCG sector (w_1),

$$\begin{aligned} y_{11}^{ca} &= \left(\frac{4139 - 3231.57}{264.16}\right)(-0.528)_{x_1} + \left(\frac{20 - 11.39}{1.938}\right)(-0.086)_{x_3} \\ &\quad + \left(\frac{1 - 3.07}{1.04}\right)(0.481)_{x_2} + \left(\frac{650 - 66841.18}{75493.08}\right)(0.491)_{x_4} \\ &\quad + \left(\frac{1000 - 152669.32}{213311.91}\right)(0.491)_{x_5} \\ &= -3.940; \end{aligned}$$

and from Equation (5.34), we have, for the FMCG sector (w_1),

$$\begin{aligned} y_{11}^{cb} &= \left(\frac{1168-3231.57}{264.16}\right)(-0.528)_{x_1} + \left(\frac{1-11.39}{1.93}\right)(-0.086)_{x_3} \\ &\quad + \left(\frac{7-3.07}{1.04}\right)(0.481)_{x_2} + \left(\frac{495310-66841.179}{75193.08}\right)(0.491)_{x_4} \\ &\quad + \left(\frac{1212300-152669.32}{213311.91}\right)(0.491)_{x_5} \\ &= 11.650. \end{aligned}$$

Hence, the first centers principal component for the FMCG sector (w_1) is

$$Y_{11}^{c} = [-3.940, 11.650].$$

The complete set of the first, second, and third centers principal components for all sectors, $u = 1, \ldots, 14$, is shown in Table 5.14.

Table 5.14 Centers principal components – finance data.

w_u	Sector	$\text{PC1} = Y_{1u}^{C}$	$\text{PC2} = Y_{2u}^{C}$	$\text{PC3} = Y_{3u}^{C}$
w_1	FMCG	[−3.940, 11.650]	[−8.297, 7.205]	[−5.879, 5.123]
w_2	Automotive	[−3.793, 6.487]	[−7.113, 6.030]	[−4.377, 2.362]
w_3	Telecommunications	[−3.843, 6.750]	[−6.908, 6.490]	[−3.222, 2.888]
w_4	Publishing	[−3.013, 1.502]	[−4.242, 4.824]	[−1.040, 0.710]
w_5	Finance	[−3.705, 5.296]	[−6.232, 6.058]	[−3.009, 2.213]
w_6	Consultants	[−3.807, 2.707]	[−2.356, 5.751]	[−1.154, 1.177]
w_7	Consumer goods	[−3.789, 3.073]	[−5.561, 5.716]	[−1.355, 1.371]
w_8	Energy	[−3.013, 2.849]	[−5.144, 5.707]	[−1.129, 1.157]
w_9	Pharmaceutical	[−3.007, 6.125]	[−6.754, 6.011]	[−1.049, 5.145]
w_{10}	Tourism	[−3.035, 1.750]	[−1.909, 5.345]	[−0.892, 0.773]
w_{11}	Textiles	[−3.093, 1.858]	[−2.308, 0.265]	[−0.771, 0.977]
w_{12}	Services	[−3.223, −1.842]	[−2.233, −1.644]	[−0.274, 0.262]
w_{13}	Durable goods	[−3.758, 1.494]	[−1.795, 5.227]	[−0.809, 0.920]
w_{14}	Others	[−3.939, −0.740]	[−5.853, 3.720]	[−0.618, 0.498]

The proportion of the total variation explained by the first centers principal component is $\lambda_1/\sum\lambda_i = 0.566$, i.e., 56.6%. Likewise, we find that the first two and three principal components together account for 80.7% and 89.3%, respectively, of the total variation.

Figure 5.9 shows the plots of the first two centers principal components for each observation. There are three distinct clusters. The first consists of the w_1 observation (i.e., the FMCG sector.) The second cluster consists of the w_2, w_3, w_5, w_9 observations (i.e., the automotive, telecommunications, finance, and pharmaceutical sectors). These are distinct clusters when viewed with respect to the

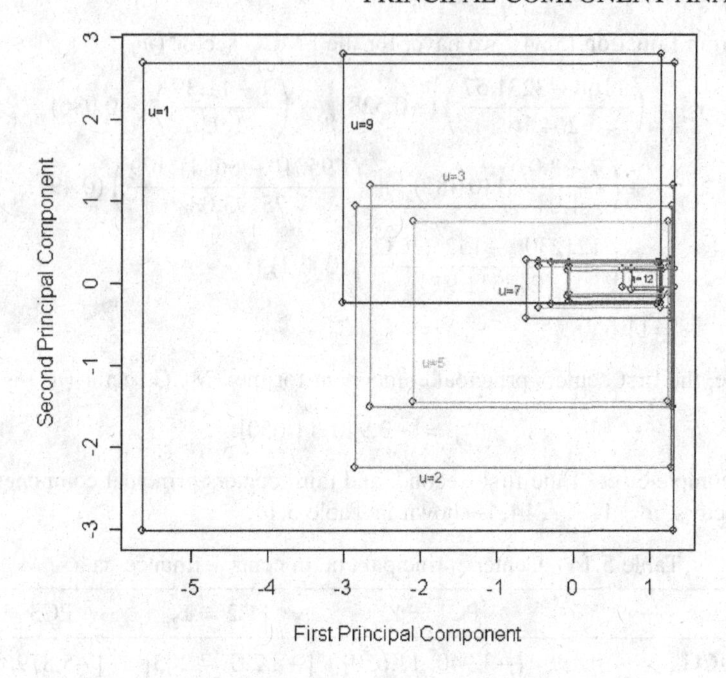

Figure 5.9(a) First/Second centers principal components.

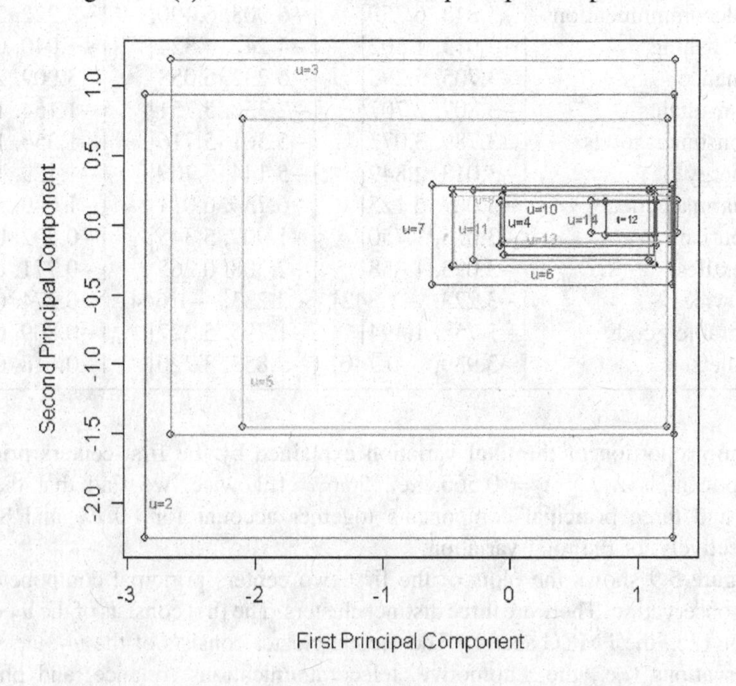

Figure 5.9(b) First/Second centers principal components.

first centers principal component. If viewed from the second centers principal component, the pharmaceutical sector (w_9) would move from the second cluster to the first. The remaining sectors clearly belong to a single third cluster. □

The analyses of the financial data in Examples 5.6, 5.7, and 5.12 all produce the same (not unreasonably) comparable clusters, with the added feature of the clusters being nested inside each other. The following example gives unnested clusters.

Example 5.13. The data of Table 5.15 provide some measurements for eigh different makes of car. The variables are Y_1 = Price ($\times 10^{-3}$, in euros), Y_2 = Maximum Velocity, Y_3 = Acceleration Time to reach a given speed, and Y_4 = Cylinder Capacity of the car.

The centered values for all four variables are displayed in Table 5.16 along with the centered variable means $\bar{X}_j^c$ and standard deviations S_j^c, $j = 1, \ldots, 4$.

Table 5.15 Cars data.

w_u	Car	Y_1 = Price	Y_2 =Max Velocity	Y_3 =Accn Time	Y_4 =Cylinder Capacity
		$[a_1, b_1]$	$[a_2, b_2]$	$[a_3, b_3]$	$[a_4, b_4]$
w_1	Aston Martin	[260.5, 460.0]	[298, 306]	[4.7, 5.0]	[5935, 5935]
w_2	Audi A6	[68.2, 140.3]	[216, 250]	[6.7, 9.7]	[1781, 4172]
w_3	Audi A8	[123.8, 171.4]	[232, 250]	[5.4, 10.1]	[2771, 4172]
w_4	BMW 7	[104.9, 276.8]	[228, 240]	[7.0, 8.6]	[2793, 5397]
w_5	Ferrari	[240.3, 391.7]	[295, 298]	[4.5, 5.2]	[3586, 5474]
w_6	Honda NSR	[205.2, 215.2]	[260, 270]	[5.7, 6.5]	[2977, 3179]
w_7	Mercedes C	[55.9, 115.2]	[210, 250]	[5.2, 11.0]	[1998, 3199]
w_8	Porsche	[147.7, 246.4]	[280, 305]	[4.2, 5.2]	[3387, 3600]

Table 5.16 Centers method statistics – cars data.

w_u	Car	X_1^c	X_2^c	X_3^c	X_5
w_1	Aston Martin	360.25	302.0	4.85	5935.0
w_2	Audi A6	104.25	233.0	8.20	2976.5
w_3	Audi A8	147.60	241.0	7.75	3471.5
w_4	BMW 7	190.85	234.0	7.80	4095.0
w_5	Ferrari	316.00	296.5	4.85	4530.0
w_6	Honda NSR	210.20	265.0	6.10	3078.0
w_7	Mercedes C	85.55	230.0	8.10	2598.5
w_8	Porsche	197.05	292.5	4.70	3493.5
	$\bar{X}_j^c$	201.47	261.75	6.54	3772.25
	S_j^c	95.86	31.21	1.58	1070.17

The principal component analysis of these centered values on all $p = 4$ random variables gave the eigenvalues

$$\boldsymbol{\lambda} = \{3.449, 0.505, 0.040, 0.005\}.$$

The first eigenvector is

$$\boldsymbol{e}_1 = \{0.512, 0.515, -0.499, 0.461\}.$$

Hence, substituting into Equations (5.33)–(5.34) for each observation, we obtain the centers principal components. For example, for Aston Martin (w_1), we have

$$PC1(\text{Aston Martin}) = Y_{11}^c = [y_{11}^{ca}, y_{11}^{cb}] = [2.339, 3.652]$$

since

$$\begin{aligned} y_{11}^{ca} = {} & \left(\frac{260.5 - 201.47}{95.86}\right)(0.512)_{x_1} + \left(\frac{298 - 261.75}{31.21}\right)(0.515)_{x_2} \\ & + \left(\frac{5.0 - 6.54}{1.58}\right)(-0.499)_{x_3} + \left(\frac{5935 - 3772.25}{1070.17}\right)(0.461)_{x_4} \\ = {} & 2.339, \end{aligned}$$

and

$$\begin{aligned} y_{11}^{cb} = {} & \left(\frac{460.0 - 201.47}{95.86}\right)(0.512)_{x_1} + \left(\frac{306 - 261.75}{31.21}\right)(0.515)_{x_2} \\ & + \left(\frac{4.7 - 6.54}{1.58}\right)(-0.499)_{x_3} + \left(\frac{5935 - 3772.25}{1070.17}\right)(0.461)_{x_4} \\ = {} & 3.652. \end{aligned}$$

The set of centers principal components for all $s = 1, \ldots, 4$ is displayed in Table 5.17. Plots based on the first and second principal components are shown in Figure 5.10.

From this figure, two (or three) clusters emerge. One cluster consists of the Aston Martin, Ferrari, and Porsche cars (i.e., w_1, w_5, and w_8), and the other

Table 5.17 Centers principal components – cars data.

w_u	Car	PC1	PC2	PC3
w_1	Aston Martin	[2.339, 3.652]	[0.429, 1.171]	[-1.030, 0.738]
w_2	Audi A6	[-3.335, -0.404]	[-1.484, 1.718]	[-1.198, 1.007]
w_3	Audi A8	[-2.467, 0.176]	[-1.104, 1.732]	[-0.793, 0.670]
w_4	BMW 7	[-2.154, 0.608]	[-0.487, 2.387]	[-1.367, 1.501]
w_5	Ferrari	[1.104, 2.013]	[-1.150, 0.783]	[-0.916, 1.321]
w_6	Honda NSR	[-0.338, 0.221]	[-0.900, -0.349]	[0.198, 0.509]
w_7	Mercedes C	[-3.818, -0.487]	[-1.868, 1.509]	[-0.910, 0.810]
w_8	Porsche	[0.266, 1.624]	[-1.720, -0.666]	[-0.880, 0.330]

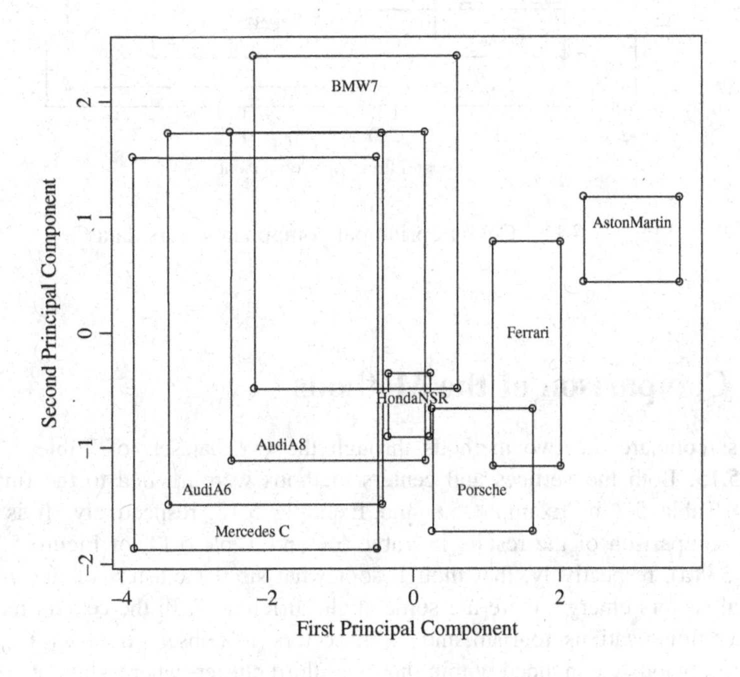

Figure 5.10 Centers principal components: cars data.

cluster consists of the Audi A6, Audi A8, BMW 7, Mercedes C, and Honda NSR cars (i.e., w_2, w_3, w_4, w_6, and w_7). Whether or not the Honda NSR should form its own third cluster is answered by studying the plots obtained by using the second and third principal components. These are shown in Figure 5.11, and do confirm the choice of the HondaNSR cars as a separate third cluster. □

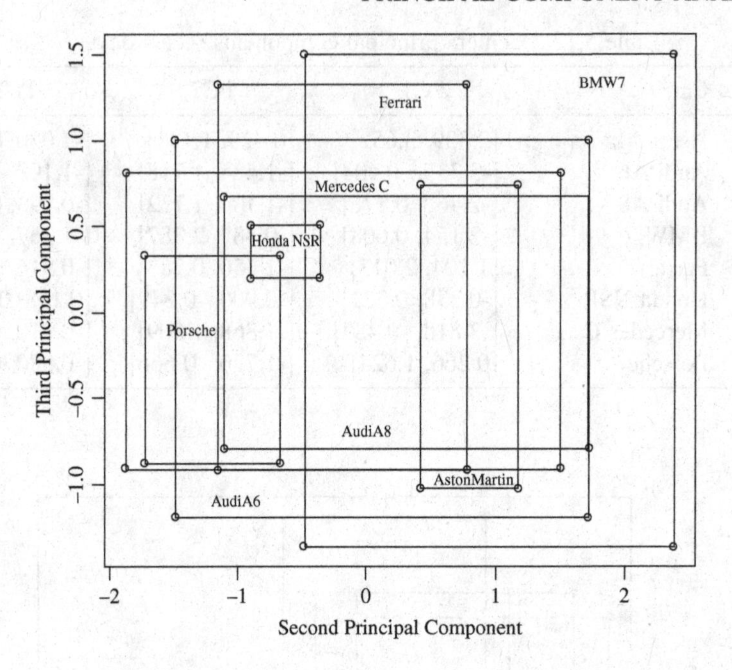

Figure 5.11 Centers principal components: cars data.

5.3 Comparison of the Methods

We first compare the two methods through the two datasets of Table 5.2 and Table 5.15. Both the vertices and centers methods were applied to the financial data of Table 5.2 in Example 5.6 and Example 5.12, respectively. It is clear from a comparison of the results in Table 5.4 and Table 5.14, or Figure 5.3 and Figure 5.9(a), respectively, that though somewhat similar clusters of the $m = 14$ financial sectors emerge, there are some slight differences. In the centers method, the telecommunications, tourism, and others sectors (i.e., observations w_4, w_{10}, w_{14}) are more obviously contained within the inner third cluster, whereas in the vertices method, these observations may be deemed to belong to a separate cluster (if attention is focused on only the first principal component, but not if the focus is the second principal component alone).

The centers method was applied to the cars dataset of Table 5.15 in Example 5.13, with the resulting principal components displayed in Table 5.17, and plotted in Figure 5.10 and Figure 5.11. Applying the vertices method to these data, we obtain the corresponding vertices principal components of Table 5.18, and plots of the first two principal components in Figure 5.12. Comparing these tables and figures, we observe that for these data, the two methods are essentially the same.

Table 5.18 Vertices principal components – cars data.

w_u	Car	PC1	PC2	PC3
w_1	Aston Martin	[2.032, 3.213]	[0.384, 0.957]	[−0.730, 0.580]
w_2	Audi A6	[−2.858, −0.394]	[−1.212, 1.404]	[−1.293, 1.117]
w_3	Audi A8	[−2.046, 0.048]	[−0.972, 1.479]	[−1.119, 0.946]
w_4	BMW 7	[−1.864, 0.470]	[−0.427, 1.916]	[−1.463, 1.159]
w_5	Ferrari	[0.998, 2.621]	[−0.920, 0.650]	[−0.711, 1.204]
w_6	Honda NSR	[−0.286, 0.186]	[−0.742, −0.289]	[0.090, 0.491]
w_7	Mercedes C	[−3.235, −0.530]	[−1.592, 1.281]	[−1.303, 1.152]
w_8	Porsche	[0.215, 1.429]	[−1.367, −0.550]	[−0.607, 0.486]

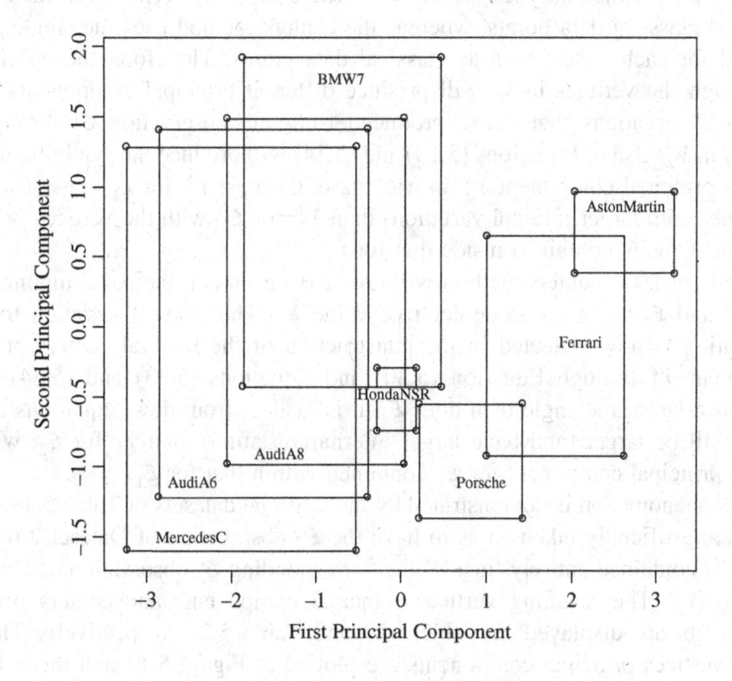

Figure 5.12 Vertices principal components: cars data.

There are, however, some fundamental differences between the two methods. Both methods are adaptations of classical principal component analysis, but they vary in how the internal variation of the symbolic observations is incorporated into the final answer.

Consider the two symbolic data points

$$\xi_1 = \{[12, 18],\ [4, 8]\},$$

$$\xi_2 = \{[14, 16],\ [5, 7]\}.$$

These have respective vertex values

$$V_1 = \{(12, 4), (12, 8), (18, 4), (18, 8)\},$$
$$V_2 = \{(14, 5), (14, 7), (16, 5), (16, 7)\},$$

and centroid values

$$\bar{X}_1^c = \bar{X}_2^c = \{(15, 6)\}.$$

Notice in particular that the data rectangle formed by $\boldsymbol{\xi}_2$ lies entirely inside the data rectangle of $\boldsymbol{\xi}_1$. This reflects the larger internal variation of $\boldsymbol{\xi}_1$ as compared to that of $\boldsymbol{\xi}_2$. However, both $\boldsymbol{\xi}_1$ and $\boldsymbol{\xi}_2$ have the same centroid values.

The vertices method uses the $m_u = 2^p$ (here $m_u = 4$) vertices for each observation as classical data points, whereas the centers method uses the single $n_c = 1$ centroid for each observation as classical data points. Therefore, the observation $\boldsymbol{\xi}_1$ through the vertices in V_1 will produce different principal components for its $m_u = 4$ observations than those produced from the observation $\boldsymbol{\xi}_2$ through the vertices in V_2. Use of Equations (5.13) and (5.14) will produce the symbolic interval vertices principal component Y_u^V in each case. Clearly, Y_1^V for ξ_1 is itself a larger rectangle (with larger internal variation) than Y_2^V for $\boldsymbol{\xi}_2$, with the vertices principal component for ξ_2 contained inside that for ξ_1.

In contrast, the centers method will produce the same principal components for both $\boldsymbol{\xi}_1$ and $\boldsymbol{\xi}_2$ using the same centroid value $\bar{X}_u^c$. The internal variation for each observation is now reflected in the construction of the interval centers principal component Y_u^c through Equation (5.30) and Equations (5.33) and (5.34). Since $\boldsymbol{\xi}_1$ spans a larger rectangle than does $\boldsymbol{\xi}_2$, it is evident from these equations that Y_1^c for $\boldsymbol{\xi}_1$ will be larger (and have larger internal variation) than Y_2^c for $\boldsymbol{\xi}_2$, with the centers principal component for $\boldsymbol{\xi}_2$ contained within that for $\boldsymbol{\xi}_1$.

This phenomenon is demonstrated by the artificial datasets of Table 5.19. These data were artificially taken so as to have the $\boldsymbol{\xi}_u$ observation of Dataset 2 (column (ii)) to be contained entirely 'inside' the corresponding $\boldsymbol{\xi}_u$ observation of Dataset 1 (column (i)). The resulting vertices principal components and centers principal components are displayed in Table 5.20 and Table 5.21, respectively. The two sets of vertices principal components are plotted in Figure 5.14 and those for the

Table 5.19 Artificial datasets.

w_u	(i) Dataset 1		(ii) Dataset 2	
	$[a_1, b_1]$	$[a_2, b_2]$	$[a_1, b_1]$	$[a_2, b_2]$
w_1	[12, 18]	[4, 8]	[14, 16]	[5, 7]
w_2	[3, 9]	[10, 14]	[4, 8]	[11, 13]
w_3	[12, 20]	[11, 19]	[14, 18]	[13, 17]
w_4	[1, 7]	[1, 9]	[3, 5]	[4, 6]

Table 5.20 Vertices principal components – artificial data.

w_u	Dataset 1		Dataset 2	
	PC1	PC2	PC1	PC2
w_1	[−0.529, 0.654]	[0.389, 1.572]	[−0.243, 0.318]	[0.856, 1.417]
w_2	[−0.730, 0.453]	[−1.386, −0.202]	[−0.537, 0.272]	[−1.322, −0.514]
w_3	[0.389, 2.316]	[−1.054, 0.873]	[1.013, 2.136]	[−0.715, 0.408]
w_4	[−2.130, −0.422]	[−0.949, 0.758]	[−1.760, −1.199]	[−0.346, 0.215]

Table 5.21 Centers principal components – artificial data.

w_u	Dataset 1		Dataset 2	
	PC1	PC2	PC1	PC2
w_1	[−0.609, 0.673]	[0.423, 1.705]	[−0.231, 0.295]	[0.801, 1.327]
w_2	[−0.763, 0.519]	[−1.500, −0.218]	[−0.500, 0.257]	[−1.237, −0.481]
w_3	[0.423, 2.525]	[−1.200, 0.903]	[0.949, 2.000]	[−0.673, 0.378]
w_4	[−2.320, −0.449]	[−0.993, 0.878]	[−1.647, −1.122]	[−0.320, 0.205]

centers principal components are shown in Figure 5.15. The distinct clusters each of size 1 corresponding to the respective w_1, w_2, w_3, w_4 data values are apparent. Notice, however, that whereas the actual data hyperrectangles do not overlap (see Figure 5.13), the principal component rectangles do overlap.

Remark: One of the distinguishing features of the financial data of Table 5.2 is that some of the interval values are wide in range. This is especially true of the FMCG sector. This particular sector involved over 600 original individual company records, including a very few dominant firms with large budgets. Recall that when aggregating individual observations into interval-valued symbolic observations, the resulting interval spans the minimum Y value to the maximum Y value. Thus, when, as in the FMCG sector, relative outlier values exist, the resulting symbolic interval is perforce wide, and as such may not truly reflect the distribution of the observations whose aggregation produced that interval. Of course, smaller aggregations such as there observed for the pharmaceutical sector (w_9, with $n_u = 31$) can also contain outlier observations.

In these situations, an aggregation producing a histogram-valued observation would more accurately reflect the set of observations making up that concept class. How vertices principal component analysis can be developed for histogram data is still an open problem. While it is possible to find the centroid values for histogram-valued observations (see Equations (3.34) and (3.35)), which in turn means that a classical analysis can be applied to these centroid values, how these can be extended (through a histogram version of Equation (5.3)) also remains an open problem.

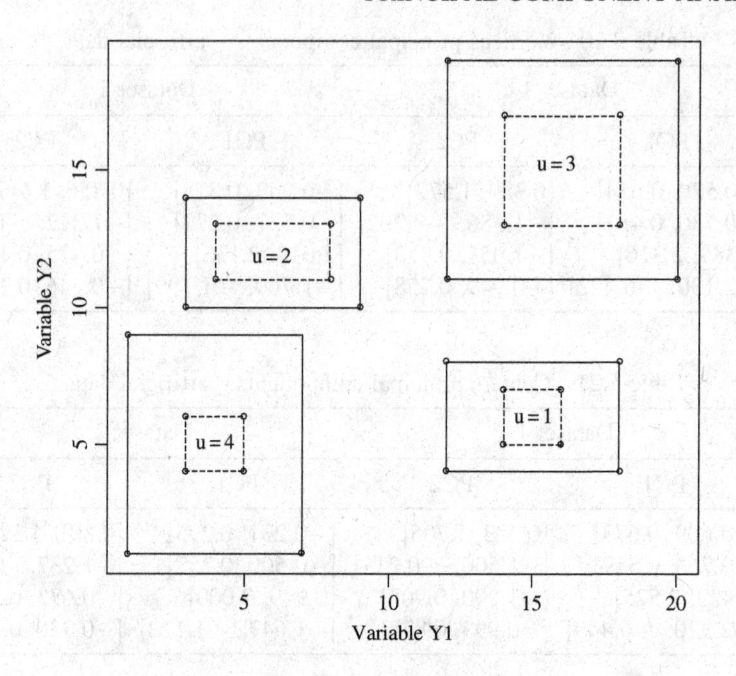

Figure 5.13 Two artificial datasets.

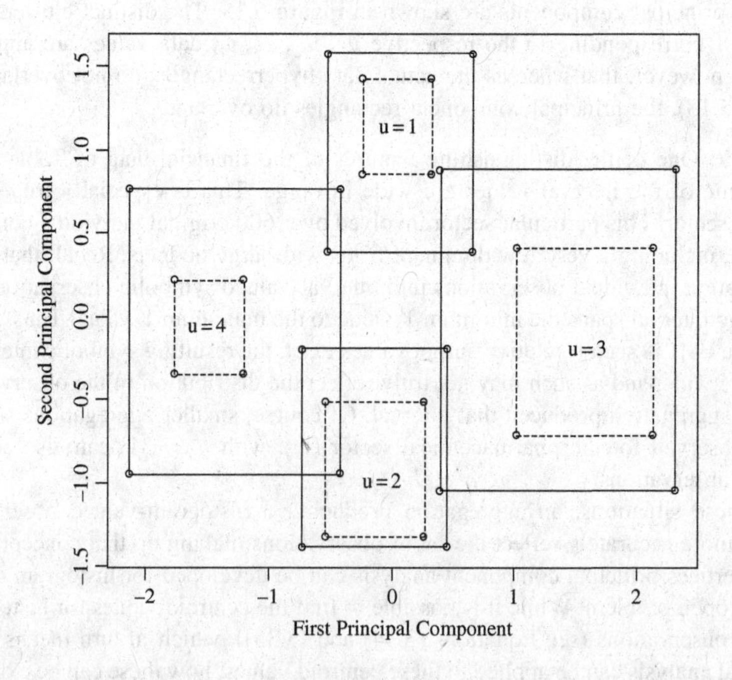

Figure 5.14 Vertices principal component: artificial data.

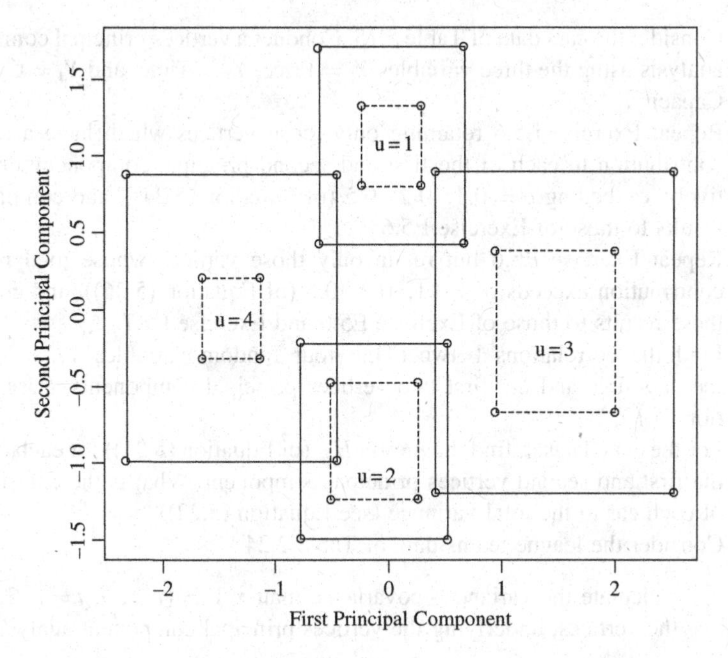

Figure 5.15 Centers principal component: artificial data.

Exercises

E5.1. Consider the financial data of Table 5.2. Do an unweighted vertices principal component analysis using the random variables Y_1 = Job Cost, Y_3 = Activity Code, and Y_4 = Annual Budget.

E5.2. For the same financial data, do an unweighted vertices principal component analysis on all five random variables.

E5.3. Carry out a weighted (with weights proportional to the number n_u) centers principal component analysis on the same financial dataset using the random variables (i) Y_1, Y_2, Y_3, and (ii) Y_1, Y_3, Y_5.

E5.4. Repeat Exercise E5.1 where now only those vertices which contribute a level exceeding α (of Equation (5.24)) on either the first or second principal component, for $\alpha = 0.1,\ 0.2,\ 0.5$, are retained. Compare these results to those for $\alpha = 0$.

E5.5. Repeat Exercise E5.4 where now the total contribution for both the first and second principal component (i.e., use Equation (5.26)) exceeds $\alpha = 0.1,\ 0.2,\ 0.5$. Compare the results of this question to those from Exercise E5.4.

E5.6. Consider the cars data of Table 5.15. Conduct a vertices principal component analysis using the three variables Y_1 = Price, Y_3 = Time, and Y_4 = Cylinder Capacity.

E5.7. Repeat Exercise E5.6 retaining only those vertices which have a relative contribution to each of the first and second principal components, respectively, exceeding $\alpha = 0.1,\ 0.2,\ 0.5$ (of Equation (5.24)), and compare the results to those of Exercise E5.6.

E5.8. Repeat Exercise E5.6 but retain only those vertices whose total relative contribution exceeds $\alpha = 0.1,\ 0.2,\ 0.5$ (of Equation (5.26)), and compare these results to those of Exercise E5.6 and Exercise E5.7.

E5.9. Find the correlations between the four random variables $Y_1, \ldots, Y_4$ of the cars data and the first two vertices principal components. (Use Equation (5.19).)

E5.10. For the cars dataset, find the inertia H_{vu} (of Equation (5.21)) for each car and the first and second vertices principal component. What is the contribution of each car to the total variance (see Equation (5.22))?

E5.11. Consider the league teams data of Table 2.24.

(i) Calculate the variance –covariance matrix $\boldsymbol{V} = (v_{ij}),\ i, j = 1, 2, 3,$ for the vertices, underlying the vertices principal component analysis.

(ii) Calculate the variance –covariance matrix $\boldsymbol{C} = (c_{ij}),\ i, j = 1, 2, 3,$ for the centers, underlying the centers principal component analysis.

(iii) Show that

$$\boldsymbol{V} = \boldsymbol{C} + \boldsymbol{E}$$

where the matrix $\boldsymbol{E} = (e_{ij}),\ i, j = 1, 2, 3,$ has elements $e_{ij} = \sum_{u \in E} w_{uj}(a_{uj} - b_{uj})^2$ where w_{uj} are constants (weights).

E5.12. Show that the relationship $\boldsymbol{V} = \boldsymbol{C} + \boldsymbol{E}$ (of Exercise E5.11) holds for $i, j = 1, \ldots, p$, in general.

References

Anderson, T. W. (1984). *An Introduction to Multivariate Statistical Analysis* (2nd ed.), John Wiley & Sons, Inc., New York.

Cazes, P., Chouakria, A., Diday, E., and Schektman, Y. (1997). Extensions de l'analyse en Composantes Principales a des Données de Type Intervalle. *Revue de Statistique Appliquée* 24, 5–24.

Chouakria, A. (1998). Extension des Méthodes d'analyse Factorielle a des Données de Type Intervalle, Doctoral Thesis, Université Paris, Dauphine.

Chouakria, A., Diday, E., and Cazes, P. (1998). An Improved Factorial Representation of Symbolic Objects. In: *Knowledge Extraction from Statistical Data.* European Commission Eurostat, Luxembourg, 301–305.

Fisher, R. A. (1936). The Use of Multiple Measurement in Taxonomic Problems. *Annals of Eugenics* 7, 179–188.

Johnson, R. A. and Wichern, D. W. (1998). *Applied Multivariate Statistical Analysis* (4th ed.), Prentice Hall, Englewood Cliffs, NJ.

Vu, T. H. T., Vu, T. M. T., and Foo, R. W. S. (2003). *Analyse de Données Symboliques sur des Projects Marketing*. Technical Report, CEREMADE, Dauphine, Université Paris IX.

6

Regression Analysis

In this chapter, linear regression methods are described for each of multi-valued (or categorical) variables, interval-valued variables, and histogram-valued variables, in turn. Then, the methodology is extended to models which include taxonomy variables as predictor variables, and to models which contain a hierarchical variable structure, as shown by Alfonso *et al.* (2005). Except for the multi-valued dependent variable case of Section 6.2.2 and a special case of interval-valued data in Example 6.6, the fitted regression models and hence their predicted values are generally single valued, even though the input variables had symbolic formats. A methodology that gives output predicted values as symbolic values is still an open question. The focus will be on fitting a linear regression model. How other types of models and model diagnostics broadly defined are handled also remains as an open problem.

6.1 Classical Multiple Regression Model

Since the symbolic regression methodology relies heavily on the classic theory, we very briefly describe the familiar classical method. The multiple classical linear regression model, for the predictor variables $X_1, \ldots, X_p$ and dependent variable Y, is defined by

$$Y = \beta_0 + \beta_1 X_1 + \ldots + \beta_p X_p + e \tag{6.1}$$

or, in vector terms,

$$\boldsymbol{Y} = \boldsymbol{X\beta} + \boldsymbol{e} \tag{6.2}$$

where the vector of observations $\boldsymbol{Y}$ is

$$\boldsymbol{Y}' = (Y_1, \ldots, Y_n),$$

Symbolic Data Analysis: Conceptual Statistics and Data Mining L. Billard and E. Diday

the regression design matrix $\boldsymbol{X}$ is the $n \times (p+1)$ matrix

$$\boldsymbol{X} = \begin{pmatrix} 1 & X_{11} & \dots & X_{1p} \\ \vdots & \vdots & & \vdots \\ 1 & X_{n1} & \dots & X_{np} \end{pmatrix},$$

the regression coefficient vector $\boldsymbol{\beta}$ is the $(p+1)$ vector

$$\boldsymbol{\beta}' = (\beta_0, \dots, \beta_p),$$

and the error vector $\boldsymbol{e}$ is

$$\boldsymbol{e}' = (e_1, \dots, e_n)$$

where the error terms satisfy $E(e_i) = 0$ and $Var(e_i) = \sigma^2$ and $Cov(e_i, e_{i'}) = 0,\ i \neq i'$. The least squares estimators of the parameters $\boldsymbol{\beta}$ are given by, if $\boldsymbol{X}$ is nonsingular,

$$\hat{\boldsymbol{\beta}} = (\boldsymbol{X}'\boldsymbol{X})^{-1}\boldsymbol{X}'\boldsymbol{Y}. \tag{6.3}$$

In the particular case when $p = 1$, Equation (6.3) simplifies to

$$\hat{\beta}_1 = \frac{\sum_{i=1}^{n}(X_i - \bar{X})(Y_i - \bar{Y})}{\sum_{i=1}^{n}(X_i - \bar{X})^2} = \frac{Cov(X, Y)}{Var(X)}, \tag{6.4}$$

$$\hat{\beta}_0 = \bar{Y} - \hat{\beta}\bar{X} \tag{6.5}$$

where

$$\bar{Y} = \frac{1}{n}\sum_{i=1}^{n} Y_i,\ \bar{X} = \frac{1}{n}\sum_{i=1}^{n} X_i. \tag{6.6}$$

An alternative formulation of the model in Equation (6.1) is

$$Y - \bar{Y} = \beta_1(X_1 - \bar{X}_1) + \dots + \beta_p(X_p - \bar{X}_p) + e$$

where

$$\bar{X}_j = \frac{1}{n}\sum_{i=1}^{n} X_{ij},\ j = 1, \dots, p.$$

When using this formulation, Equation (6.3) becomes

$$\hat{\boldsymbol{\beta}} = [(\boldsymbol{X} - \bar{\boldsymbol{X}})'(\boldsymbol{X} - \bar{\boldsymbol{X}})]^{-1}(\boldsymbol{X} - \bar{\boldsymbol{X}})'(\boldsymbol{Y} - \bar{\boldsymbol{Y}})$$

where now there is no column of ones in the matrix $\boldsymbol{X}$. It follows that β_0 in Equation (6.1) is

$$\beta_0 \equiv \bar{Y} - (\beta_1\bar{X}_1 + \dots + \beta_p\bar{X}_p);$$

hence the two formulations are equivalent.

Given the nature of the definitions of the symbolic covariance for interval-valued and histogram-valued observations (see Equation (4.16) and Equation (4.19), respectively), this latter formulation is preferable when fitting $p \geq 2$ predictor variables in these cases (as illustrated in subsequent sections below).

The model of Equation (6.1) assumes there is only one dependent variable Y for the set of p predictor variables $\boldsymbol{X} = (X_1, \ldots, X_p)$. When the variable Y is itself of dimension $q > 1$, we have a multivariate multiple regression model, given by

$$Y_k = \beta_{0k} + \beta_{1k}X_1 + \ldots + \beta_{pk}X_p + e_k, \; k = 1, \ldots, q, \tag{6.7}$$

or, in vector form,

$$\boldsymbol{Y}^* = \boldsymbol{X}\boldsymbol{\beta}^* + \boldsymbol{e}^* \tag{6.8}$$

where the observation matrix $\boldsymbol{Y}^*$ is the $n \times q$ matrix

$$\boldsymbol{Y}^* = (\boldsymbol{Y}_{(1)}, \ldots, \boldsymbol{Y}_{(q)})$$

with

$$\boldsymbol{Y}'_{(k)} = (Y_{1k}, \ldots, Y_{nk}),$$

the regression design matrix $\boldsymbol{X}$ is the same $n \times (p+1)$ matrix $\boldsymbol{X}$ used in Equation (6.2), the regression parameter matrix β^* is the $(p+1) \times q$ matrix

$$\boldsymbol{\beta}^* = (\boldsymbol{\beta}_{(1)}, \ldots, \boldsymbol{\beta}_{(q)})$$

with

$$\boldsymbol{\beta}'_{(k)} = (\beta_{0k}, \ldots, \beta_{pk}),$$

and where the error terms $\boldsymbol{e}^*$ form the $n \times q$ matrix

$$\boldsymbol{e}^* = (\boldsymbol{e}_{(1)}, \ldots, \boldsymbol{e}_{(q)})$$

with

$$\boldsymbol{e}'_{(k)} = (e_{1k}, \ldots, e_{nk})$$

and where the error terms satisfy

$$E(\boldsymbol{e}_{(k)}) = 0, \; Cov(\boldsymbol{e}_{(k_1)}, \boldsymbol{e}_{(k_2)}) = \sigma_{k_1 k_2}\boldsymbol{I}.$$

Note that these errors from different responses on the same trials are correlated in general but that observations from different trials are not. Then, the least squares estimators of the parameters $\boldsymbol{\beta}^*$ are

$$\hat{\boldsymbol{\beta}}^* = (\boldsymbol{X}'\boldsymbol{X})^{-1}\boldsymbol{X}'\boldsymbol{Y}^*. \tag{6.9}$$

In the special case when $p = 1$, the model of Equation (6.7) becomes, for $i = 1, \dots, n$,

$$Y_{ik} = \beta_{0k} + \beta_{1k} X_i + e_{ik}, \; k = 1, \dots, q, \tag{6.10}$$

and hence we can show that

$$\hat{\beta}_{1k} = \frac{\sum_{i=1}^{n} (X_i - \bar{X})(Y_{ik} - \bar{Y}_k)}{\sum_{i=1}^{n} (X_i - \bar{X})^2} \tag{6.11}$$

and

$$\hat{\beta}_{0k} = \bar{Y}_k - \hat{\beta}_{1k} \bar{X} \tag{6.12}$$

where

$$\bar{Y}_k = \frac{1}{n} \sum_{i=1}^{n} Y_{ik}, \; k = 1, \dots, q, \tag{6.13}$$

and $\bar{X}$ is as defined in Equation (6.6). There is a plethora of standard texts covering the basics of classical regression analysis, including topics beyond the model fitting stages touched on herein, such as model diagnostics, non-linear and other types of models, variable selection, and the like. See, for example, Montgomery and Peck (1992) and Myers (1986) for an applications-oriented introduction. As for the univariate Y case, there are numerous texts available for details on multivariate regression; see, for example, Johnson and Wichern (2002).

6.2 Multi-Valued Variables

6.2.1 Single dependent variable

Recall from Definition 2.2 that a multi-valued random variable is one whose value is a list of categorical values, i.e., each observation w_u takes a value

$$\xi_u = \{\eta_k, \; k = 1, \dots, s_u\}, \; u = 1, \dots, m. \tag{6.14}$$

A modal multi-valued random variable, from Definition 2.4, is one whose value takes the form

$$\xi_u = \{\eta_{uk}, \; \pi_{uk}; \; k = 1, \dots, s_u\}, \; u = 1, \dots, m. \tag{6.15}$$

Clearly, the non-modal multi-valued variable of Equation (6.14) is a special case of the modal multi-valued variable where each value η_k in this list is equally likely. In the sequel, we shall take the measures π_k to be relative frequencies p_k, $k = 1, \dots, s_u$; adjustments to other forms of measures π_k can readily be made (see Section 2.2).

For clarity, let us denote the independent predictor variables by $X = (X_1, \ldots, X_p)$ taking values in $\Re^p$, where X_j has realizations

$$X_j(w_u) = \{\eta_{ujk}, p_{ujk};\ k = 1, \ldots, s_{uj}\},\ j = 1, \ldots, p, \tag{6.16}$$

on the observation w_u, $u = 1, \ldots, m$. Let the possible values of X_j in $\mathcal{X}_j$ be $\{X_{j1}, \ldots, X_{js_j}\}$. These X_{jk}, $k = 1, \ldots, s_j$, $j = 1, \ldots, p$, are a type of indicator variable taking relative frequency values p_{jk}, respectively. Without loss of generality in Equation (6.16), we take $s_{uj} = s_j$ for all observations w_u and set the observed $p_{ujk} \equiv 0$ for those values of $X_j \subset \mathcal{X}_j$ not appearing in Equation (6.16). Let the dependent variable be denoted by Y taking values in the space $\mathcal{Y} \equiv \Re$. Initially, let Y take a single (quantitative, or qualitative) value in $\Re$; this is extended to the case of several Y values in Section 6.2.2.

The regression analysis for these multi-valued observations proceeds as for the multiple classical regression model of Equation (6.1) in the usual way, where the X_j variables in Equation (6.1) take values equal to the $\{p_{ujk}\}$ of the symbolic observations. Note that, since the X are a type of indicator variable, only $(k_j - 1)$ are included in the model to enable inversion of the associated $(X'X)$ matrix.

Example 6.1. Table 6.1 lists the average fuel costs (Y, in coded dollars) paid per annum and the types of fuel used (X) for heating in several different regions of a country. The complete set of possible X values is $X = \{X_1, \ldots, X_4\}$, i.e., $\mathcal{X} =$ {gas, oil, electric, other (forms of heating)}. Thus, we see, for instance, that in region1 ($w_u = w_1$), 83% of the customers used gas, 3% used oil, and 14% used electricity for heating their homes, at an average expenditure of $Y = 28.1$.

Table 6.1 Fuel consumption.

w_u	Region	Y = Expenditure	X = Fuel Type
w_1	region1	28.1	{gas, 0.83; oil, 0.03; electric, 0.14}
w_2	region2	23.4	{gas, 0.69; oil, 0.13; electric, 0.12; other, 0.06}
w_3	region3	33.2	{oil, 0.40; electric, 0.15; other, 0.45}
w_4	region4	25.1	{gas, 0.61; oil, 0.07; electric, 0.16; other, 0.16}
w_5	region5	21.7	{gas, 0.67; oil, 0.15; electric, 0.18}
w_6	region6	32.5	{gas, 0.40; electric, 0.45; other, 0.15}
w_7	region7	26.6	{gas, 0.83; oil, 0.01; electric, 0.09; other, 0.07}
w_8	region8	19.9	{gas, 0.66; oil, 0.15; electric, 0.19}
w_9	region9	28.4	{gas, 0.86; electric, 0.09; other, 0.05}
w_{10}	region10	25.5	{gas, 0.77; electric, 0.23}

The resulting regression equation is, from Equation (6.3),

$$\hat{Y} = 61.31 - 36.73(\text{gas}) - 61.34(\text{oil}) - 32.74(\text{electric}). \tag{6.17}$$

Note that although the $X_4 =$ 'other' variable does not enter directly into Equation (6.17), it is present indirectly by implication when the corresponding values for X_{1k}, $k = 1, 2, 3$, are entered into the model.

By substituting observed X values into Equation (6.17), predicted expenditures $\hat{Y}$ and the corresponding residuals $R = Y - \hat{Y}$ are obtained. For example, for region1,

$$\hat{Y}(w_1) = 61.31 - 36.73(0.83) - 61.34(0.03) - 32.74(0.14) = 24.40,$$

and hence the residual is

$$R(w_1) = Y(w_1) - \hat{Y}(w_1) = 28.10 - 24.40 = 3.70.$$

The predicted $\hat{Y}$ values and the residuals R for all regions are displayed in Table 6.2.

Table 6.2 Fuel – regression parameters.

w_u	Region	Y	gas	oil	electric	other	$\hat{Y}$	R
w_1	region1	28.1	0.83	0.03	0.14	0.00	24.40	3.70
w_2	region2	23.4	0.69	0.13	0.12	0.06	24.07	−0.67
w_3	region3	33.2	0.00	0.40	0.15	0.45	31.87	1.33
w_4	region4	25.1	0.61	0.07	0.16	0.16	29.37	−4.27
w_5	region5	21.7	0.67	0.15	0.18	0.00	21.61	0.09
w_6	region6	32.5	0.40	0.00	0.45	0.15	31.89	0.61
w_7	region7	26.6	0.83	0.01	0.09	0.07	27.26	−0.66
w_8	region8	19.9	0.66	0.15	0.19	0.00	21.65	−1.75
w_9	region9	28.4	0.86	0.00	0.09	0.05	26.78	1.62
w_{10}	region10	25.5	0.77	0.00	0.23	0.00	25.50	0.00

□

Example 6.2. Suppose there are $p = 2$ predictor variables. The data of Table 6.3 extend the analysis of Table 6.1 by relating the expenditure Y on heating to two predictors $X_1 =$ Type of heating fuel (as in Table 6.1) and $X_2 =$ Size of household with $\mathcal{X}_2 = \{\text{small, large}\}$. Thus, the third column of Table 6.3 gives the observation values for X_2. The X_2 possible values are recoded to $X_{21} =$ small and $X_{22} =$ large. The fourth column in Table 6.3 gives the corresponding recoded values. The X_1 and the related X_{1k}, $k = 1, \ldots, 4$, values are as given in Table 6.1 and Table 6.2, and the X_2 and related X_{2k}, $k = 1, 2$, values are as in Table 6.3.

Table 6.3 Fuel consumption – two variables.

w_u	Region	X_2	X_{21}	X_{22}	$\hat{Y}$	R
w_1	region1	{small, 0.57; large, 0.43}	0.57	0.43	25.93	2.17
w_2	region2	{small, 0.65; large, 0.35}	0.65	0.35	23.97	−0.57
w_3	region3	{small, 0.38; large, 0.62}	0.38	0.62	32.72	0.48
w_4	region4	{small, 0.82; large, 0.18}	0.82	0.18	26.58	−1.48
w_5	region5	{small, 0.74; large, 0.26}	0.74	0.26	20.35	1.35
w_6	region6	{small, 0.55; large, 0.45}	0.55	0.45	32.28	0.22
w_7	region7	{small, 0.81; large, 0.19}	0.81	0.19	25.40	1.20
w_8	region8	{small, 0.63; large, 0.37}	0.63	0.37	21.86	−1.96
w_9	region9	{small, 0.47; large, 0.53}	0.47	0.53	29.66	−1.26
w_{10}	region10	{small, 0.66; large, 0.34}	0.66	0.34	25.66	−0.16

The regression model that fits these data is, from Equation (6.3),

$$\hat{Y} = 66.40 - 32.08(\text{gas}) - 59.77(\text{oil}) - 30.69(\text{electric}) - 13.60(\text{small}). \quad (6.18)$$

Then, from Equation (6.18), we predict, for example, for region1,

$$\hat{Y}(w_1) = 66.40 - 32.08(0.83) - 59.77(0.03) - 30.69(0.14) - 13.60(0.57) = 25.93,$$

and hence the residual is

$$R(w_1) = 28.1 - 25.93 = 2.17.$$

The predicted and residual values for all regions are shown in the last two columns, respectively, of Table 6.3. □

6.2.2 Multi-valued dependent variable

Suppose now that the dependent variable Y can take any of a number of values for a list of possible values $\mathcal{Y} = \{\zeta_1, \zeta_2, \ldots, \zeta_q\}$. Let us write this as, for observation w_u,

$$Y(w_u) = \{\zeta_{uk},\ q_{uk};\ k = 1, \ldots, t_u\}$$

where for observation w_u the outcome ζ_{uk} occurred with relative frequency q_{uk}, and where t_u is the number of components in $\mathcal{Y}$ that were observed in $Y(w_u)$. Without loss of generality we let $t_u = q$ for all u, i.e., those ξ_k in $\mathcal{Y}$ not actually occurring in $Y(w_u)$ take relative frequency $q_{uk} \equiv 0$.

Then, in the same manner as the predictor variables $\{X_j,\ j = 1, \ldots, p\}$ were coded in Section 6.2.1, we also code the dependent variable to the q-dimensional dependent variable $Y = (Y_1, \ldots, Y_q)$ with Y_k representing the possible outcome ζ_k with the observed relative frequency q_k, $k = 1 \ldots, q$. Our linear regression model between the coded Y and X variables can be viewed as analogous to the classical multivariate multiple regression model of Equation (6.7).

Therefore, the techniques of multivariate multiple regression models are easily extended to our symbolic multi-valued variable observations. This time we obtain predicted fits $\hat{Y}$ which have symbolic modal-multi-valued values.

Example 6.3. A study was undertaken to investigate the relationship if any between Gender (X_1) and Age (X_2) on the types of convicted Crime (Y) reported in $m = 15$ areas known to be populated by gangs. The dependent random variable Y took possible values in $\mathcal{Y} = \{\text{violent, non-violent, none}\}$. The Age X_1 of gang members took values in $\mathcal{X}_1 = \{< 20, \geq 20 \text{ years}\}$ and Gender X_2 took values in $\mathcal{X}_2 = \{\text{male, female}\}$. These random variables were coded, respectively, to $Y = (Y_1, Y_2, Y_3)$, $X_1 = (X_{11}, X_{12})$, and $X_2 = (X_{21}, X_{22})$. The coded data are as shown in Table 6.4. Thus, for example, for gang1, the original data had

$$\xi(\text{gang1}) = (Y(w_1), X_1(w_1), X_2(w_1)) = (\{\text{violent, 0.73; non-violent, 0.16, none, 0.11}\},$$
$$\{\text{male}, 0.68; \text{female}, 0.32\}, \{< 20, 0.64; \geq 20, 0.36\}),$$

i.e., in this gang, 64% were under 20 years of age and 36% were 20 or older, 68% were male and 32% were female, and 73% had been convicted of a violent crime, 16% a non-violent crime, and 11% had not been convicted.

Table 6.4 Crime demographics.

w_u	Crime			Gender		Age	
	Y_1 violent	Y_2 non-violent	Y_3 none	X_{11} male	X_{12} female	X_{21} < 20	X_{22} ≥ 20
gang1	0.73	0.16	0.11	0.68	0.32	0.64	0.36
gang2	0.40	0.20	0.40	0.70	0.30	0.80	0.20
gang3	0.20	0.20	0.60	0.50	0.50	0.50	0.50
gang4	0.10	0.20	0.70	0.60	0.40	0.40	0.60
gang5	0.20	0.40	0.40	0.35	0.65	0.55	0.45
gang6	0.48	0.32	0.20	0.53	0.47	0.62	0.38
gang7	0.14	0.65	0.21	0.40	0.60	0.33	0.67
gang8	0.37	0.37	0.26	0.51	0.49	0.42	0.58
gang9	0.47	0.32	0.21	0.59	0.41	0.66	0.34
gang10	0.18	0.15	0.77	0.37	0.63	0.22	0.78
gang11	0.35	0.35	0.30	0.41	0.59	0.44	0.56
gang12	0.18	0.57	0.25	0.39	0.61	0.45	0.55
gang13	0.74	0.16	0.10	0.70	0.30	0.63	0.37
gang14	0.33	0.45	0.22	0.37	0.64	0.29	0.71
gang15	0.35	0.39	0.26	0.50	0.50	0.44	0.56

The multivariate regression model is, from Equation (6.7), for $u = 1, \ldots, m$,

$$
\begin{aligned}
Y_{u1} &= \beta_{01} + \beta_{11} X_{u11} + \beta_{31} X_{u21} + e_{u1}, \\
Y_{u2} &= \beta_{02} + \beta_{12} X_{u11} + \beta_{32} X_{u21} + e_{u2}, \\
Y_{u3} &= \beta_{03} + \beta_{13} X_{u11} + \beta_{33} X_{u21} + e_{u3},
\end{aligned}
\tag{6.19}
$$

where for observation w_u the dependent variable is written as $Y(w_u) = \{Y_{u1}, Y_{u2}, Y_{u3}\}$.

Note that since the X_1 and X_2 variables are a type of coded variable, one of the X_{1k} and one of the X_{2k} variables is omitted in the regression model of Equation (6.9). However, all the coded Y variables are retained, so in this case there are $q = 3$ equations, one for each of the coded Y variables. Then, from Equation (6.19), we can estimate the parameters $\boldsymbol{\beta}$ by $\hat{\boldsymbol{\beta}}^*$ to give the regression model as $\hat{Y} = \{\hat{Y}_1, \hat{Y}_2, \hat{Y}_3\}$ where

$$
\begin{aligned}
\hat{Y}_1 &= -0.202 + 0.791(\text{male}) + 0.303(\text{under } 20) \\
\hat{Y}_2 &= 0.711 - 0.9800(\text{male}) + 0.226(\text{under } 20) \\
\hat{Y}_3 &= 0.531 + 0.213(\text{male}) - 0.622(\text{under } 20).
\end{aligned}
\tag{6.20}
$$

Substituting the observed X values into Equation (6.20), we can calculate $\hat{Y}$. Thus, for example, for the first gang,

$$
\begin{aligned}
\hat{Y}_1(\text{gang1}) &= -0.202 + 0.791(0.68) + 0.303(0.64) = 0.53, \\
\hat{Y}_2(\text{gang1}) &= 0.711 - 0.980(0.68) + 0.226(0.64) = 0.19, \\
\hat{Y}_3(\text{gang1}) &= 0.531 + 0.213(0.68) - 0.622(0.64) = 0.28.
\end{aligned}
$$

That is, the predicted crime for a gang with the age and gender characteristics of gang1 is

$$
\hat{Y}(\text{gang1}) = \hat{Y}(w_1) = \{\text{violent, } 0.53\text{; non-violent, } 0.19\text{; none, } 0.28\},
$$

i.e., 53% are likely to be convicted of a violent crime, 19% of a non-violent crime, and 28% will not be convicted of any crime. The predicted crime rates $\hat{Y}$ for each of the gangs along with the corresponding residuals ($R = Y - \hat{Y}$) are shown in Table 6.5.

Table 6.5 Crime predictions.

w_u	$\hat{Y}$ = Predicted Crime			Residuals		
	$\hat{Y}_1$ violent	$\hat{Y}_2$ non-violent	$\hat{Y}_3$ none	R_1	R_2	R_3
gang1	0.53	0.19	0.28	0.20	−0.03	−0.17
gang2	0.59	0.21	0.18	−0.19	−0.01	0.22
gang3	0.34	0.33	0.33	−0.14	−0.13	0.27
gang4	0.39	0.21	0.41	−0.29	−0.01	0.29
gang5	0.24	0.49	0.26	−0.04	−0.09	0.14
gang6	0.41	0.33	0.26	0.08	−0.01	−0.06
gang7	0.21	0.39	0.41	−0.07	0.26	−0.20
gang8	0.33	0.31	0.38	0.04	0.06	−0.12
gang9	0.46	0.28	0.25	0.01	0.04	−0.04
gang10	0.16	0.40	0.47	0.02	−0.25	0.30
gang11	0.26	0.41	0.34	0.09	−0.06	−0.04
gang12	0.24	0.43	0.33	−0.06	0.14	−0.08
gang13	0.54	0.17	0.29	0.20	−0.01	−0.19
gang14	0.18	0.41	0.43	0.15	0.04	−0.21
gang15	0.33	0.32	0.36	0.02	0.07	−0.10

□

6.3 Interval-Valued Variables

For interval-valued observations, the dependent variable is $Y = [a, b]$, and the predictor variables are $\boldsymbol{X} = (X_1, \ldots, X_p)$ with $X_j = [c_j, d_j]$, $j = 1, \ldots, p$. The regression analysis proceeds by using Equations (6.1)–(6.3) of the classical methodology but with symbolic counterparts for the terms in the $(\boldsymbol{X}'\boldsymbol{X})$ and $(\boldsymbol{X}'\boldsymbol{Y})$ matrices. Thus, for $p = 1$, we have

$$\hat{\beta}_1 = Cov(Y, X)/S_x^2 \tag{6.21}$$

and

$$\hat{\beta}_0 = \bar{Y} - \hat{\beta}_1 \bar{X} \tag{6.22}$$

where

$$\bar{Y} = \sum_{u=1}^{m} (a_u + b_u)/2m, \quad \bar{X} = \sum_{u=1}^{m} (c_u + d_u)/2m, \tag{6.23}$$

and where $Cov(Y, X)$ is calculated from Equation (4.16) and S_x^2 is calculated from Equation (3.23).

Example 6.4. The data of Table 6.6 give interval-valued observations for a predictor variable X = Age and two dependent random variables Y_1 = Cholesterol and Y_2 = Weight for a certain population. The concept here is the age group represented by the age decade of those individuals making up the original dataset.

Table 6.6 Age, Cholesterol, and Weight variables.

w_u	X Age	Y_1 Cholesterol	Y_2 Weight	n_u
w_1	[20, 30)	[114, 192]	[108, 141]	43
w_2	[30, 40)	[103, 189]	[111, 150]	66
w_3	[40, 50)	[120, 191]	[127, 157]	75
w_4	[50, 60)	[136, 223]	[130, 166]	43
w_5	[60, 70)	[149, 234]	[139, 161]	59
w_6	[70, 80)	[142, 229]	[143, 169]	35
w_7	[80, 90)	[140, 254]	[140, 176]	18

Suppose we fit Y_1 = Cholesterol as a single dependent variable to the predictor variable X = Age. From Equations (6.21)–(6.23), we obtain

$$\hat{\beta}_1 = 1.387, \quad \hat{\beta}_0 = 96.278;$$

hence, the regression equation modeling cholesterol and age is

$$\hat{Y}_1 = \text{Cholesterol} = 96.28 + 1.39(\text{Age}). \tag{6.24}$$

Suppose we now fit the dependent variable Y_2 = Weight, and suppose we conduct a weighted analysis with weights proportional to the number of individuals (n_u, in Table 6.6) that were aggregated to form the respective classes (w_u). In this case, we obtain

$$\hat{\beta}_1 = 0.588, \quad \hat{\beta}_0 = 111.495,$$

and hence

$$\hat{Y}_2 = \text{Weight} = 111.50 + 0.59(\text{Age}). \tag{6.25}$$

□

An alternative regression methodology is to calculate the variable midpoints for each observation w_u, i.e.,

$$y^c(w_u) = (a_u + b_u)/2, \; x^c_j(u) = (c_{uj} + d_{uj})/2, \; j = 1, \ldots, p; \tag{6.26}$$

and then to use the regression methodology of Equations (6.6)–(6.9) on these (y^c, x^c_j) values.

Example 6.5. Take the data of Table 6.6, and fit a regression model by using the centered values of Equation (6.26). Let the dependent variable be $Y_1 =$ Cholesterol, and the predictor variable be $X =$ Age. We obtain

$$\hat{Y}_1 = \text{Cholesterol} = 124.05 + 0.88(\text{Age}). \tag{6.27}$$

When we fit a model with weights proportional to class size, we obtain

$$\hat{Y}_1 = \text{Weight} = 118.62 + 0.98(\text{Age}). \tag{6.28}$$

□

A disadvantage of this approach is that, unlike the first approach which incorporates the interval lengths into the analysis (through Equations (6.21)–(6.23) when $p = 1$, say), this alternative approach using the single midpoint value does not. When the intervals are short, the two methods will give reasonably comparable results. A suggestion by De Carvalho *et al.* (2004) and Lima Neto *et al.* (2005) is to use the interval midpoints along with the length of the intervals as an additional variable.

Example 6.6. Consider the $Y_1 =$ Cholesterol and $X =$ Age interval-valued data of Table 6.6. Setting the midpoints as one variable and the interval length as another variable gives the data in the form shown in Table 6.7. Note that there is now a multivariate dependent variable (Y_1, Y_2), with

$$Y_1(w_u) = (a_u + b_u)/2, \ Y_2(w_u) = b_u - a_u,$$

and two predictor variables (X_1, X_2) with

$$X_1(w_u) = (c_u + d_u)/2, \ X_2(w_u) = d_u - c_u.$$

Table 6.7 Range as added variables.

w_u	X_1 Age	Y_1 Cholesterol	X_2 Range Age	Y_2 Range Cholesterol	$\hat{Y}_1$	$\hat{Y}_2$	Symbolic $\hat{Y}$
w_1	[20, 30)	[114, 192]	10	78	146.05	66.90	[112.60, 179.45]
w_2	[30, 40)	[103, 189]	10	86	154.93	78.00	[115.93, 193.93]
w_3	[40, 50)	[120, 191]	10	71	163.75	82.43	[122.54, 204.96]
w_4	[50, 60)	[136, 223]	10	87	172.57	86.86	[129.14, 216.00]
w_5	[60, 70)	[149, 234]	10	85	181.39	91.29	[135.75, 227.04]
w_6	[70, 80)	[142, 229]	10	87	190.21	95.71	[142.36, 238.07]
w_7	[80, 90)	[140, 254]	10	114	199.04	100.14	[148.97, 249.11]

Then, using the results from Equations (6.10)–(6.13), we obtain the unweighted regression equations

$$\begin{aligned}\hat{Y}_1 &= \text{Cholesterol} = 124.05 + 0.88(\text{Age}),\\ \hat{Y}_2 &= \text{Range} = 62.50 + 0.44(\text{Age}).\end{aligned} \tag{6.29}$$

Note that the age range X_2 did not appear in Equations (6.29) as its value in this dataset was 10 years for all observations. Calculating the estimated predictors $\hat{Y}_1$ and $\hat{Y}_2$ for different midpoint age values gives the predictions shown in Table 6.7. For example, for the first observation w_1, when $X_1 = 25$, we obtain from Equations (6.29),

$$\hat{Y}_1(w_1) = 124.05 + 0.88(25) = 146.05,$$

$$\hat{Y}_2(w_1) = 62.50 + 0.44(10) = 66.90.$$

Hence, for those with age $X = [20, 30]$, the predicted cholesterol is $\hat{Y} = [112.60, 179.45]$. Thus, the prediction for the dependent variable cholesterol now takes interval-valued values. These are plotted in Figure 6.1; also plotted are the original interval-valued Y_1 (cholesterol) and X (age) values. The interval-valued predictions for each X value are evident. □

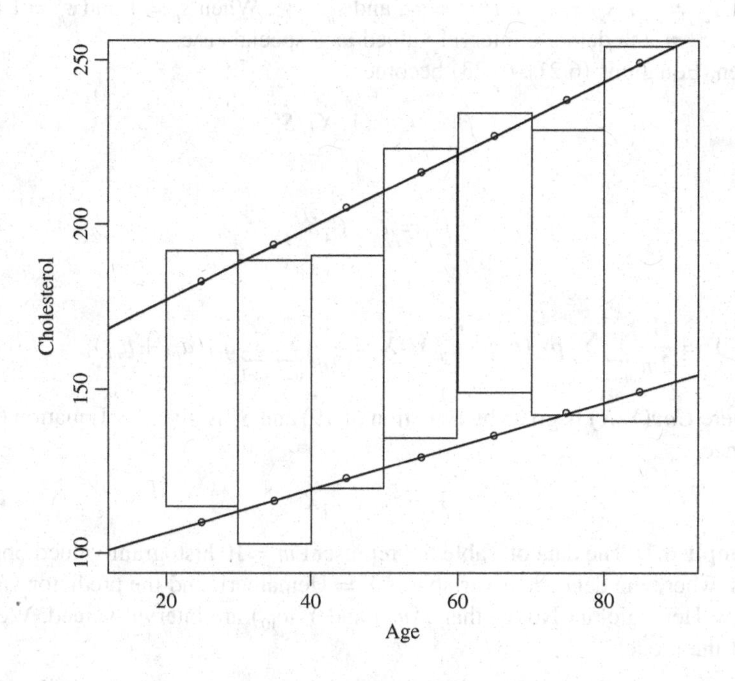

Figure 6.1 Prediction intervals based on range.

6.4 Histogram-Valued Variables

The principles invoked for interval-valued observations in Section 6.2 can be extended to histogram-valued observations in a manner analogous to the extensions developed in Section 3.6 and Section 4.6 of the mean, variance, and covariance functions to histogram-valued data. For notational simplicity, let us assume that there is a single $p = 1$ predictor variable X, and the dependent variable is Y; the case for $p > 1$ follows analogously and is left as an exercise for the reader. When we considered histogram multi-valued observations in Section 6.2, we were able to assume that all possible values could occur, albeit some with zero probability. This generalization is not possible for modal interval-valued (i.e., quantitative histogram-valued) observations as the histogram subintervals are not the same for all observation w_u. Different methods are required for the present case of quantitative interval-based histograms.

Following Section 3.6, we let, for each observation w_u,

$$Y(w_u) = \{[a_{u1}, b_{u1}), p_{u1}; \ldots ; [a_{us_u^y}, b_{us_u^y}), p_{us_u^y}\}$$

where s_u^y is the number of histogram subintervals for the observed Y values, and

$$X(w_u) = \{[c_{u1}, d_{u1}), q_{u1}; \ldots ; [c_{us_u^x}, d_{us_u^x}), q_{us_u^x}\}$$

where s_u^x is the number of histogram subintervals for the observed X values. In general $s_{u_1}^y \neq s_{u_2}^y$, $s_{u_1}^x \neq s_{u_2}^x$ for $u_1 \neq u_2$, and $s_u^y \neq s_u^x$. When $s_u^y = 1$ and $s_u^x = 1$ for all $u = 1, \ldots, m$, the data are interval-valued as a special case.

Then, Equations (6.21)–(6.23) become

$$\hat{\beta}_1 = Cov(Y, X)/S_x^2 \tag{6.30}$$

and

$$\hat{\beta}_0 = \bar{Y} - \hat{\beta}_1 \bar{X}, \tag{6.31}$$

where

$$\bar{Y} = \frac{1}{2m} \sum_{u=1}^{m} \sum_{k=1}^{s_u^y} p_{uk}(c_{uk} + d_{uk}), \quad \bar{X} = \frac{1}{2m} \sum_{u=1}^{m} \sum_{k=1}^{s_u^x} q_{uk}(a_{uk} + b_{uk}), \tag{6.32}$$

and where $Cov(Y, X)$ is given by Equation (4.19) and S_x^2 is given by Equation (3.40).

Hence,

$$\hat{Y} = \hat{\beta}_0 + \hat{\beta}_1 X. \tag{6.33}$$

Example 6.7. The data of Table 6.8 represent $m = 10$ histogram-valued observations, where the dependent variable is Y = Hematocrit and the predictor variable is X = Hemoglobin. Notice that $Y(w_5)$ and $Y(w_{10})$ are interval-valued. We want to fit the model

$$Y = \beta_0 + \beta_1 X$$

to these data. Using Equations (6.30)–(6.32), we obtain

$$\hat{Y} = -1.34 + 3.10X. \tag{6.34}$$

Substituting into Equation (6.34) gives the relevant predicted values.

Table 6.8 Histogram-valued regression variables.

w_u	Y = Hematocrit	X = Hemoglobin
w_1	{[33.29, 37.52), 0.6; [37.52, 39.61], 0.4}	{[11.54, 12.19), 0.4; [12.19, 12.80], 0.6}
w_2	{[36.69, 39.11), 0.3; [39.11, 45.12], 0.7}	{[12.07, 13.32), 0.5; [13.32, 14.17], 0.5}
w_3	{[36.69, 42.64), 0.5; [42.64, 48.68], 0.5}	{[12.38, 14.20), 0.3; [14.20, 16.16], 0.7}
w_4	{[36.38, 40.87), 0.4; [40.87, 47.41], 0.6}	{[12.38, 14.26), 0.5; [14.26, 15.29], 0.5}
w_5	{[39.19, 50.86]}	{[13.58, 14.28), 0.3; [14.28, 16.24], 0.7}
w_6	{[39.70, 44.32), 0.4; [44.32, 47.24], 0.6}	{[13.81, 14.50), 0.4; [14.50, 15.20], 0.6}
w_7	{[41.56, 46.65), 0.6; [46.65, 48.81], 0.4}	{[14.34, 14.81), 0.5; [14.81, 15.55], 0.5}
w_8	{[38.40, 42.93), 0.7; [42.93, 45.22], 0.3}	{[13.27, 14.00), 0.6; [14.00, 14.60], 0.4}
w_9	{[28.83, 35.55), 0.5; [35.55, 41.98], 0.5}	{[9.92, 11.98), 0.4; [11.98, 13.80], 0.6}
w_{10}	{[44.48, 52.53]}	{[15.37, 15.78), 0.3; [15.78, 16.75], 0.7}

Had the midpoint values of each histogram subinterval been used instead of classical values, Equations (6.3)–(6.6) would give

$$\hat{Y} = -2.16 + 3.16X. \tag{6.35}$$

□

As discussed in Section 6.2 for interval-valued observations, use of the subinterval midpoints alone has the effect of losing the internal variations within observations, whereas using the symbolic covariance and variance formula in Equation (6.30) incorporates both the internal variations and the variations between observations. This distinction is highlighted in the following example.

Example 6.8. For the hematocrit–hemoglobin histogram-valued data of Table 6.8, use of the full symbolic intervals gives a covariance value, from Equation (4.19), of

$$Cov(X, Y) = 5.699,$$

whereas using the subinterval midpoints gives a covariance value of

$$Cov(X, Y) = 4.328.$$

Thus, by using only the midpoints, some of the variation in the data is lost. □

Adding a range variable along the lines suggested by De Carvalho *et al.* (2004) and Lima Neto *et al.* (2005) for interval-valued data is not easily achieved for histogram-valued data, not least because there are several subintervals for each variable with these subintervals not necessarily the same for each observation. In the unlikely but particular case that the subintervals for the dependent variable were always the same for all observations with $s_u^y = s^*$ say, then a multivariate regression could perhaps be performed on $(Y_1, \ldots, Y_{s^*})$. How these issues are to be resolved remains an open problem.

6.5 Taxonomy Variables

Recall from Definition 2.12 that taxonomy variables are variables that impose on the data a tree-like structure. These arise by the nature of the dataset such as in Example 2.23 and Figure 2.1. They can arise by virtue of logical rules necessary to maintain data coherence after aggregation of an original dataset. The regression methodology of the previous sections is extended to include taxonomy variables as follows. For ease of presentation, we first assume that the predictor variables $\mathbf{X}$ and the dependent variable Y are all classical-valued variables. It is also assumed that all the predictor variables are taxonomy variables. We shall later add non-taxonomy predictor variables to the model (see Examples 6.12 and 6.13).

The taxonomy variables are typically categorical or multi-valued variables. Therefore, following the principles of Section 6.1.1, for each X_j taking possible values in $\mathcal{X}_j$, we code these possibilities as $X_j \equiv (X_{j1}, \ldots, X_{js_j})$ with X_{jk}, $k = 1, \ldots, s_j$, occurring with probability $p_{jk} = 1$, or 0, depending on whether $X_j = X_{jk}$ does, or does not, hold.

It is often the case that information is not known for all levels of the tree for all observations. That is, the value for a variable X_j is known for a given observation but its value for the immediately following lower branch X_{j+1} is not. Neither is it known for all subsequent lower levels of the tree; that is, $X_{j'}$, $j' = j+1, \ldots, t$, are unknown. When this occurs, the probability measures take the form

$$p_{j'k} = 1/b_{jk}, \quad j' = j+1, \ldots, t, \tag{6.36}$$

where b_{jk} is the number of branches that extend from X_{jk} to the level(s) j' below.

Example 6.9. Table 6.9 gives the observed values for a taxonomy tree relating to color (of birds, say). The tree structure can be seen in Figure 6.2.

There are $t = 2$ levels in the tree. At the top of the tree, the variable $X_1 =$ Tone takes possible values dark or light which are recoded to $X_1(w_u) \equiv (X_{11}(w_u), X_{12}(w_u))$, respectively. At the base of the tree, the variable $X_2 =$ Shade has possible values purple or red (if $X_1 =$ dark), or white or yellow (if $X_1 =$ light). The X_2 variable is recoded to $X_2 \equiv (X_{21}, \ldots, X_{24})$.

Table 6.9 Bird color.

Species	Density	Color	Species	Density	Color
w_1	230	red	w_{11}	305	red
w_2	301	dark	w_{12}	207	purple
w_3	373	yellow	w_{13}	242	white
w_4	280	white	w_{14}	345	white
w_5	281	dark	w_{15}	199	dark
w_6	374	red	w_{16}	248	red
w_7	330	yellow	w_{17}	311	yellow
w_8	215	purple	w_{18}	256	white
w_9	340	white	w_{19}	388	light
w_{10}	159	purple	w_{20}	260	red

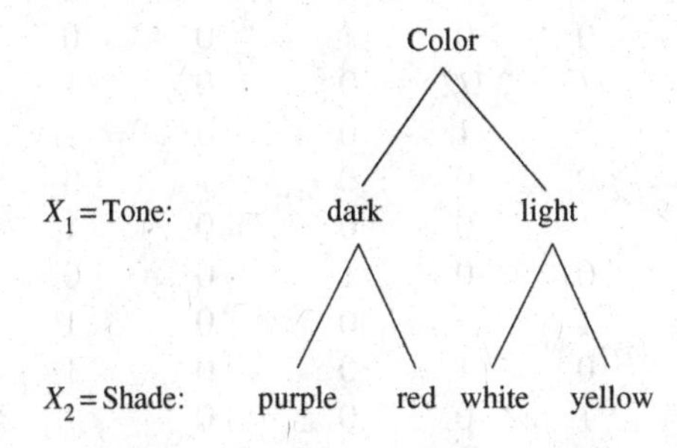

Figure 6.2 Color taxonomy tree.

Thus, for example, the third observation at $(w_3) = \{\text{yellow}\}$ tells us the taxonomy values at all levels of the tree, with $X_2 = \text{yellow}$ and hence $X_1 = \text{light}$. The recoded X_1 value is therefore

$$X_1(w_3) = (X_{11}(w_3), X_{12}(w_3)) = \{\text{dark}, 0; \text{light}, 1\}$$

and that for X_2 is

$$X_2(w_3) = (X_{21}(w_3), \ldots, X_{24})(w_3)) = \{\text{purple}, 0; \text{red}, 0; \text{white}, 0; \text{yellow}, 1\},$$

where we write $X_{jk} \equiv (X_{jk}, p_{jk})$ where p_{jk} is the probability that X_j takes the value X_{jk}. Thus, in this example, for the observation w_3, X_1 takes the value 'dark' with probability 0 and 'light' with probability 1.

In contrast, the observation w_5 is such that

$$X(w_5) = \{\text{dark}\},$$

i.e., we know its tone is dark but we do not know its shade. Therefore, the recoded X_2 variable becomes

$$X_2(w_5) = \{\text{purple}, 0.5; \text{red}, 0.5; \text{white}, 0; \text{yellow}, 0\}.$$

The complete set of weights after recoding is shown in Table 6.10.

Table 6.10 Weights color – birds.

Species	Level 1				Level 2	
	purple	red	white	yellow	dark	light
w_1	0	1	0	0	1	0
w_2	1/2	1/2	0	0	1	0
w_3	0	0	0	1	0	1
w_4	0	0	1	0	0	1
w_5	1/2	1/2	0	0	1	0
w_6	0	1	0	0	1	0
w_7	0	0	0	1	0	1
w_8	1	0	0	0	1	0
w_9	0	0	1	0	0	1
w_{10}	1	0	0	0	1	0
w_{11}	0	1	0	0	1	0
w_{12}	1	0	0	0	1	0
w_{13}	0	0	1	0	0	1
w_{14}	0	0	1	0	0	1
w_{15}	1/2	1/2	0	0	1	0
w_{16}	0	1	0	0	1	0
w_{17}	0	0	0	1	0	1
w_{18}	0	0	1	0	0	1
w_{19}	0	0	1/2	1/2	0	1
w_{20}	0	1	0	0	1	0

□

The regression analysis proceeds at each level of the tree using these recoded taxonomy variables as predictor variables in the model. Since these recoded variables are indicator variables, one (X_{jk}) in each set is dropped in order to invert the corresponding $(X'X)$ matrix.

Example 6.10. Table 6.9 also includes a dependent Y = Density (as a measure of prevalence) for each observation. There are $t = 2$ levels. Therefore, there will be two regression models. At the bottom of the tree, at level 1, we fit the model

$$Y^{(1)} = \beta_0^{(1)} + \beta_1^{(1)} X_{21} + \beta_2^{(1)} X_{22} + \beta_3^{(1)} X_{23}. \tag{6.37}$$

For these data, we obtain, from Equations (6.3)–(6.6),

$$\hat{Y}^{(1)} = 348.69 - 147.34(\text{purple}) - 60.62(\text{red}) - 49.68(\text{white}).$$

Notice that the X_{24} = yellow variable is only indirectly entered into this equation. In this case, each of X_{2k}, $k = 1, 2, 3$, takes the value 0, so that the predicted $\hat{Y} = 348.69$ when X_2 = yellow.

Likewise, at the second level, we fit the model

$$Y^{(2)} = \beta_0^{(2)} + \beta_1^{(2)} X_{11} \tag{6.38}$$

and obtain

$$\hat{Y}^{(2)} = 318.33 - 65.70(\text{dark}).$$ □

There can be more than one taxonomy variable with differing levels t, as the following example shows.

Example 6.11. A second taxonomy variable Region is added to that of Example 6.10; see Table 6.11 and Figure 6.3. There are $t = 3$ levels. At the

Table 6.11 Region taxonomy.

Species	Population type	
w_1	South Urban, ≥1m	pop6
w_2	South Rural, <50k	pop7
w_3	South Rural, ≥50k	pop8
w_4	South Rural, <50k	pop7
w_5	North Rural, ≥50k	pop4
w_6	North Urban, <1m	pop1
w_7	North Urban, ≥1m	pop2
w_8	North Rural	.
w_9	South Rural, <50k	pop7
w_{10}	North Rural, ≥50k	pop4
w_{11}	North Urban, <1m	pop1
w_{12}	North Rural, <50m	pop3
w_{13}	North Rural, <50m	pop3
w_{14}	North	.
w_{15}	North Urban	.
w_{16}	North	.
w_{17}	South Urban, <1m	pop5
w_{18}	South Urban	.
w_{19}	South Urban, <1m	pop5
w_{20}	South Rural, <50k	pop7

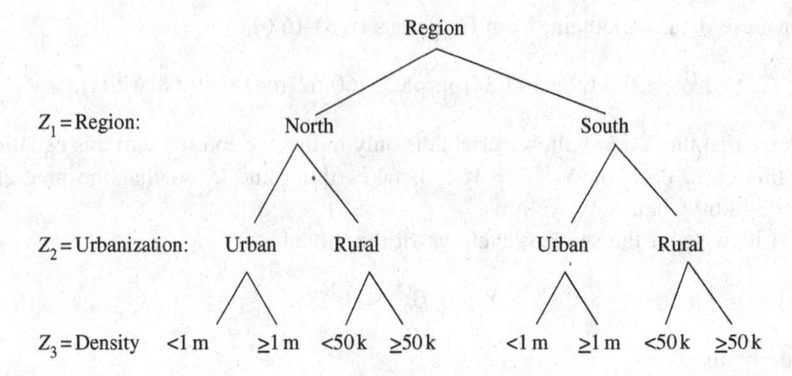

Figure 6.3 Region taxonomy tree.

top of the tree, define Z_1 = Region with two possible values, North or South. For each region, let Z_2 = Urbanization be divided into Urban or Rural branches; and at the base of the tree at level 1, let the population density be measured as $\{<1\text{m}, \geq 1\text{m}\}$ if it is an urban area and as $\{<50,000, \geq 50,000\}$ if it is a rural area. Let us refer to these eight possible populations at the base of this tree as 'pop$k$', $k = 1, \ldots, 8$.

The recoded variables are therefore $Z_1 \equiv (Z_{11}, Z_{12})$, $Z_2 \equiv (Z_{21}, \ldots, Z_{24})$, and $Z_3 \equiv (Z_{31}, \ldots, Z_{38})$ with $Z_{3k} \equiv \text{pop}k$, $k = 1, \ldots, 8$. Then, we have, for example,

$$Z(w_1) = \{\text{South Urban}, \geq 1\text{ m population}\}.$$

Since we know the Z_j values for all levels of tree, the recoded values or weights become for the observation w_1, e.g.,

$$Z_3(w_1) = \{\text{pop6}, 1; \text{pop}k, 0, k \neq 6\}.$$

For the observation w_{14}, however, we have only

$$Z_1(w_{14}) = \{\text{North}\}.$$

Here, both the Urbanization (Z_2) and population levels (Z_1) are unknown. Therefore, the recoded variables become

$$Z_3(w_{14}) = \{\text{pop}k, 1/4, k = 1, \ldots, 4; \quad \text{pop}k, 0, k = 5, \ldots, 8\}.$$

The complete set of weights for this Region taxonomy variable for the three levels are shown in Tables 6.12a, b, and c, respectively. The Color taxonomy

Table 6.12a Level 1 weights for region.

Species	pop1	pop2	pop3	pop4	pop5	pop6	pop7	pop8
w_1	0	0	0	0	0	1	0	0
w_2	0	0	0	0	0	0	1	0
w_3	0	0	0	0	0	0	0	1
w_4	0	0	0	0	0	0	1	0
w_5	0	0	0	1	0	0	0	0
w_6	1	0	0	0	0	0	0	0
w_7	0	1	0	0	0	0	0	0
w_8	0	0	1/2	1/2	0	0	0	0
w_9	0	0	0	0	0	0	1	0
w_{10}	0	0	0	1	0	0	0	0
w_{11}	1	0	0	0	0	0	0	0
w_{12}	0	0	1	0	0	0	0	0
w_{13}	0	0	1	0	0	0	0	0
w_{14}	1/4	1/4	1/4	1/4	0	0	0	0
w_{15}	1/2	1/2	0	0	0	0	0	0
w_{16}	1/4	1/4	1/4	1/4	0	0	0	0
w_{17}	0	0	0	0	1	0	0	0
w_{18}	0	0	0	0	1/2	1/2	0	0
w_{19}	0	0	0	0	1	0	0	0
w_{20}	0	0	0	0	0	0	1	0

Table 6.12b, c Level 2, 3 weights for region.

Species	(b) Level 2				(c) Level 3	
	North Urban	North Rural	South Urban	South Rural	North	South
w_1	0	0	1	0	0	1
w_2	0	0	0	1	0	1
w_3	0	0	0	1	0	1
w_4	0	0	0	1	0	1
w_5	0	1	0	0	1	0
w_6	1	0	0	0	1	0
w_7	1	0	0	0	1	0
w_8	0	1	0	0	1	0
w_9	0	0	0	1	0	1
w_{10}	0	1	0	0	1	0
w_{11}	1	0	0	0	1	0
w_{12}	0	1	0	0	1	0
w_{13}	0	1	0	0	1	0
w_{14}	1/2	1/2	0	0	1	0
w_{15}	1	0	0	0	1	0
w_{16}	1/2	1/2	0	0	1	0
w_{17}	0	0	1	0	0	1
w_{18}	0	0	1	0	0	1
w_{19}	0	0	1	0	0	1
w_{20}	0	0	1	0	0	1

variable weights are as given in Table 6.10. Using these recoded variables, a regression model is fitted at each level of the tree.

The regression at the bottom of the tree takes the X_2 weights for Color and the Z_3 weights for Region, i.e., it takes level 1 values for all taxonomy variables. Thus, we fit

$$Y^{(1)} = \beta_0^{(1)} + \sum_{k=1}^{3} \beta_{1k}^{(1)} X_{2k} + \sum_{k=1}^{7} \beta_{2k}^{(1)} Z_{3k}.$$

For these data, the fitted model at level 1 is, from Equation (6.3) with $p = 10$,

$$\begin{aligned}\hat{Y}^{(1)} =& 373.00 - 146.05(\text{purple}) - 87.65(\text{red}) - 55.21(\text{white}) \\ &+ 45.07(\text{pop1}) - 61.17(\text{pop2}) - 40.77(\text{pop3}) - 14.45(\text{pop4}) \\ &- 15.02(\text{pop5}) - 65.99(\text{pop6}) - 0.98(\text{pop7}). \end{aligned} \quad (6.39)$$

The regression model at level 2 fits the X_1 Color variables and the Z_2 Region variables, i.e., we fit

$$Y^{(2)} = \beta_0^{(2)} + \beta_{11}^{(2)} X_{11} + \sum_{k=1}^{3} \beta_{2k}^{(2)} Z_{2k}.$$

For these data, the regression model is, from Equation (6.3) with $p = 4$,

$$\begin{aligned}\hat{Y}^{(2)} =& 334.11 - 58.28(\text{dark}) + 15.49(\text{North Urban}) \\ &- 63.57(\text{North Rural}) - 23.29(\text{South Urban}). \end{aligned} \quad (6.40)$$

Finally, we fit a regression model to the top level of the trees. Here, we take the X_1 Color variables and the Z_1 Region variables. Therefore, we fit

$$Y^{(3)} = \beta_0^{(3)} + \beta_{11}^{(3)} X_{21} + \beta_{21}^{(3)} Z_{11},$$

which becomes, from Equation (6.3) with $p = 2$,

$$\hat{Y}^{(3)} = 324.00 - 59(\text{dark}) - 17(\text{North}). \quad (6.41)$$

□

Non-taxonomy variables can be included in the regression model. Further, these variables can be classical or symbolic variables. It is now clear that fitting regression models to predictor variables that include both taxonomy and symbolic-valued variables is executed by simultaneously applying the methods described above (for the taxonomy variables) and those in Sections 6.2–6.4 for symbolic-valued and histogram-valued ones, respectively. This is illustrated in the following three examples.

Example 6.12. Suppose the Color taxonomy tree and Example 6.9 pertain, and suppose the dependent variable Y = Density takes the interval-valued observations of Table 6.13.

Table 6.13 Taxonomy regression: interval-valued variables.

w_u	Y = Density	X_3 = Size	w_u	Y = Density	X_3 = Size
w_1	[225, 235]	[16, 18]	w_{11}	[300, 310]	[26, 28]
w_2	[299, 303]	[23, 28]	w_{12}	[206, 208]	[14, 17]
w_3	[368, 378]	[24, 33]	w_{13}	[238, 246]	[16, 19]
w_4	[270, 290]	[22, 25]	w_{14}	[340, 350]	[28, 33]
w_5	[277, 285]	[20, 24]	w_{15}	[196, 202]	[11, 16]
w_6	[371, 377]	[31, 37]	w_{16}	[244, 252]	[17, 19]
w_7	[322, 338]	[29, 33]	w_{17}	[305, 317]	[26, 31]
w_8	[214, 216]	[15, 19]	w_{18}	[255, 257]	[18, 19]
w_9	[335, 345]	[27, 31]	w_{19}	[380, 396]	[29, 36]
w_{10}	[155, 163]	[12, 15]	w_{20}	[258, 262]	[19, 24]

Recoding the taxonomy variables $X_1 \equiv (X_{11}, X_{12})$ and $X_2 \equiv (X_{21}, \ldots, X_{24})$, we obtain the weights of Table 6.10. We use Equation (4.16) to obtain the elements of the $\boldsymbol{X'X}$ and $\boldsymbol{X'Y}$ matrices, and follow the procedure described above for the taxonomy tree levels.

Hence, at the base of the tree (level = 1), we obtain from Equation (6.3) with $p = 3$,

$$\hat{Y}^{(1)} = 348.71 - 147.26(\text{purple}) - 60.63(\text{red}) - 49.68(\text{white}).$$

At the top of the tree (level = 2) we obtain

$$\hat{Y}^{(2)} = 318.35 - 65.71(\text{dark}).$$

We observe the similarity between the regression equations $\hat{Y}^{(1)}$ and $\hat{Y}^{(2)}$ obtained here with those obtained in Example 6.10. We also observe that the single classical Y values used in Example 6.10 are in fact the midpoints of the intervals used in this Example 6.12; compare Table 6.9 and Table 6.13. □

Example 6.13. Let us add the second taxonomy tree Region used in Example 6.10, and also add a non-taxonomy interval-valued random variable X_3 = Size, as given in Table 6.13. The weights for the Region taxonomy tree are as shown in Table 6.12. Overall, there are now three taxonomy levels to be considered.

At the base of the tree, we fit

$$Y^{(1)} = \beta_0^{(1)} + \sum_{k=1}^{3} \beta_{1k}^{(1)} X_{2k} + \sum_{k=1}^{7} \beta_{2k}^{(1)} Z_{3k} + \beta_3^{(1)} X_3. \tag{6.42}$$

For these data, from Equation (6.3) with $p = 11$, we find

$$\begin{aligned}\hat{Y}^{(1)} &= 86.93 + 60.19(\text{purple}) + 88.83(\text{red}) + 70.43(\text{white}) - 137.41(\text{pop1})\\ &\quad - 63.35(\text{pop2}) - 90.80(\text{pop3}) - 108.42(\text{pop4}) - 56.41(\text{pop5})\\ &\quad - 111.68(\text{pop6}) - 114.60(\text{pop7}) + 9.86X_3.\end{aligned}$$

At level = 2, we fit

$$Y^{(2)} = \beta_0^{(2)} + \beta_{11}^{(2)} X_{21} + \sum_{k=1}^{3} \beta_{2k}^{(2)} Z_{2k} + \beta_3^{(2)} X_3. \tag{6.43}$$

Then, from Equation (6.3) with $p = 5$, we obtain

$$\begin{aligned}\hat{Y}^{(2)} &= 85.24 - 4.92(\text{dark}) - 10.04(\text{North Urban}) - 8.22(\text{North Rural})\\ &\quad - 1.86(\text{South Urban}) + 8.84X_3.\end{aligned}$$

At the top of the tree, we fit

$$Y^{(3)} = \beta_0^{(3)} + \beta_{11}^{(3)} X_{21} + \beta_{21}^{(3)} Z_{11} + \beta_3^{(3)} X_3 \tag{6.44}$$

to give, from Equation (6.3) with $p = 3$,

$$\hat{Y}^{(3)} = 85.34 - 5.25(\text{dark}) - 8.04(\text{North}) + 8.80X_3.$$ □

Example 6.14. Let us take the same two taxonomy trees Color and Region as in Example 6.9 and Example 6.13 with respective level weights as given in Table 6.10 and Table 6.12. Let us suppose that the dependent random variable Y = Density and the non-taxonomy variable X_3 = Size have the histogram-valued observations of Table 6.14.

The components of the $\boldsymbol{X}'\boldsymbol{X}$ and $\boldsymbol{X}'\boldsymbol{Y}$ matrices are now calculated from Equation (4.19). At level = 1, we fit a regression of the form of Equation (6.42), to obtain

$$\begin{aligned}\hat{Y}^{(1)} &= 86.02 + 56.66(\text{purple}) + 82.40(\text{red}) + 68.19(\text{white}) - 130.44(\text{pop1})\\ &\quad - 63.41(\text{pop2}) - 89.30(\text{pop3}) - 105.86(\text{pop4}) - 56.55(\text{pop5})\\ &\quad - 107.33(\text{pop6}) - 114.30(\text{pop7}) + 9.44X_3.\end{aligned}$$

Table 6.14 Taxonomy regression: histogram-valued variables.

w_u	Y = Density	X_3 = Size
w_1	{[225, 226), 0.2; [226, 231), 0.5; [231, 235], 0.3}	{[16, 17), 0.5; [17, 18], 0.5}
w_2	{[299, 301), 0.5; [301, 303], 0.5}	{[23, 26), 0.4; [26, 28], 0.6}
w_3	{[368, 370), 0.1; [370, 374), 0.5; [374, 376), 0.2; [376, 378], 0.2}	{[24, 29), 0.5; [29, 33], 0.5}
w_4	{[270, 277), 0.2; [277, 283), 0.6; [283, 290], 0.2}	{[22, 23), 0.4; [23, 25], 0.6}
w_5	{[277, 280), 0.3; [280, 284), 0.5; [284, 285], 0.2}	{[20, 22), 0.5; [22, 24], 0.5}
w_6	{[371, 374), 0.5; [374, 377], 0.5}	{[31, 34), 0.6; [34, 37], 0.4}
w_7	{[322, 326), 0.2; [326, 330), 0.2; [330, 334), 0.4; [334, 338], 0.2}	{[29, 31), 0.5; [31, 33], 0.5}
w_8	{[214, 216], 1.0}	{[15, 17), 0.5; [17, 19], 0.5}
w_9	{[335, 340), 0.4; [340, 345], 0.6}	{[27, 29), 0.4; [29, 31], 0.6}
w_{10}	{[155, 157), 0.3; [157, 161), 0.4; [161, 163], 0.3}	{[12, 13), 0.3; [13, 15], 0.7}
w_{11}	{[300, 302), 0.1; [302, 308), 0.8; [308, 310], 0.1}	{[26, 28], 1.0}
w_{12}	{[206, 208], 1.0}	{[14, 16), 0.6; [16, 17], 0.4}
w_{13}	{[238, 242), 0.5; [242, 246], 0.5}	{[16, 17), 0.3; [17, 19], 0.7}
w_{14}	{[340, 343), 0.3; [343, 347), 0.4; [347, 350), 0.3}	{[28, 30), 0.5; [30, 33], 0.5}
w_{15}	{[196, 199), 0.5; [199, 202], 0.5}	{[11, 13), 0.4; [13, 16], 0.6}
w_{16}	{[244, 246), 0.1; [246, 250), 0.8; [250, 252], 0.1}	{[17, 19], 1.0}
w_{17}	{[305, 308), 0.2; [308, 314), 0.6; [314, 317], 0.2}	{[26, 29), 0.7; [29, 31], 0.3}
w_{18}	{[255, 257], 1.0}	{[18, 19], 1.0}
w_{19}	{[380, 383), 0.1; [383, 388), 0.4; [388, 393), 0.3; [393, 396], 0.2}	{[29, 33), 0.5; [33, 36], 0.5}
w_{20}	{[258, 260), 0.5; [260, 262], 0.5}	{[19, 22), 0.4; [22, 24], 0.6}

At the second level, we fit a regression of the form of Equation (6.43) to give

$$\hat{Y}^{(2)} = 83.29 - 6.89(\text{dark}) - 8.09(\text{North Urban}) - 7.28(\text{North Rural}) - 2.40(\text{South Urban}) + 8.93X_3;$$

and at the top of the tree, we fit Equation (6.44) to obtain

$$\hat{Y}^{(3)} = 85.40 - 5.25(\text{dark}) - 8.04(\text{North}) + 8.80X_3. \quad \square$$

6.6 Hierarchical Variables

Recall from Definition 2.13 that for a hierarchical tree structure which particular variables (the daughter variables) are present depends on the values obtained from the mother variable at the previous level of the tree. The difference between a taxonomy and hierarchy structure can be illustrated in the following example.

Example 6.15. Figure 6.4 represents a hierarchy tree where at the top of the tree the random variable X_1 is the question 'What am I holding in my hand?' The possible answers are $X_1 = \text{bird}$ or $X_1 = \text{coin}$. At the next level the question then is 'If $X_1 = \text{bird}$, what is its color X_2?' with possible answers on $\mathcal{X}_2 = \{\text{purple}, \text{red}\}$, or 'If $X_1 = \text{coin}$, what is its face X_3?' with possible answers in $\mathcal{X}_3 = \{\text{head}, \text{tail}\}$.

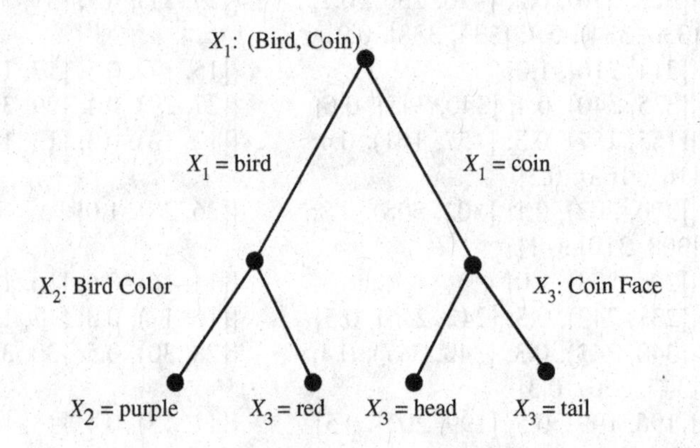

Figure 6.4 Hierarchy – bird or coin.

If we compare this structure to that of Figure 6.2 giving a taxonomy for bird color, at the base of the tree there is a color regardless of which branch was followed. In contrast, the base of the hierarchy tree has a color variable only when we sit on the branch that says we have a bird. □

Regression analysis starts by fitting the dependent variable Y to the predictor variables X at the top of the tree and then successively moves down by fitting the relevant predictor variables X to the residuals obtained at the previous level. At each level, we fit the respective daughter variables, in turn. The final complete regression model is a single equation containing a series of indicator subcomponents relating to which hierarchy branch pertains. Thus, for the structure of Figure 6.4 and Example 6.15, the regression model has the form

$$Y = \beta_0 + \beta_1 X_1 + \delta(X_1 = \text{bird})(\beta_{01} + \beta_{11} X_2) + \delta(X_1 = \text{coin})(\beta_{02} + \beta_{12} X_3) \quad (6.45)$$

where $\delta(z) = 1$ or 0 if z is true or not true.

We illustrate the methodology first with classical-valued variables and then with symbolic-valued variables. For symbolic variables, the methodology uses both the classical hierarchy techniques and that from Sections 6.2–6.4 for symbolic-valued data.

Example 6.16. The dataset of Table 6.15 represents outcomes for a dependent variable Y and five predictor variables $X_1, \ldots, X_5$ which are defined by the hierarchy tree of Figure 6.5. At the top of the tree $X_1 = \{a, b, c\}$. If $X_1 = a$, then the daughter variable is X_2; if $X_1 = b$, the daughter variable is X_5. Notice that if $X_1 = c$, there are no branches (i.e., no daughter variables).

Table 6.15 Hierarchy dataset.

w_u	Y	X_1	X_2	X_3	X_4	X_5
w_1	415	a	$a1$	$a11$	NA	NA
w_2	390	a	$a1$	$a11$	NA	NA
w_3	381	a	$a1$	$a11$	NA	NA
w_4	385	a	$a1$	$a11$	NA	NA
w_5	324	a	$a1$	$a12$	NA	NA
w_6	355	a	$a1$	$a12$	NA	NA
w_7	260	a	$a2$	NA	10	NA
w_8	255	a	$a2$	NA	12	NA
w_9	244	a	$a2$	NA	12	NA
w_{10}	200	a	$a2$	NA	20	NA
w_{11}	238	a	$a2$	NA	15	NA
w_{12}	182	b	NA	NA	NA	40
w_{13}	165	b	NA	NA	NA	50
w_{14}	158	b	NA	NA	NA	55
w_{15}	150	b	NA	NA	NA	60
w_{16}	139	b	NA	NA	NA	70
w_{17}	120	b	NA	NA	NA	80
w_{18}	35	c	NA	NA	NA	NA
w_{19}	45	c	NA	NA	NA	NA
w_{20}	50	c	NA	NA	NA	NA

We first fit the X_1 variable at the top of the tree to the Y variable, i.e., we fit

$$Y = \beta_0 + \beta_1 a + \beta_2 b + e \tag{6.46}$$

where $a = 1$ (0) if $X_1 = a$ ($\neq a$) and likewise $b = 1$ (0) if $X_1 = b$ ($\neq b$). Since the a and b are indicator variables, the $X_1 = c$ option does not appear explicitly in Equation (6.43); it is present indirectly with $a = b = 0$ when $X_1 = c$. The resulting regression model is found to be

$$\hat{Y} = 43.33 + 270.03a + 109.00b. \tag{6.47}$$

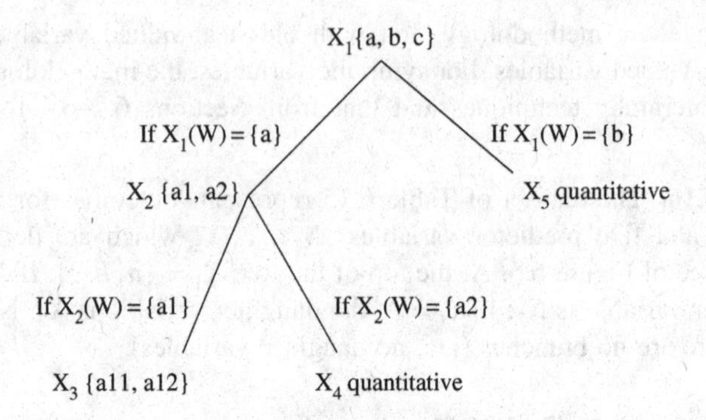

Figure 6.5 Hierarchically dependent variables.

Substituting in Equation (6.44), we obtain the predicted values $\hat{Y}$ and the corresponding residuals

$$R_1 = Y - \hat{Y}.$$

Therefore, for the first observation

$$\hat{Y}(w_1) = 43.33 + 270.03(1) + 109.00(0) = 313.36;$$

hence

$$R_1(w_1) = 415 - 313.36 = 101.64.$$

For the $u = 20$th observation,

$$\hat{Y}(w_{20}) = 43.33 + 270.03(0) + 109.00(0) = 43.33,$$

and hence

$$R_1(w_{20}) = 50 - 43.33 = 6.67.$$

The complete set of predicted values $\hat{Y}$ and residuals R_1, at the top of the tree from fitting X_1, is shown in Table 6.16.

The next step is to fit the daughter variables of the mother variable X_1 (here X_2 and X_5) of the hierarchy tree to the residuals R_1. We first fit X_2 to the residuals R_1 for the observations w_u, $u = 1, \ldots, 11$. That is, we fit

$$R_1 = \beta_0 + \beta_1 a1 + e \tag{6.48}$$

where, with $X_2 = a1$ or $a2$, being indicator variables, only one of $a1$ or $a2$ is needed in Equation (6.45). The analysis gives

$$\hat{R}_1 = -73.96 + 135.60a1. \tag{6.49}$$

Table 6.16 Hierarchy regression fits.

w_u		Fit of X_1		Fit of X_2 and X_5		Fit of X_3 and X_4		Final fit	
	Y	$\hat{Y}$	R_1	$\hat{R}_1$	R_2	$\hat{R}_2$	R_3	Y^*	R
w_1	415	313.4	101.6	61.6	40.0	17.75	22.25	392.75	22.25
w_2	390	313.4	76.6	61.6	15.0	17.75	−2.75	392.75	−2.75
w_3	381	313.4	67.6	61.6	6.0	17.75	−11.75	392.75	−11.75
w_4	385	313.4	71.6	61.6	10.0	17.75	−7.75	392.75	−7.75
w_5	324	313.4	10.6	61.6	−51.0	−35.5	−15.5	339.50	−15.50
w_6	355	313.4	41.6	61.6	−20.0	−35.5	15.5	339.50	15.50
w_7	260	313.4	−53.4	−74.0	20.6	22.5	−1.9	261.90	−1.90
w_8	255	313.4	−58.4	−74.0	15.6	10.7	4.9	250.04	4.96
w_9	244	313.4	−69.4	−74.0	4.6	10.7	−6.1	250.04	−6.04
w_{10}	200	313.4	−113.4	−74.0	−39.4	−36.8	−2.6	202.60	−2.60
w_{11}	238	313.4	−75.4	−74.0	−1.4	−7.1	5.7	232.25	5.75
w_{12}	182	152.3	29.7	28.7	1.0	28.7	1.0	180.91	1.09
w_{13}	165	152.3	12.7	13.7	−1.1	13.7	−1.1	165.91	−0.91
w_{14}	158	152.3	5.7	6.2	−0.6	6.2	−0.6	158.41	−0.41
w_{15}	150	152.3	−2.3	−1.2	−1.1	−1.2	−1.1	150.91	−0.91
w_{16}	139	152.3	−13.3	−16.2	2.9	−16.2	2.9	135.91	3.09
w_{17}	120	152.3	−32.3	−31.2	−1.1	−31.2	−1.1	120.91	−0.91
w_{18}	35	43.3	−8.3	43.3	−8.3	43.3	−8.3	43.33	−8.33
w_{19}	45	43.3	1.7	43.3	1.7	43.3	1.7	43.33	1.67
w_{20}	50	43.3	6.7	43.3	6.7	43.3	6.7	43.33	6.67

Hence, the predicted $\hat{R}_1$ values and the new residuals $R_2 = R_1 - \hat{R}_1$ can be found from Equation (6.46). For example, for observation w_6,

$$\hat{R}_1(w_6) = -73.96 + 135.60(1) = 61.6,$$

and hence

$$R_2(w_6) = 41.6 - 61.6 = -20.0.$$

The other branch at this level of the tree involves X_5. Therefore, a regression model with X_5 as the predictor variable and R_1 as the dependent variable is fitted using the observations w_u, $u = 12, \ldots, 17$. That is, we fit the model

$$R_1 = \beta_0 + \beta_1 X_5 + e \tag{6.50}$$

to these observations. Note that this X_5 is not an indicator variable. We obtain

$$R_1 = 88.55 - 1.50X_5. \tag{6.51}$$

Again, we calculate the predicted residual $\hat{R}_1$ and the corresponding residual $R_2 = R_1 - \hat{R}_1$. For example, for observation w_{12}, from Equation (6.48),

$$\hat{R}_1(w_{12}) = 88.55 - 1.50(40) = 28.7,$$

and hence

$$R_2(w_{12}) = 29.7 - 28.7 = 1.0.$$

The complete set of fitted residuals $\hat{R}_1$ and their residuals R_2 at this second level of the hierarchy tree is provided in Table 6.16.

The next step is to proceed to the daughter variables at the immediately succeeding level of the hierarchy tree, here X_3 and X_4. Thus, we want to fit the models

$$R_2 = \beta_{03} + \beta_{13}a11 + e \tag{6.52}$$

to the observations w_u, $u = 1, \dots, 6$; and

$$R_2 = \beta_{04} + \beta_{14}X_4 + e \tag{6.53}$$

to the observations w_u, $u = 7, \dots, 11$. We obtain

$$\hat{R}_2 = -35.50 + 53.25a11 \tag{6.54}$$

and

$$\hat{R}_2 = 81.85 - 5.93X_4, \tag{6.55}$$

respectively. The corresponding predicted residuals $\hat{R}_2$ and their residuals $R_3 = R_2 - \hat{R}_2$ are shown in Table 6.16.

Finally, the complete regression model obtained from fitting these data to these hierarchical predictor variables is, from Equations (6.47), (6.49), (6.51), (6.54), and (6.55),

$$\begin{aligned} Y^* =& 43.33 + 270.03a + 109.00b + \delta(X_1 = a)(-73.96 + 135.60a1) \\ &+ \delta(X_1 = b)(88.55 - 1.50X_5) + \delta(X_2 = a1)(-35.50 + 53.25a11) \\ &+ \delta(X_2 = a2)(81.85 - 5.93X_4). \end{aligned} \tag{6.56}$$

Then, by substituting the relevant $X_1, \dots, X_5$ values into Equation (6.56), we obtain the final predicted value Y^* and the corresponding residual $R = Y - Y^*$. For example, for observation w_5,

$$\begin{aligned} Y^*(w_5) =& 43.33 + 270.03(1) + 109.00(0) + (1)[-73.96 + 135.60(1)] \\ &+ (0)[88.55 - 1.50(\text{NA})] + (1)[-35.00 + 53.25(0)] \\ &+ (0)[81.85 - 5.93(\text{NA})] \\ =& 339.5, \end{aligned}$$

and hence

$$R(w_5) = 324 - 339.5 = -15.5.$$

The complete set of predicted values Y^* and residuals R is given in Table 6.16. □

We now apply this technique to symbolic-valued variables. Example 6.17 considers interval-valued variables for the hierarchy tree. This is followed by Example 6.18 which considers histogram-valued variables and also adds another non-hierarchy predictor variable.

Example 6.17. Consider the data of Table 6.17 where the dependent variable Y = Sales, X_4 = Volume and X_5 = Costs are interval-valued variables, and where X_1, X_2, X_3 are as in Example 6.15 with X_3 being a classical-valued variable taking values in $\mathcal{Y}_3 = \{a11, a12\}$ and with X_1 and X_2 being hierarchy-tree variables as displayed in Figure 6.5.

Table 6.17 Hierarchy: interval-valued variables.

w_u	Y	X_1	X_2	X_3	X_4	X_5
w_1	[410, 420]	a	$a1$	$a11$	NA	NA
w_2	[388, 392]	a	$a1$	$a11$	NA	NA
w_3	[378, 384]	a	$a1$	$a11$	NA	NA
w_4	[381, 389]	a	$a1$	$a11$	NA	NA
w_5	[319, 330]	a	$a1$	$a12$	NA	NA
w_6	[352, 358]	a	$a1$	$a12$	NA	NA
w_7	[255, 265]	a	$a2$	NA	[9, 11]	NA
w_8	[251, 259]	a	$a2$	NA	[10, 14]	NA
w_9	[240, 248]	a	$a2$	NA	[11, 13]	NA
w_{10}	[197, 203]	a	$a2$	NA	[17, 23]	NA
w_{11}	[232, 244]	a	$a2$	NA	[13, 17]	NA
w_{12}	[180, 184]	b	NA	NA	NA	[38, 42]
w_{13}	[162, 168]	b	NA	NA	NA	[47, 53]
w_{14}	[156, 160]	b	NA	NA	NA	[50, 60]
w_{15}	[149, 151]	b	NA	NA	NA	[58, 62]
w_{16}	[136, 142]	b	NA	NA	NA	[66, 74]
w_{17}	[119, 121]	b	NA	NA	NA	[78, 82]
w_{18}	[33, 37]	c	NA	NA	NA	NA
w_{19}	[41, 49]	c	NA	NA	NA	NA
w_{20}	[48, 52]	c	NA	NA	NA	NA

Then, as in Example 6.16, we fit a regression to the variable X_1 at the top of the tree, i.e., Equation (6.4), where now the interval-valued counterparts of the components of the $\boldsymbol{X}'\boldsymbol{X}$ and $\boldsymbol{X}'\boldsymbol{Y}$ matrices are used through Equation (4.16). We obtain

$$\hat{Y}_1 = 43.29 + 270.16a + 109.00b. \tag{6.57}$$

We can then calculate the predictor $\hat{Y}$ from Equation (6.57) and hence the residuals. Since X_1 takes classical values, the predicted $\hat{Y}$ will be classical valued. However, since the observed Y is interval-valued, then the residuals $R_1 = Y - \hat{Y}$ will be interval-valued. For example, for the first observation,

$$\hat{Y}(w_1) = 43.29 + 270.16(1) + 109.00(0) = 313.4.$$

Hence, the residual is the interval

$$R_1(w_1) = (R_{1a}, R_{1b}) = [96.6, 106.6]$$

since

$$R_{1a} = 410 - 313.4 = 96.6,$$

$$R_{1b} = 420 - 313.4 = 106.6.$$

The complete set of residuals R_1 obtained by fitting this X_1 variable at the top of the hierarchy tree is shown in Table 6.18. Note that for observations w_{18}, w_{19}, w_{20}, the predictor $\hat{Y}$ is

$$\hat{Y} = 43.29 + 270.16(0) + 109.00(0) = 43.29.$$

We now proceed to the second level of the hierarchy tree and fit the daughter variables X_2 and X_5 in turn to the residuals R_1. To fit X_2, we calculate the regression equation in the form of Equation (6.48) where now the components of the $\boldsymbol{X}'\boldsymbol{X}$ and $\boldsymbol{X}'\boldsymbol{Y}$ matrices are calculated using the symbolic form of Equation (4.16), on the observations $w_1, \ldots, w_{11}$ only. Thus, we obtain the regression equation

$$\hat{R}_1 = -74.09 + 135.83a1. \tag{6.58}$$

We can now obtain the residuals $R_2 = R_1 - \hat{R}_1$, where $\hat{R}_1$ is obtained from Equation (6.58).

For example, for w_1, we have

$$\hat{R}_1(w_1) = -74.09 + 135.83(1) = 61.7.$$

Hence, the residual is the interval

$$R_2(w_1) = [R_{2a}, R_{2b}] = [34.8, 44.8]$$

where

$$R_{2a} = 96.55 - 61.74 = 34.8$$

$$R_{2b} = 106.55 - 61.74 = 44.8.$$

Table 6.18 Hierarchy: residuals.

w_u	Y	Fit of X_1		Fit of X_2 and X_5		Fit of X_3 and X_4	
		$\hat{Y}$	R_1	$\hat{R}_1$	R_2	$\hat{R}_2$	R_3
w_1	[410, 420]	313.4	[96.6, 106.6]	61.8	[34.8, 44.8]	18.2	[20.7, 26.7]
w_2	[388, 392]	313.4	[74.6, 78.6]	61.8	[12.8, 16.8]	18.2	[−5.3, −1.3]
w_3	[378, 384]	313.4	[64.6, 70.6]	61.8	[2.8, 8.8]	18.2	[−15.3, −9.3]
w_4	[381, 389]	313.4	[67.6, 75.6]	61.8	[5.8, 13.8]	18.2	[−12.3, −4.3]
w_5	[319, 330]	313.4	[5.6, 16.6]	61.8	[−56.2, −45.2]	−35.6	[−20.6, −9.6]
w_6	[352, 358]	313.4	[38.6, 44.6]	61.8	[−23.2, −17.2]	−35.6	[12.4, 18.4]
w_7	[255, 265]	313.4	[−58.4, −48.4]	−74.1	[15.7, 25.7]	21.8	[−6.1, 3.9]
w_8	[251, 259]	313.4	[−62.4, −54.4]	−74.1	[11.7, 19.7]	10.4	[1.3, 9.3]
w_9	[240, 248]	313.4	[−73.4, −65.4]	−74.1	[0.7, 8.7]	10.4	[−9.7, −1.7]
w_{10}	[197, 203]	313.4	[−116.4, −110.4]	−74.1	[−42.3, −36.3]	−35.3	[−7.0, −1.0]
w_{11}	[232, 244]	313.4	[−81.4, −69.4]	−74.1	[−7.3, 4.7]	−6.8	[−0.5, 11.5]
w_{12}	[180, 184]	152.3	[27.7, 31.7]	28.4	[−0.8, 3.2]	28.4	[−0.8, 3.2]
w_{13}	[162, 168]	152.3	[9.7, 15.7]	13.6	[−3.9, 2.1]	13.6	[−3.9, 2.1]
w_{14}	[156, 160]	152.3	[3.7, 7.7]	6.2	[−2.5, 1.5]	6.2	[−2.5, 1.5]
w_{15}	[149, 151]	152.3	[−3.3, −1.3]	−1.2	[−2.1, −0.1]	−1.2	[−2.1, −0.1]
w_{16}	[136, 142]	152.3	[−16.3, −10.3]	−16.1	[−0.3, 5.7]	−16.1	[−0.3, 5.7]
w_{17}	[119, 121]	152.3	[−33.3, −31.3]	−30.9	[−2.4, −0.4]	−30.9	[−2.4, −0.4]
w_{18}	[33, 37]	43.3	[−10.3, −6.3]	43.3	[−10.3, −6.3]	43.4	[−10.4, −6.4]
w_{19}	[41, 49]	43.3	[−2.3, 5.7]	43.3	[−2.3, 5.7]	43.4	[−2.4, 5.6]
w_{20}	[48, 52]	43.3	[4.7, 8.7]	43.3	[4.7, 8.7]	43.4	[4.6, 8.6]

Likewise we obtain all R_2 residuals for $w_1, \ldots, w_{11}$ as shown in Table 6.18.

We fit the daughter variable X_5 to the residuals R_1 using the observations $w_{12}, \ldots, w_{17}$. We obtain

$$\hat{R}_1 = 87.81 - 1.48X_5. \tag{6.59}$$

Again, the residuals $R_2 = R_1 - \hat{R}_1$ are calculated from Equation (6.59). Since X_5 is interval-valued the value used in this prediction is taken to be the midpoint value. For example, for the observation w_{12}, we have

$$\hat{R}_1(w_{12}) = 87.81 - 1.48(38+42)/2 = 28.5.$$

Hence,

$$R_2 = (R_{2a}, R_{2b}) = [-0.8, 3.2]$$

where

$$R_{2a} = 27.7 - 28.5 = -0.8$$
$$R_{2b} = 31.7 - 28.5 = 3.2.$$

The residuals R_2 for $w_{12}, \ldots, w_{17}$ after fitting X_5 are shown in Table 6.18.

The residuals R_2 for the remaining observations $w_{18}, \ldots, w_{20}$ are unchanged from R_1, since the X_2 and X_5 variables are not applicable (NA) here.

At the third level, we fit the daughter variables X_3 and X_4. Thus, using observations $w_1, \ldots, w_6$, we fit X_3 to the residuals R_2, and obtain

$$\hat{R}_2 = -35.57 + 53.74a11. \tag{6.60}$$

Hence, we obtain the residuals R_3 as shown in Table 6.18. Likewise, using observations $w_7, \ldots, w_{11}$, we fit X_4 to the residuals R_2, and obtain

$$\hat{R}_2 = 78.91 - 5.71X_4. \tag{6.61}$$

The residuals R_3 after fitting X_4 are given in Table 6.18.

Therefore, the complete regression equation is obtained by combining Equations (6.57)–(6.61). We have

$$\begin{aligned}\hat{Y} = 43.29 + 270.16a + 108.99b + \delta(X_1 = a)(-74.09 + 135.83a1)\\ + \delta(X_1 = b)(87.81 - 1.48X_5) + \delta(X_2 = a1)(-35.57 + 53.74a11)\\ + \delta(X_2 = a2)(78.91 - 5.71X_4).\end{aligned} \tag{6.62}$$

Substituting the $X_1, \ldots, X_5$ values from Table 6.17 into Equation (6.62), we obtain the final predicted value $\hat{Y}$.

In Table 6.19 column (i), the predicted Y value, $\hat{Y}$, is calculated by using the midpoint value for each of the $X_1, \ldots, X_5$ variables. For example, from Equation (6.62),

$$\begin{aligned}\hat{Y}(w_8) &= 43.29 + 270.16(1) + 108.99(0) + (1)[-74.09 + 135.83(0)]\\ &\quad + (1)[87.81 - 1.48(10 + 14)/2]\\ &= 249.71.\end{aligned}$$

Hence,

$$R_b(w_8) = 251 - 249.71 = 1.29$$

$$R_a(w_8) = 259 - 249.71 = 9.29,$$

to give the residual interval

$$R(w_8) = [1.29, 9.29].$$

The predicted $\hat{Y}$ and residual values for all observations are shown in column (i). This approach produces a single prediction equation for $\hat{Y}$.

Table 6.19 Final hierarchy fit.

w_u	Y	(i) First approach		(ii) Second approach	
		$\hat{Y}$	$[R_a, R_b]$	$[\hat{Y}_a, \hat{Y}_b]$	$[R_a, R_b]$
w_1	[410, 420]	393.36	[16.64, 26.64]	[393.36, 393.36]	[16.26, 26.64]
w_2	[388, 392]	393.36	[−5.36, −1.36]	[393.36, 393.36]	[5.36, −1.36]
w_3	[378, 384]	393.36	[−15.36, −9.36]	[393.36, 393.36]	[15.36, −9.36]
w_4	[381, 389]	393.36	[−12.36, −4.36]	[393.36, 393.36]	[12.36, −4.36]
w_5	[319, 330]	339.62	[−20.62, −9.62]	[339.62, 339.62]	[20.62, −9.62]
w_6	[352, 358]	339.62	[12.38, 18.38]	[339.62, 339.62]	[12.38, 18.38]
w_7	[255, 265]	261.14	[−6.14, 3.86]	[255.42, 266.85]	[1.85, −0.42]
w_8	[251, 259]	249.71	[1.29, 9.29]	[238.29, 261.4]	[2.14, 12.71]
w_9	[240, 248]	249.71	[−9.71, −1.71]	[244.00, 255.42]	[7.42, −4.00]
w_{10}	[197, 203]	204.01	[−7.01, −1.01]	[186.87, 221.15]	[18.15, 10.13]
w_{11}	[232, 244]	232.57	[−0.57, 11.43]	[221.15, 244.00]	[0.00, 10.85]
w_{12}	[180, 184]	180.73	[−0.72, 3.28]	[177.77, 183.69]	[0.31, 2.24]
w_{13}	[162, 168]	165.88	[−3.88, 2.12]	[161.43, 170.34]	[2.34, 0.57]
w_{14}	[156, 160]	158.46	[−2.46, 1.54]	[151.04, 165.88]	[5.88, 4.96]
w_{15}	[149, 151]	151.04	[−2.04, −0.04]	[148.07, 154.01]	[3.01, 0.93]
w_{16}	[136, 142]	136.20	[−0.20, 5.80]	[130.26, 142.14]	[0.14, 5.74]
w_{17}	[119, 121]	121.36	[−2.36, −0.36]	[118.39, 124.33]	[3.33, 0.61]
w_{18}	[33, 37]	43.29	[−10.29, −6.29]	[43.29, 43.29]	[−10.29, −6.29]
w_{19}	[41, 49]	43.29	[−2.29, 5.71]	[43.29, 43.29]	[−2.29, 5.71]
w_{20}	[48, 52]	43.29	[4.71, 8.71]	[43.29, 43.29]	[4.71, 8.71]

An alternative approach is to calculate $\hat{Y}_a$ using the lower limits of X_j, i.e., the a_{uj}, in Equation (6.62) and likewise calculate $\hat{Y}_b$ using the upper limits b_{uj}. Thus, for example,

$$\begin{aligned}\hat{Y}_b(w_8) &= 43.29 + 270.16(1) + 108.99(0) \\ &\quad + (1)[-74.09 + 135.83(0)] + (1)[87.81 - 1.48(10)] \\ &= 261.14;\end{aligned}$$

similarly,

$$\hat{Y}_a(w_8) = 238.29.$$

Note that the lower and upper limits for the predictor $\hat{Y}_a$, $\hat{Y}_b$ are interchanged when, as in this case, the regression slope is negative. Then, we can calculate the residuals $R = (R_a, R_b)$ with $R_k = Y_k - \hat{Y}_k$, $k = a, b$. Thus, for example,

$$R_b(w_8) = 251 - 238.29 = 12.7$$
$$R_a(w_8) = 259 - 261.14 = -2.1,$$

to give

$$R(w_8) = [-2.1, 12.7],$$

where again the limits are interchanged as necessary. The complete set of interval-valued predictions and the residual intervals from this approach are given in column (ii) of Table 6.19.

Notice that the $\hat{Y}(w_8)$ value is the midpoint of the interval-valued $[\hat{Y}_a(w_8), \hat{Y}_b(w_8)]$. This result holds in general. When all the X_j values pertaining to a particular $Y(w_u)$ are non-intervals, the two approaches will give the same results; see, for example, $\hat{Y}(w_1)$ and $[\hat{Y}_a(w_1), \hat{Y}_b(w_1)]$. □

Example 6.17 brings to light some unresolved issues in symbolic regression analysis. The two prediction fits and residual calculations of Table 6.19 are but two of the approaches that might have been used at the prediction stage (be it the interim predictions of the interim residuals $\hat{R}_1, \hat{R}_2, \ldots$, or the final predictions and final residuals). Substituting the midpoint $X_j(w_u)$ values into the prediction equation, e.g., Equation (6.62), perforce produces a single regression line based on $\hat{Y}(w_u)$. If, however, an interval-valued prediction equation is desired, then an approach such as that used here to give $[\hat{Y}_a(w_u), \hat{Y}_b(w_u)]$ is required. How these approaches compare in general is still an open problem; the question of whether a single prediction equation or an interval prediction equation is preferable is also open to debate.

When the data contain histogram-valued variables, this dilemma is further exacerbated. End points of an interval are clearly defined; their counterparts for use in calculating residual histograms from histogram-based data are less obvious to define. At this stage, we confine ourselves to using the weighted 'midpoint' of each histogram-valued observation (as in Example 6.18 below). There is considerable scope for more attention and work in these areas.

Example 6.18. Consider the data of Table 6.20. The predictor variables X_1 and X_2 are the same hierarchy variables used in Example 6.17 and displayed in the hierarchy tree of Figure 6.5.

Patients come to a hospital emergency room and are treated. The hierarchy variable X_1 takes three possible values $X_1 = a$ if the patient is admitted to this hospital, $X_1 = b$ if sent to a second hospital, or $X_1 = c$ if sent home. The second hierarchy variable X_2 takes values $X_2 = a1$ if the patient is admitted to the cardiology unit, or $X_2 = a2$ if admitted to a (non-cardiology) ward. The third hierarchy branch is X_3 with values $X_3 = a11$ if the patient dies, or $X_3 = a12$ if the patient is sent home after recovery. Suppose that for patients moved to a ward, one of the variables measured is $X_4 =$ Glucose level, and suppose that for patients sent to the second hospital, measurements include $X_5 =$ Pulse Rate. Finally, suppose that all patients are observed for $X_6 =$ Cholesterol. Suppose the dependent variable $Y =$ Duration Time (coded) before exiting the system. The random variables Y, X_4, X_5, and X_6 are histogram-valued random variables. The variables X_1, X_2, and X_3 are hierarchy-tree structural variables, and so are indicator variables. As

Table 6.20 Hierarchy: histogram-valued variables.

w_u	Y	X_1	X_2	X_3	X_4	X_5	X_6
w_1	{[5.3, 6.2), 0.3; [6.2, 8.3], 0.7}	a	$a1$	$a11$	NA	NA	[223, 264]
w_2	{[5.5, 6.7), 0.4; [6.7, 9.0], 0.6}	a	$a1$	$a11$	NA	NA	[220, 250]
w_3	{[5.1, 6.6), 0.4; [6.6, 7.8], 0.6}	a	$a1$	$a12$	NA	NA	[219, 241]
w_4	{[3.7, 5.8), 0.6; [5.8, 6.3], 0.4}	a	$a2$	NA	{[87, 110), 0.4; [110, 140], 0.6}	NA	[209, 239]
w_5	{[4.5, 5.9), 0.4; [5.9, 6.2], 0.6}	a	$a2$	NA	{[90, 120), 0.5; [120, 150], 0.5}	NA	[211, 237]
w_6	{[4.1, 6.1), 0.5; [6.1, 6.9], 0.5}	a	$a2$	NA	{[100, 120), 0.3; [120, 155], 0.7}	NA	[219, 245]
w_7	{[2.4, 4.8), 0.3; [4.8, 6.2], 0.7}	b	NA	NA	NA	{[70, 80), 0.4; [80, 90], 0.6}	[215, 228]
w_8	{[2.1, 5.4), 0.2; [5.4, 6.9], 0.8}	b	NA	NA	NA	{[65, 80), 0.4; [80, 95], 0.6}	[208, 233]
w_9	{[4.8, 6.5), 0.5; [6.5, 8.2], 0.5}	b	NA	NA	NA	{[64, 82), 0.5; [82, 98], 0.5}	[227, 253]
w_{10}	{[0.2, 1.1), 0.6; [1.1, 3.7], 0.4}	c	NA	NA	NA	NA	[190, 210]

before, we write $X_1 \equiv (X_{11}, X_{12}, X_{13})$, $X_2 \equiv (X_{21}, X_{22})$, $X_3 = (X_{31}, X_{32})$ with each taking values 0, 1 as appropriate.

At the top of the hierarchy tree, we fit the predictor variables X_1 and X_6 to the variable Y. That is, we fit

$$Y = \beta_0 + \beta_1 X_{11} + \beta_2 X_{12} + \beta_3 X_6 + e$$

using all $m = 10$ observations. Hence, we obtain

$$\hat{Y} = -9.752 + 3.270a + 3.200b + 0.055X_6. \tag{6.63}$$

To calculate $\hat{Y}(w_u)$ for a given observation, we substitute the weighted midpoint value for each predictor variable for w_u. Therefore, for the first observation w_1, for example, we have, from Equation (6.63),

$$\begin{aligned}\hat{Y}(w_1) &= -9.752 + 3.270(1) + 3.200(0) + 0.055(243.5)\\ &= 6.832.\end{aligned}$$

The residuals are calculated from $R_1 = Y_m - \hat{Y}$ where Y_m is the weighted midpoint of Y. Thus, for example,

$$\begin{aligned}R_1(w_1) &= Y_m(w_1) - \hat{Y}(w_1)\\ &= 6.80 - 6.832 = -0.032.\end{aligned}$$

The predicted values $\hat{Y}(w_u)$ and the residuals $R_1(w_u)$ are shown in Table 6.21, along with the weighted $Y_m(w_u)$ values.

Table 6.21 Fits of histogram hierarchy.

w_u	Y_m	Fit of X_1 and X_6		Fit of X_2 and X_5		Fit of X_3 and X_4		Final fit	
		$\hat{Y}$	R_1	$\hat{R}_1$	R_2	$\hat{R}_2$	R_3	$\hat{Y}$	R
w_1	6.80	6.832	−0.032	0.439	−0.471	−0.064	−0.407	7.28	−0.48
w_2	7.15	6.368	0.782	0.439	0.343	−0.064	0.407	6.82	0.33
w_3	6.66	6.094	0.566	0.439	0.127	0.127	−0.000	6.73	−0.07
w_4	5.27	5.766	−0.496	−0.319	−0.177	0.028	−0.205	5.55	−0.28
w_5	5.71	5.766	−0.056	−0.319	0.263	0.005	0.258	5.53	0.18
w_6	5.80	6.204	−0.404	−0.319	−0.085	−0.032	−0.053	5.93	−0.13
w_7	4.93	5.556	−0.626	−0.186	−0.440	−0.186	−0.440	5.48	−0.55
w_8	5.67	5.502	0.168	−0.170	0.338	−0.170	0.338	5.44	0.23
w_9	6.50	6.568	−0.068	−0.170	0.102	−0.170	0.102	6.52	−0.02
w_{10}	1.35	1.184	0.166	0.166	0.000	0.166	0.000	1.25	0.10

At the second level of the hierarchy tree, the predictor variables X_2 and X_5 are fitted to the $R_1(w_u)$ residuals. Thus, fitting X_2 to the residuals for w_u, $u = 1, \ldots, 6$, we find

$$R_1 = -0.319 + 0.757a1. \tag{6.64}$$

Fitting X_5 to the residuals for w_u, $u = 7, 8, 9$, gives

$$R_1 = -2.819 + 0.033X_5. \tag{6.65}$$

Hence, the residuals $R_2 = R_1 - \hat{R}_1$ can be calculated and are shown in Table 6.21.

At the third hierarchy level, the variable X_3 is fitted to the $R_2(w_u)$ residuals for $w_u = 1, 2, 3$, and the variable X_4 is fitted to these residuals for w_u, $u = 4, 5, 6$. We obtain

$$R_2 = 0.127 - 0.191X_3 \tag{6.66}$$

and

$$R_2 = 0.490 - 0.004X_4, \tag{6.67}$$

respectively. Hence, the residuals at this level are $R_3 = R_2 - \hat{R}_2$ and are given in Table 6.21.

From Equations (6.63)–(6.67), the complete hierarchy regression equation is therefore

$$\begin{aligned}\hat{Y} = &-9.752 + 3.270a + 3.197b + 0.055X_6 + \delta(a = 1)(-0.319 + 0.757a1)\\ &+ \delta(b = 1)(-2.819 + 0.033X_5) + \delta(a = 1)\delta(a1 = 1)(0.127 - 0.191a11)\\ &+ \delta(a = 1)\delta(a2 = 1)(0.490 - 0.004X_4).\end{aligned} \tag{6.68}$$

The final fits $\hat{Y}(w_u)$ and their residuals $R(w_u) = Y_m(w_u) - \hat{Y}(w_u)$ are provided in Table 6.21. For example, for w_5, we calculate

$$Y_m(w_5) = 5.71,\ X_{4m}(w_5) = 120.00,\ X_{6m}(w_5) = 224.00.$$

Hence, from Equation (6.68),

$$\begin{aligned}\hat{Y}(w_5) &= -9.752 + 3.270(1) + 3.197(0) + 0.055(224.00)\\ &\quad + (1)[-0.319 + 0.757(0)] + (1)(1)[0.490 - 0.004(120.00)]\\ &= 5.53,\end{aligned}$$

and

$$R(w_5) = 5.71 - 5.53 = 0.18.$$

□

Exercises

E6.1. Fit a linear regression equation (of the form of Equation (6.1)) through the data of Table 2.2b where the dependent variables is $Y =$ Cholesterol ($\equiv Y_{13}$) and the predictor variable is $X =$ Weight ($\equiv Y_9$).

E6.2. Take the airline data of Table 2.7. Consider Y_1 = Airtime to be a multi-valued random variable taking possible categorical values $\eta_1 = [Y_1 < 120]$, $\eta_2 = [120 \leq Y_1 \leq 220]$, and $\eta_3 = [Y_1 > 220]$. For the random variable Y_3 = Arrival Delay, suppose you know only the sample mean arrival delay time, $\bar{Y}_3$ say, for each airline. Fit a linear regression line with $\bar{Y}_3$ as the dependent variable and Y_1 as the predictor variable.

E6.3. For the setup in Exercise E6.2, model a regression line to $\bar{Y}_3$ as a function of the two predictor variables Y_1 and Y_6 (= Weather Delay).

E6.4. Let Y_3 = Arrival Delay be a multi-valued random variable with categories $\eta_1 = [Y_3 < 0]$, $\eta_2 = [0 \leq Y_3 \leq 60]$, $\eta_3 = [Y_3 > 60]$. Fit a linear regression model with this Y_3 as the dependent variable and Y_1 as the predictor variable.

E6.5. Repeat Exercise E6.3 using the multi-valued form of the random variable Y_3 as described in Exercise E6.4.

E6.6. Fit a linear regression model to the blood pressure data of Table 3.5 where Y_2 = Systolic Pressure is the dependent variable and Y_3 = Diastolic Pressure is the predictor variable.

E6.7. Take the hemoglobin data of Table 4.6 and the hematocrit data of Table 4.14. Let Y = Hematocrit be the dependent variable and X = Hemoglobin be the predictor variable. What is the relationship between Y and X for women?

E6.8. Repeat Exercise E6.7 but for men.

E6.9. Do the regression equations for women and men obtained in Exercises E6.7 and E6.8 differ? How might any such difference(s) be tested?

E6.10. Consider the data of Table 6.22. The two taxonomy variables are Y_1 = Type of Work and Y_2 = Profession with the possible Y_2 values being dependent

Table 6.22 Profession taxonomy tree.

w_u	Y_1 = Type of Work	Y_2 = Profession	Y_3 = Salary	Y_4 = Duration
w_1	White Collar	Manager	[94500, 97400]	[46, 51]
w_2	White Collar	Teacher	[58600, 62100]	[41, 44]
w_3	Blue Collar	Plumber	[72300, 81000]	[43, 47]
w_4	Services	Health	[48350, 52200]	[38, 40]
w_5	Services	Health	[67000, 74500]	[45, 46]
w_6	Blue Collar	Plumber	[65860, 72500]	[44, 46]
w_7	White Collar	Teacher	[63100, 65000]	[40, 42]
w_8	White Collar	Manager	[92000, 98400]	[48, 51]
w_9	Blue Collar	Electrician	[73000, 78000]	[42, 45]
w_{10}	Services	Food	[26000, 32000]	[38, 42]
w_{11}	Blue Collar	Electrician	[64600, 67200]	[39, 43]
w_{12}	Services	Food	[21000, 24000]	[35, 39]
w_{13}	Services	Services	[25000, 27500]	[39, 41]
w_{14}	Blue Collar	Blue Collar	[73200, 74800]	[43, 46]
w_{15}	White Collar	White Collar	[88000, 90000]	[47, 50]

on the actual realization of Y_1 (see Example 2.24). Suppose the random variables are $Y_3 =$ Salary and $Y_4 =$ Duration (of training, for that profession). Fit a taxonomy regression model where Y_3 is the dependent variable and Y_1 and Y_2 are the regression variables, using the first 12 observations $w_1, \ldots, w_{12}$, only.

E6.11. Refer to Exercise E6.10. Fit a regression model where Y_3 is the dependent variable and Y_1 and Y_2 are taxonomy regressor variables and Y_4 is a non-taxonomy regressor variable.

E6.12. Redo Exercise E6.10 but use all 15 observations.

E6.13. Redo Exercise E6.11 but use all 15 observations.

E6.14. Consider the first 17 observations $w_1, \ldots, w_{17}$ of Table 6.17. Fit a (hierarchy) regression model to these data using Y as the dependent variable.

E6.15. Take the data of Table 6.20. Fit a hierarchy regression model where Y is the dependent variable and $X_1, \ldots, X_5$ are predictor variables.

References

Alfonso, F., Billard, L., and Diday, E. (2005). Symbolic Linear Regression in the Presence of Taxonomy and Hierarchical Variables. Technical Report, University of Georgia, USA.

De Carvalho, F. A. T., Lima Neto, E. A., and Tenorio, C. P. (2004). A New Method to Fit a Linear Regression Model for Interval-valued Data. In: *Advances in Artificial Intelligence: Proceedings of the Twenty Seventh German Conference on Artificial Intelligence* (eds. S. Biundo, T. Frühwirth, and G. Palm). Springer-Verlag, Berlin, 295–306.

Johnson, R. A. and Wichern, D. W. (2002). *Applied Multivariate Statistical Analysis*. Prentice Hall, Englewood Cliffs, NJ.

Lima Neto, E. A., de Carvalho, F. A. T., and Freire, E. S. (2005). Applying Constrained Linear Regression Models to Predict Interval-Valued Data. In: *Lecture Notes on Artificial Intelligence*, LNAI 3238 (ed. U. Furbach). Springer-Verlag, Berlin, 92–106.

Myers, R. H. (1986). *Classical and Modern Regression with Applications*. Duxbury Press, Boston, MA.

Montgomery, D. C. and Peck, E. A. (1992). *Introduction to Linear Regression Analysis* (2nd ed.). John Wiley & Sons, Inc., New York.

7

Cluster Analysis

An important aspect of analyzing data is the formation of groups or clusters of observations, with those observations within a cluster being similar and those between clusters being dissimilar according to some suitably defined dissimilarity or similarity criteria. These clusters, often referred to as classes, are not necessarily those considered in previous chapters. Up to now, a class was the aggregation of observations satisfying a category (e.g., age $\times$ gender). The clusters that result satisfy a similarity (or dissimilarity) criterion. However, these cluster outputs can, but need not, be a category.

After describing briefly the basics of partitioning, hierarchical, and pyramidal clustering, we then delve more deeply into each of these in turn. Since these structures are based on dissimilarity measures, we first describe these measures as they pertain to symbolic data.

7.1 Dissimilarity and Distance Measures

7.1.1 Basic definitions

The formation of subsets $(C_1, \ldots, C_r)$ of E into a partition, hierarchy, or pyramid is governed by similarity $s(a, b)$ or dissimilarity $d(a, b)$ measures between two objects a and b, say. These measures take a variety of forms. Since typically a similarity measure is an inverse functional of its corresponding dissimilarity measure (e.g., $d(b, b) = 1 - s(a, b)$, and the like), we consider just dissimilarity measures. Distance measures are important examples of dissimilarity measures. We will define these entities in terms of an 'object' (which has a description D). These relate to an 'observation' (which has a realization ξ); see Chapter 2.

Symbolic Data Analysis: Conceptual Statistics and Data Mining L. Billard and E. Diday

Definition 7.1: Let a and b be any two objects in E. Then, a **dissimilarity measure** $d(a, b)$ is a measure that satisfies

(i) $d(a, b) = d(b, a)$;

(ii) $d(a, a) = d(b, b) > d(a, b)$ for all $a \neq b$;

(iii) $d(a, a) = 0$ for all $a \in E$. □

Definition 7.2: A **distance measure**, also called a **metric**, is a dissimilarity measure as defined in Definition 7.1 which further satisfies

(iv) $d(a, b) = 0$ implies $a = b$;

(v) $d(a, b) \leq d(a, c) + d(c, b)$ for all $a, b, c \in E$. □

Dissimilarity measures are symmetric from property (i), and property (v) is called the triangular inequality.

Definition 7.3: An **ultrametric** measure is a distance measure as defined in Definition 7.2 which also satisfies

(vi) $d(a, b) \leq Max\{d(a, c), d(c, b)\}$ for all $a, b, c \in E$. □

We note that it can be shown that ultrametrics are in one-to-one correspondence with hierarchies. Therefore, to compare hierarchies, we can compare their associated ultrametrics.

Definition 7.4: For the collection of objects $a_1, \ldots, a_m$ in E, the **dissimilarity matrix** (or **distance matrix**) is the $m \times m$ matrix $\boldsymbol{D}$ with **elements** $d(a_i, a_j)$, $i, j = 1, \ldots, m$. □

Example 7.1. Consider the observations represented by a, b, c (or, equivalently, w_1, w_2, w_3) whose dissimilarity matrix is

$$\boldsymbol{D} = \begin{pmatrix} 0 & 2 & 1 \\ 2 & 0 & 3 \\ 1 & 3 & 0 \end{pmatrix}.$$

That is, the dissimilarity measure between a and b is $d(a, b) = 2$, while $d(a, c) = 1$ and $d(b, c) = 3$. Since property (v) holds for all a, b, c, this matrix $\boldsymbol{D}$ is also a distance matrix, e.g.,

$$d(b, c) = 3 \leq d(b, a) + d(a, c) = 2 + 1 = 3. \quad □$$

Definition 7.5: A dissimilarity (or distance) matrix whose elements monotonically increase as they move away from the diagonal (by column and by row) is called a **Robinson matrix**. □

Example 7.2. The matrix $\boldsymbol{D}$ in Example 7.1 is not a Robinson matrix since, for example,

$$d(a, c) = 1 < d(a, b) = 2.$$

However, the matrix

$$\boldsymbol{D} = \begin{pmatrix} 0 & 1 & 3 \\ 1 & 0 & 2 \\ 3 & 2 & 0 \end{pmatrix}$$

is a Robinson matrix. □

Robinson matrices play a crucial role in pyramidal clustering as it can be shown that they are in one-to-one correspondence with indexed pyramids (see Diday, 1989); see Section 7.5. Each clustering process requires a dissimilarity or distance matrix $\boldsymbol{D}$ of some form. Some distance and dissimilarity measures that arise frequently in symbolic clustering methodologies will be presented here. We pay special attention to those measures developed by Gowda and Diday (1991) and Ichino and Yaguchi (1994) and their extensions to Minkowski distances. In many instances, these measures are extensions of a classical counterpart, and in all cases reduce to a classical equivalent as a special case. For excellent reviews of these measures, see Gordon (1999), Esposito *et al.* (2000), and Gower (1985).

There are a number of dissimilarity and distance measures based on the notions of the Cartesian operators 'join' and 'meet' between two sets A and B, introduced by Ichino (1988). We denote $A = (A_1, \ldots, A_p)$ and $B = (B_1, \ldots, B_p)$. In the context of symbolic data, the set A corresponds to the observed value of a p-dimensional observation $\boldsymbol{\xi} = (Y_1, \ldots, Y_p)$, i.e., $A \equiv \boldsymbol{\xi}$. Thus, when A is a multi-valued symbolic object, A_j takes single values in the space $\mathcal{Y}_j^A$ (e.g., $A_j = \{\text{green, red}\}$); and when A is interval valued, A_j takes interval values $[a_j, b_j]$ in $\mathcal{Y}_j \subset \Re$ (e.g., $A_j = [3, 7]$). The possible values for B_j are defined analogously. How these notions, and their application to the relevant dissimilarity measures, pertain for modal-valued objects in general is still an open problem (though we shall see below that there are some instances where modal multi-valued objects are covered).

Definition 7.6: The Cartesian **join** $A \oplus B$ between two sets A and B is their componentwise union,

$$A \oplus B = (A_1 \oplus B_1, \ldots, A_p \oplus B_p)$$

where $A_j \oplus B_j =$ '$A_j \cup B_j$'. When A and B are multi-valued objects with $A_j = \{a_{j1}, \ldots, a_{js_j}\}$ and $B_j = \{b_{j1}, \ldots, b_{t_j}\}$, then

$$A_j \oplus B_j = \{a_{j1}, \ldots, b_{t_j}\},\ j = 1, \ldots, p, \tag{7.1}$$

is the set of values in A_j, B_j, or both. When A and B are interval-valued objects with $A_j = [a_j^A, b_j^A]$ and $B_j = [a_j^B, b_j^B]$, then

$$A_j \oplus B_j = [Min(a_j^A, a_j^B),\ Max(b_j^A, b_j^B)]. \tag{7.2}$$

□

Definition 7.7: The Cartesian **meet** $A \otimes B$ between two sets A and B is their componentwise intersection,

$$A \otimes B = (A_1 \otimes B_1, \ldots, A_p \otimes B_p)$$

where

$$A_j \otimes B_j = A_j \cap B_j, \ j = 1, \ldots, p.$$

When A and B are multi-valued objects, then $A_j \otimes B_j$ is the list of possible values from $\mathcal{Y}_j$ common to both. When A and B are interval-valued objects forming overlapping intervals on Y_j,

$$A_j \otimes B_j = [Max(a_j^A, a_j^B), \ Min(b_j^A, b_j^B)], \tag{7.3}$$

and when $A_j \cap B_j = \phi$, then $A_j \otimes B_j = 0$. □

Example 7.3. Let A and B be multi-valued symbolic objects with

$$A = (\{\text{red, green}\}, \ \{\text{small}\})$$
$$B = (\{\text{red, blue}\}, \ \{\text{small, medium}\}).$$

Then, their join is, from Equation (7.1),

$$A \oplus B = (\{\text{red, green, blue}\}, \ \{\text{small, medium}\})$$

and their meet is, from Equation (7.3),

$$A \otimes B = (\{\text{red}\}, \{\text{small}\}).$$ □

Example 7.4. Let A and B be the interval-valued symbolic objects

$$\begin{bmatrix} A = ([3,7],[21,25],[5,9]), \\ B = ([5,8],[19,24],[6,11]). \end{bmatrix}$$

Then, their join is, from Equation (7.2),

$$A \oplus B = ([3,8],[19,25],[5,11])$$

and from Equation (7.3) their meet is

$$A \otimes B = ([5,7],[19,24],[6,9]).$$ □

These same definitions apply to mixed variables, as shown in the following example.

Example 7.5. Let A and B be the symbolic objects

$$A = ([3, 7], \{\text{red, green}\}),$$
$$B = ([5, 8], \{\text{red, blue}\}).$$

Then, their join and meets are, respectively,

$$\begin{bmatrix} A \oplus B = ([3, 8], \{\text{red, green,blue}\}), \\ A \otimes B = ([5, 7], \{\text{red}\}). \end{bmatrix}$$ □

Two important distance measures are those developed by Gowda and Diday (1991) and Ichino and Yaguchi (1994). We present these distances here in their general form. Their applications to multi-valued variables and to interval-valued variables are presented in Section 7.1.2 and Section 7.1.3, respectively.

Definition 7.8: The **Gowda–Diday dissimilarity measure** between two sets A and B is

$$D(A, B) = \sum_{j=1}^{p} [D_{1j}(A_j, B_j) + D_{2j}(A_j, B_j) + D_{3j}(A_j, B_j)] \tag{7.4}$$

where $D_{1j}(A_j, B_j)$ is a measure that relates to the relative sizes (or span) of A_j and B_j, $D_{2j}(A_j, B_j)$ relates to the relative content of A_j and B_j, and $D_{3j}(A_j, B_j)$ measures their relative position. □

Definition 7.9: The **Ichino–Yaguchi dissimilarity measure** on the variable Y_j component of the two sets A and B is

$$\phi_j(A, B) = |A_j \oplus B_j| - |A_j \otimes B_j| + \gamma(2|A_j \otimes B_j| - |A_j| - |B_j|) \tag{7.5}$$

where $|A|$ is the number of elements in A if Y_j is multi-valued and is the length of the interval A_j if Y_j is interval-valued, and where $0 \leq \gamma \leq 0.5$ is a prespecified constant. □

Definition 7.10: The **generalized Minkowski distance** of order $q \geq 1$ between two sets A and B is

$$d_q(A, B) = \left(\sum_{j=1}^{p} w_j^* [\phi_j(A, B)]^q \right)^{1/q} \tag{7.6}$$

where w_j^* is an appropriate weight for the distance component $\phi_j(A, B)$ on Y_j, $j = 1, \ldots, p$. □

As the number of variables p increases, the Minkowski distance in Equation (7.6) can rise. To account for any inordinate increase in this distance, Equation (7.6) can be replaced by

$$d_q(A, B) = \left(\frac{1}{p} \sum_{j=1}^{p} w_j^* [\phi_j(A, B)]^q \right)^{1/q}. \tag{7.7}$$

Distances weighted to account for different scales of measurement take the form

$$\phi'_j(A, B) = \phi_j(A, B)/|\mathcal{Y}_j|, \tag{7.8}$$

where $|\mathcal{Y}_j| =$ number of possible values if Y_j is multi-valued, or $|\mathcal{Y}_j| =$ total length spanned by observations if Y_j is interval-valued. It is easy to show that scale-weighted distances of Equation (7.8) when used in Equation (7.7) are such that $0 \leq d_q(A, B) \leq 1$. There are two special cases.

Definition 7.11: A generalized Minkowski distance of order $q = 1$ is called a **city block distance** when

$$d(A, B) = \sum_{j=1}^{p} c_j \phi_j(A, B) \tag{7.9}$$

where city weights are such that $\sum c_j = 1$. □

Definition 7.12: A generalized Minkowski distance of order $q = 2$ is a **Euclidean distance** when

$$d(A, B) = \left(\sum_{j=1}^{p} [\phi_j(A, B)]^2 \right)^{1/2}. \tag{7.10}$$

□

The Euclidean distances of Equation (7.10) can be normalized to account for differences in scale and for large values of p. This gives us the following definition.

Definition 7.13: A **normalized Euclidean distance** is such that

$$d(A, B) = \left(\frac{1}{p} \sum_{j=1}^{p} [|\mathcal{Y}_j|^{-1} \phi_j(A, B)]^2 \right)^{1/2}. \tag{7.11}$$

□

These Definitions 7.8–7.13 all involve the concepts of join and meet in some manner, with the precise nature depending on whether or not a given random variable Y_j is multi-valued or interval-valued. They can also contain, or not contain, weight factors such as those of Equation (7.6). Examples of their use are provided in the respective Sections 7.1.2–7.1.4 below.

7.1.2 Multi-valued variables

Let $\mathbf{Y} = (Y_1, \ldots, Y_p)$ be a p-dimensional multi-valued random variable with Y_j taking values in $\mathcal{Y}_j = \{Y_{j1}, \ldots, Y_{js_j}\}$. Without loss of generality, let us first assume that each Y_j is modal-valued and that all possible values in $\mathcal{Y}_j$ occur. The observation w_u in E therefore can be written in the form, for each $u = 1, \ldots, m$,

$$\boldsymbol{\xi}(w_u) = \left(\{Y_{u1k_1}, p_{u1k_1}; k_1 = 1, \ldots, s_1\}; \ldots; \{Y_{upk_p}, p_{upk_p}; k_p = 1, \ldots, s_p\}\right) \tag{7.12}$$

where p_{ujk_j} is the relative frequency of Y_{ujk_j}. Then, those values Y_{jk_j} in $\mathcal{Y}_j$ which do not occur in the observation $\boldsymbol{\xi}$ have corresponding values of $p_{jk_j} \equiv 0$. Likewise, for a non-modal multi-valued random variable, these $p_{jk_j} \equiv 1/n_{uj}$ where n_{uj} is the number of values from $\mathcal{Y}_j$ which do occur.

Example 7.6. Let $\mathcal{Y}_1 = \{\text{blue, green, red}\}$. Then the observation

$$\xi_1 = \{\text{blue, green}\}$$

is rewritten as

$$\xi_1 = \{\text{blue, 0.5; green, 0.5; red, 0}\}.$$

Likewise, the observation

$$\xi_2 = \{\text{blue, 0.4; red, 0.6}\}$$

becomes

$$\xi_2 = \{\text{blue, 0.4; green, 0.0; red, 0.6}\}.$$

□

Definition 7.14: For multi-valued modal data of the form of Equation (7.12), a **categorical distance measure** between any two observations $\xi(w_{u_1})$ and $\xi(w_{u_2})$ is $d(w_{u_1}, w_{u_2})$ where

$$d^2(w_{u_1}, w_{u_2}) = \sum_{j=1}^{p} \sum_{k_j=1}^{s_j} \left(p \sum_{u=1}^{m} p_{ujk_j}\right)^{-1} \left(p_{u_1jk_j} - p_{u_2jk_j}\right)^2. \tag{7.13}$$

□

Example 7.7. Consider the observations

$$\xi(w_1) = (\{\text{blue, 0.4; green, 0.6}\}, \{\text{urban}\})$$

$$\xi(w_2) = (\{\text{blue, 0.3; green, 0.4; red, 0.3}\}, \{\text{urban, rural}\})$$

$$\xi(w_3) = (\{\text{blue}\}, \{\text{urban, 0.4; rural, 0.6}\}).$$

When transformed to the format of Equation (7.12), the relative frequencies $\{p_{ujk_j},\ u = 1, 2, 3;\ k_1 = 1, 2, 3,\ k_2 = 1, 2,\ j = 1, 2\}$ take the values as shown in Table 7.1.

Table 7.1 Relative frequencies p_{ujk_j} – Example 7.7 data.

w_u	Y_1			Y_2		$p_{u..}$
	blue	green	red	urban	rural	
w_1	0.4	0.6	0.0	1.0	0.0	2
w_2	0.3	0.4	0.3	0.5	0.5	2
w_3	1.0	0.0	0.0	0.4	0.6	2
$p_{.jk_j}$	1.7	1.0	0.3	1.9	1.1	6

Then, from Equation (7.13), the squared distance between the observations $\xi(w_1)$ and $\xi(w_2)$ is

$$d^2(w_1, w_2) = \frac{1}{2}\left[\frac{(0.4-0.3)^2}{1.7} + \frac{(0.6-0.4)^2}{1.0} + \frac{(0.0-0.3)^2}{0.3} + \frac{(1.0-0.5)^2}{1.9} + \frac{(0.0-0.5)^2}{1.1}\right] = 0.352.$$

Likewise, we obtain

$$d^2(w_1, w_3) = 0.544, \; d^2(w_2, w_3) = 0.381.$$

Hence, the distance matrix is

$$\boldsymbol{D} = \begin{pmatrix} 0 & 0.593 & 0.738 \\ 0.593 & 0 & 0.617 \\ 0.738 & 0.617 & 0 \end{pmatrix}.$$

Notice that this $\boldsymbol{D}$ is a Robinson matrix. □

There are two measures that apply to non-modal multi-valued observations. For these data, let the format be written as

$$\xi(w_u) = (\{Y_{u1k_1^u}, k_1^u = 1, \ldots, s_1\}; \ldots; \{Y_{upk_u^p}, k_p^u = 1, \ldots, s_p\}). \tag{7.14}$$

Definition 7.15: The **Gowda–Diday dissimilarity measure** between two multi-valued observations $\xi(w_1)$ and $\xi(w_2)$ of the form of Equation (7.14) is

$$D(w_1, w_2) = \sum_{j=1}^{p}[D_{1j}(w_1, w_2) + D_{2j}(w_1, w_2)]$$

where

$$D_{1j}(w_1, w_2) = (|k_j^1 - k_j^2|)/k_j, \; j = 1, \ldots, p, \tag{7.15}$$

$$D_{2j}(w_1, w_2) = (k_j^1 + k_j^2 - k_j^*)/k_j, \; j = 1, \ldots, p, \tag{7.16}$$

where k_j is the number of values from $\mathcal{Y}_j$ in the join $\xi(w_1) \oplus \xi(w_2)$ and k_j^* is the number of values in the meet $\xi(w_1) \otimes \xi(w_2)$ for the variable Y_j. (The relative position component of Definition 7.8 is zero for multi-valued data.) □

Definition 7.16: The **Ichino–Yaguchi dissimilarity measure (or distance)** between the two multi-valued observations $\xi(w_1)$ and $\xi(w_2)$ of the form of Equation (7.14) for the variable Y_j is

$$\phi_j(w_1, w_2) = k_j - k_j^* + \gamma(2k_j^* - k_j^1 - k_j^2) \tag{7.17}$$

where k_j, k_j^*, k_j^1, and k_j^2 have the same meaning as in Definition 7.15 and where $0 \leq \gamma \leq 0.5$ is a prespecified constant. □

Example 7.8. The data of Table 7.2 give values for Color Y_1 from $\mathcal{Y}_1 =$ {red, black, blue} and Habitat Y_2 from $\mathcal{Y}_2 =$ {urban, rural} for $m = 4$ species of birds. The components for the Gowda–Diday distances for each variable (Y_1, Y_2) and for each pair of species (w_{u_1}, w_{u_2}) are displayed in Table 7.3. For example, consider the pair (species1, species2). From Equation (7.15), the span distance component for Y_1 is

$$D_1(w_1, w_2) = |2 - 1|/2 = 1/2,$$

and the content distance component from Equation (7.16) for Y_1 is

$$D_2(w_1, w_2) = (2 + 1 - 2 \times 1)/2 = 1/2.$$

Table 7.2 Birds – Color and Habitat.

w_u	Species	(Color, Habitat)
w_1	species1	({red, black}; {urban, rural})
w_2	species2	({red}; {urban})
w_3	species3	({red, black, blue}; {rural})
w_4	species4	({red, black, blue}; {urban, rural}

Hence, for Y_1, the distance measure is

$$\phi_1(w_1, w_2) = 1/2 + 1/2 = 1.$$

Likewise, for the variable Y_2, we obtain

$$\phi_2(w_1, w_2) = 1/2 + 1/2 = 1.$$

Hence, the unweighted Gowda–Diday distance between species1 and species2 is

$$d(w_1, w_2) = \phi_1(w_1, w_2) + \phi_2(w_1, w_2) = 2.$$

Table 7.3 Gowda–Diday distances: birds – Color and Habitat.

(w_{u_1}, w_{u_2})	$Y_1 = \text{Color}$			$Y_2 = \text{Habitat}$			(Y_1, Y_2)
	$D_1(w_1, w_2)$	$D_2(w_1, w_2)$	$\phi_1(w_1, w_2)$	$D_1(w_1, w_2)$	$D_2(w_1, w_2)$	$\phi_2(w_1, w_2)$	$\phi(w_1, w_2)$
(w_1, w_2)	1/2	1/2	1	1/2	1/2	1	2
(w_1, w_3)	1/3	1/3	2/3	1/2	1/2	1	5/3
(w_1, w_4)	1/3	1/3	2/3	0	0	0	2/3
(w_2, w_3)	2/3	2/3	4/3	0	1	1	7/3
(w_2, w_4)	0	2/3	2/3	1/2	1/2	1	7/3
(w_3, w_4)	0	0	0	1/2	1/2	1	1

The Gowda–Diday distance matrix becomes

$$\boldsymbol{D} = \begin{pmatrix} 0 & 2 & 5/3 & 2/3 \\ 2 & 0 & 7/3 & 7/3 \\ 5/3 & 7/3 & 0 & 1 \\ 2/3 & 7/3 & 1 & 0 \end{pmatrix}.$$

When these distances are normalized to adjust for scale, the weights for the variables Y_1 and Y_2 are $|\mathcal{Y}_1| = 3$ and $|\mathcal{Y}_2| = 2$, respectively. Thus, for example, the weighted distance between species1 and species2 is

$$d(w_1, w_2) = 1/3 + 1/2 = 5/6.$$

The weighted distance matrix is

$$\boldsymbol{D} = \begin{pmatrix} 0 & 5/6 & 13/18 & 2/9 \\ 5/6 & 0 & 17/18 & 17/18 \\ 13/18 & 17/18 & 0 & 1/2 \\ 2/9 & 17/18 & 1/2 & 0 \end{pmatrix}.$$

□

Example 7.9. We can obtain the Ichino–Yaguchi distances for the data of Table 7.2 by substitution into Equation (7.17). For example, for species1 and species4 we have, for the variables Y_1 and Y_2, respectively,

$$\phi_1(w_1, w_4) = 3 - 2 + \gamma(2 \times 2 - 3 - 2) = 1 + \gamma(-1),$$

and

$$\phi_2(w_1, w_4) = 2 - 2 + \gamma(2 \times 2 - 2 - 2) = 0.$$

These distances for each pair of species are collected in Table 7.4.

These distances are now substituted into Equation (7.6) to obtain the Minkowski distance of order q. For example, taking $\gamma = 1/2$, we have the unweighted Minkowski distance between species2 and species4 as

$$d_q(w_2, w_4) = (1^q + 0.5^q)^{1/q}. \tag{7.18}$$

Table 7.4 Ichino–Yaguchi distances: birds – Color and Habitat.

w_{u_1}, w_{u_2}	$\phi_j(w_{u_1}, w_{u_2})$		Unweighted		Weighted	
	Y_1	Y_2	$q=1$	$q=2$	$q=1$	$q=2$
(w_1, w_2)	$1+\gamma(-1)$	$1+\gamma(-1)$	0.500	0.707	0.208	0.300
(w_1, w_3)	$1+\gamma(-1)$	$1+\gamma(-1)$	0.500	0.707	0.208	0.300
(w_1, w_4)	$1+\gamma(-1)$	0	0.250	0.500	0.083	0.167
(w_2, w_3)	$2+\gamma(-2)$	$2+\gamma(-2)$	1.000	1.414	0.417	0.601
(w_2, w_4)	$2+\gamma(-2)$	$1+\gamma(-1)$	0.750	1.118	0.181	0.417
(w_3, w_4)	0	$1+\gamma(-1)$	0.250	0.500	0.125	0.250

Were these distances to be weighted to account for scale, this would become

$$d_q(w_2, w_4) = [(1/3)^q + (0.5/2)^q]^{1/q}. \tag{7.19}$$

Then the city block distances are obtained by substituting $q = 1$ into Equation (7.18) or Equation (7.19) for the unweighted or weighted distances, respectively. The resulting distances for $c_1 = c_2$ are summarized in Table 7.4. Hence, the unweighted city block distance matrix is

$$\boldsymbol{D} = \begin{pmatrix} 0 & 0.50 & 0.50 & 0.25 \\ 0.50 & 0 & 1.00 & 0.75 \\ 0.50 & 1.0 & 0 & 0.25 \\ 0.25 & 0.75 & 0.25 & 0 \end{pmatrix}.$$

The Euclidean distances are found by substituting $q = 2$ into Equation (7.18) or Equation (7.19) for the unweighted or weighted distances, respectively. These values are given in Table 7.4. Therefore, we see that the weighted Euclidean distance matrix for these data is

$$\boldsymbol{D} = \begin{pmatrix} 0 & 0.30 & 0.30 & 0.17 \\ 0.30 & 0 & 0.60 & 0.42 \\ 0.30 & 0.60 & 0 & 0.25 \\ 0.17 & 0.42 & 0.25 & 0 \end{pmatrix}.$$

□

7.1.3 Interval-valued variables

There are a number of dissimilarity and distance measures for interval-valued data, some of which have a kind of analogous counterpart to those for multi-valued observations and some of which are unique to interval-valued data. Let us denote an interval-valued realization $\boldsymbol{\xi}_u$ for the observation w_u by

$$\boldsymbol{\xi}_u = ([a_{uj}, b_{uj}],\ j = 1, \ldots, p),\ u = 1, \ldots, m. \tag{7.20}$$

Definition 7.17: The **Gowda–Diday dissimilarity measure** between two interval-valued observations w_{u_1} and w_{u_2} of the form of Equation (7.20) is given by

$$D(w_{u_1}, w_{u_2}) = \sum_{j=1}^{p} D_j(w_{u_1}, w_{u_2}) \tag{7.21}$$

where for the variable Y_j, the distance is

$$D_j(w_{u_1}, w_{u_2}) = \sum_{k=1}^{3} D_{jk}(w_{u_1}, w_{u_2}) \tag{7.22}$$

with

$$D_{j1}(w_{u_1}, w_{u_2}) = ||b_{u_1j} - a_{u_1j}| - |b_{u_2j} - a_{u_2j}||/k_j \tag{7.23}$$

where

$$k_j = |Max(b_{u_1j}, b_{u_2j}),\ Min(a_{u_1j}, a_{u_2j})|$$

is the length of the entire distance spanned by w_{u_1} and w_{u_2}; with

$$D_{j2}(w_{u_1}, w_{u_2}) = (|b_{u_1j} - a_{u_1j}| + |b_{u_2j} - a_{u_2j}| - 2I_j)/k_j. \tag{7.24}$$

where I_j is the length of the intersection of the intervals $[a_{u_1j}, b_{u_1j}]$ and $[a_{u_2j}, b_{u_2j}]$, i.e.,

$$I_j = |Max(a_{u_1j}, a_{u_2j}) - Min(b_{u_1j}, b_{u_2j})|$$

if the intervals overlap, and $I_j = 0$ if not; and with

$$D_{j3}(w_{u_1}, w_{u_2}) = |a_{u_1j} - a_{u_2j}|/|\mathcal{Y}_j| \tag{7.25}$$

where $\mathcal{Y}_j$ is the total length in $\mathcal{Y}$ covered by the observed values of Y_j, i.e.,

$$|\mathcal{Y}_j| = \max_u(b_{uj}) - \min_u(a_{uj}). \tag{7.26}$$

□

The components $D_{jk}(w_{u_1}, w_{u_2})$, $k = 1, 2$, of Equations (7.23)–(7.24) are counterparts of the span and content components of the Gowda–Diday distance for multi-valued variables given in Definition 7.15. The third component $D_{j3}(w_{u_1}, w_{u_2})$ is a measure of the relative positions of the two observations.

Example 7.10. We consider the veterinary clinic data of Table 7.5, and take the first three observations only. We first calculate the Gowda–Diday distances for the random variable Y_1 = Height. Then, substituting the data values from

Table 7.5 into Equation (7.23), we obtain the distance component between the male and female horses as

$$D_{11}(\text{HorseM, HorseF}) = \frac{|(180-120)-(160-158)|}{|Max(180,160)-Min(120,158)|}$$

$$= |60-2|/|180-120| = 0.967.$$

Table 7.5 Veterinary data.

w_u	Animal	Y_1 Height	Y_2 Weight
w_1	HorseM	[120.0, 180.0]	[222.2, 354.0]
w_2	HorseF	[158.0, 160.0]	[322.0, 355.0]
w_3	BearM	[175.0, 185.0]	[117.2, 152.0]
w_4	DeerM	[37.9, 62.9]	[22.2, 35.0]
w_5	DeerF	[25.8, 39.6]	[15.0, 36.2]
w_6	DogF	[22.8, 58.6]	[15.0, 51.8]
w_7	RabbitM	[22.0, 45.0]	[0.8, 11.0]
w_8	RabbitF	[18.0, 53.0]	[0.4, 2.5]
w_9	CatM	[40.3, 55.8]	[2.1, 4.5]
w_{10}	CatF	[38.4, 72.4]	[2.5, 6.1]

Likewise, by substituting into Equation (7.24), we obtain

$$D_{12}(\text{HorseM, HorseF}) = \frac{|180-120|+|160-158|-2|160-158|}{|180-120|}$$

$$= 0.967;$$

and by substituting into Equation (7.25), we find

$$D_{13}(\text{HorseM, HorseF}) = \frac{|120-158|}{Max(180,165,185)-Min(120,158,175)}$$

$$= 38/(185-120) = 0.584.$$

Hence, the Gowda–Diday distance for the variable Y_1 is

$$D_1(\text{HorseM, HorseF}) = 0.967+0.967+0.584 = 2.518.$$

For the random variable Y_2 = Weight, we likewise obtain

$$D_{21}(\text{HorseM, HorseF}) = 0.744,$$

$$D_{22}(\text{HorseM, HorseF}) = 0.759,$$

$$D_{23}(\text{HorseM, HorseF}) = 0.419,$$

and hence

$$D_2(\text{HorseM, HorseF}) = 1.922.$$

Therefore, the Gowda–Diday distance between the male and female horses over all variables is

$$D(\text{HorseM}, \text{HorseF}) = 2.518 + 1.922 = 4.440.$$

Each distance component for each variable and each animal pair is shown in Table 7.6. From these results, the Gowda–Diday distance matrix for the male and female horses and the female bears is

$$\boldsymbol{D} = \begin{pmatrix} 0 & 4.44 & 3.67 \\ 4.44 & 0 & 2.13 \\ 3.67 & 2.13 & 0 \end{pmatrix}. \qquad \square$$

Table 7.6 Gowda–Diday distances – horses and bears.

(w_{u_1}, w_{u_2})	$Y_1 =$ Height				$Y_2 =$ Weight				(Y_1, Y_2)
	D_{11}	D_{12}	D_{13}	D_1	D_{21}	D_{22}	D_{23}	D_2	D
(HorseM, HorseF)	0.967	0.967	0.584	2.518	0.744	0.759	0.419	1.922	4.440
(HorseM, BearM)	0.769	0.923	0.846	2.538	0.409	0.703	0.021	1.133	3.671
(HorseF, BearM)	0.296	0.444	0.231	0.971	0.008	0.285	0.861	1.154	2.125

Definition 7.18: The **Ichino–Yaguchi dissimilarity measure** between the two interval-valued observations w_{u_1} and w_{u_2} of the form of Equation (7.20) for the interval-valued variable Y_j is

$$\phi_j(w_{u_1}, w_{u_2}) = |w_{u_1j} \oplus w_{u_2j}| - |w_{u_1j} \otimes w_{u_2j}| + \gamma(2|w_{u_1j} \otimes w_{u_2j}| - |w_{u_1j}| - |w_{u_2j}|) \tag{7.27}$$

where as before $|A|$ is the length of the interval $A = [a, b]$, i.e., $|A| = b - a$, and $0 \leq \gamma \leq 0.5$ is a prespecified constant. $\square$

Definition 7.19: The **generalized (weighted) Minkowski distance** or order q for interval-valued objects w_{u_1} and w_{u_2} is

$$d_q(w_{u_1}, w_{u_2}) = \{\sum_{j=1}^{p} w_j^*[\phi_j(w_{u_1}, w_{u_2})]^q\}^{1/q} \tag{7.28}$$

where $\phi_j(w_{u_1}, w_{u_2})$ is the Ichino–Yaguchi distance of Equation (7.27) and where w_j^* is a weight function associated with Y_j.

The **city block distance** is a Minkowski distance of order $q = 1$, namely,

$$d(w_{u_1}, w_{u_2}) = \sum_{j=1}^{p} c_j w_j^* \phi_j(w_{u_1}, w_{u_2}) \tag{7.29}$$

where $c_j > 0$, $\sum c_j = 1$, and the **normalized Euclidean distance** is a Minkowski distance of order $q = 2$, i.e.,

$$d(w_{u_1}, w_{u_2}) = \left(\frac{1}{p} \sum_{j=1}^{p} w_j^* [\phi_j(w_{u_1}, w_{u_2})]^2 \right)^{1/2}. \tag{7.30}$$

□

These definitions for a Minkowski distance are completely analogous with those for multi-valued objects; the difference is in the nature of the Ichino–Yaguchi dissimilarity measure $\phi_j(w_{u_1}, w_{u_2})$.

A weight function which takes account of the scale of measurements on Y_j is $w_j^* = 1/|\mathcal{Y}_j|$ where $|\mathcal{Y}_j|$ is the span of the real line $\Re^1$ covered by the observations as defined in Equation (7.26).

Example 7.11. We consider again the veterinary clinic data of Table 7.5 and take the first three classes, male and female horses and male bears. The Ichino–Yaguchi dissimilarity measures for each variable Y_j, obtained from Equation (7.27) are shown in Table 7.7. For example,

$$\begin{aligned} \phi_1(\text{HorseM}, \text{BearM}) &= (185 - 120) - (180 - 175) + \gamma(2 \times 5 - 60 - 10) \\ &= 60 + \gamma(-60) \\ \phi_2(\text{HorseM}, \text{BearM}) &= (354 - 117.2) - 0 + \gamma(2 \times 0 - 131.8 - 34.8) \\ &= 236.8 + \gamma(-166.6). \end{aligned}$$

Taking $\gamma = 0.5$ and substituting into Equation (7.28) gives the Minkowski distances. Weighted distances obtained by taking $w_j^* = 1/|\mathcal{Y}_j|$ can also be found. Here,

$$w_1^* = 1/[Max(180, 160, 185) - Min(120, 158, 175)] = 1/65$$

and similarly,

$$w_2^* = 1/(355 - 117.2) = 1/237.8.$$

The weighted (by w_j^*) and unweighted distances for $q = 1, 2$ are also shown in Table 7.7. □

Table 7.7 Ichino–Yaguchi distances – horses and bears.

	$\phi_j(w_{u_1}, w_{u_2})$		Unweighted		Weighted	
	$j=1$	$j=2$	$q=1$	$q=2$	$q=1$	$q=2$
(HorseM, HorseF)	$58+\gamma(-58)$	$100.8+\gamma(-100.8)$	79.4	58.1	0.658	0.494
(HorseM, BearM)	$60+\gamma(-60)$	$236.8+\gamma(-166.6)$	183.5	156.4	1.107	0.794
(HorseF, BearM)	$27+\gamma(-12)$	$237.8+\gamma(-67.8)$	214.9	195.0	1.138	0.877

Extensions of Minkowski distances of order $p = 2$ for interval-valued data play an important role in the divisive clustering methodology (of Section 7.4). Given this role these extensions are described herein; the general case of order q follows readily.

Definition 7.20: The **Hausdorff distance** for the variable Y_j between two interval-valued observations ξ_{u_1} and ξ_{u_2} is

$$\phi_j(w_{u_1}, w_{u_2}) = Max[|a_{u_1j} - a_{u_2j}|, |b_{u_1j} - b_{u_2j}|]. \tag{7.31}$$

□

Example 7.12. For the horses and male bears observations of Table 7.5, we can see that the Hausdorff distances are

$$\phi_1(\text{HorseM}, \text{HorseF}) = Max[|120 - 158|, |180 - 160|] = 38,$$

$$\phi_2(\text{HorseM}, \text{HorseF}) = Max[|222.2 - 322|, |354 - 355|] = 99.8.$$

Likewise,

$$\phi_1(\text{HorseM}, \text{BearM}) = 55, \phi_2(\text{HorseM}, \text{BearM}) = 202;$$

$$\phi_1(\text{HorseF}, \text{BearM}) = 27, \phi_2(\text{HorseF}, \text{BearM}) = 204.8.$$

□

Definition 7.21: The **Euclidean Hausdorff distance matrix** for two interval-valued observations ξ_{u_1} and ξ_{u_2} is $\boldsymbol{D} = (d(w_{u_1}, w_{u_2}))$, $u_1, u_2 = 1, \ldots, m$, where

$$d(w_{u_1}, w_{u_2}) = \{\sum_{j=1}^{p} [\phi_j(w_{u_1}, w_{u_2})]^2\}^{1/2} \tag{7.32}$$

and $\phi_j(w_{u_1}, w_{u_2})$ are the Hausdorff distances defined in Equation (7.31). □

Definitions 7.22: A **normalized Euclidean Hausdorff distance matrix** has elements

$$d(w_{u_1}, w_{u_2}) = \left(\sum_{j=1}^{p} \left[\frac{\phi_j(w_{u_1}, w_{u_2})}{H_j} \right]^2 \right)^{1/2} \tag{7.33}$$

where

$$H_j^2 = \frac{1}{2m^2} \sum_{u_1=1}^{m} \sum_{u_2=1}^{m} [\phi_j(w_{u_1}, w_{u_2})]^2 \tag{7.34}$$

and where $\phi_j(w_{u_1}, w_{u_2})$ is the Hausdorff distance of Equation (7.31). □

Example 7.13. From the Hausdorff distances found in Example 7.12 between the male and female horses and the male bears, it follows by substituting into Equation (7.32) that the Euclidean Hausdorff distance matrix is

$$\boldsymbol{D} = \begin{pmatrix} 0 & 106.79 & 209.35 \\ 106.79 & 0 & 206.57 \\ 209.35 & 206.57 & 0 \end{pmatrix}$$

where, for example,

$$d(\text{HorseM}, \text{HorseF}) = (38^2 + 99.8^2)^{1/2} = 106.79.$$

To find the normalized distance matrix, we first need, from Equation (7.34),

$$H_1^2 = \frac{1}{2 \times 3^2}[38^2 + 55^2 + 27^2] = 288.778$$

and similarly,

$$H_2^2 = 5150.39.$$

Hence, from Equation (7.33) we obtain the normalized Euclidean Hausdorff distance matrix as

$$\boldsymbol{D} = \begin{pmatrix} 0 & 2.63 & 4.29 \\ 2.63 & 0 & 3.27 \\ 4.29 & 3.27 & 0 \end{pmatrix}.$$

For example,

$$d\,(\text{HorseM}, \text{HorseF}) = \left[\left(\frac{38}{16.99}\right)^2 + \left(\frac{99.8}{71.77}\right)^2\right]^{1/2} = 2.633.$$

□

Definition 7.23: A **span normalized Euclidean Hausdorff distance matrix** has elements

$$d(w_{u_1}, w_{u_2}) = \left(\sum_{j=1}^{p}\left[\frac{\phi_j(w_{u_1}, w_{u_2})}{|\mathcal{Y}_j|}\right]^2\right)^{1/2} \tag{7.35}$$

where $|\mathcal{Y}_j|$ is the span of the observations Y_j defined in Equation (7.26). □

Example 7.14. For our female and male horses and male bears observations, the spans for Y_1 and Y_2 are, respectively,

$$|\mathcal{Y}_1| = 65, \ |\mathcal{Y}_2| = 237.8.$$

Then, for example,

$$d\,(\text{HorseM}, \text{HorseF}) = \left[\left(\frac{38}{65}\right)^2 + \left(\frac{99.8}{237.8}\right)^2\right]^{1/2} = 0.720.$$

We obtain the span normalized Euclidean Hausdorff distance matrix as

$$D = \begin{pmatrix} 0 & 0.72 & 1.20 \\ 0.72 & 0 & 0.94 \\ 1.20 & 0.94 & 0 \end{pmatrix}. \qquad \square$$

The first normalization of Definition 7.22 is a type of standard deviation corresponding to symbolic data. If the data are classical data, then the Hausdorff distances are equivalent to Euclidean distances on $\Re^2$, in which case H_j corresponds exactly to the standard deviation of Y_j. We refer to this as the dispersion normalization. The normalization of Definition 7.23 is based on the length of the domain of $\mathcal{Y}_j$ actually observed, i.e., the maximum deviation. We refer to this as the span normalization.

7.1.4 Mixed-valued variables

Dissimilarity and distance measures can be calculated between observations that have a mixture of multi-valued and interval-valued variables. In these cases, formulas for these measures provided in Section 7.1.2 are used on the Y_j which are multi-valued and those formulas of Section 7.1.3 are used on the interval-valued Y_j variables. The distances over all Y_j variables are then calculated using the appropriate summation formula, e.g., Equation (7.6) for a generalized Minkowski distance matrix. These principles are illustrated in the following example.

Example 7.15. Table 7.8 provides the oils dataset originally presented by Ichino (1988). The categories are 6 types of oil (linseed, perilla, cottonseed, sesame, camellia, and olive) and 2 fats (beef and hog); $m = 8$. There are four interval-valued variables, namely, Y_1 = Specific Gravity (in g/cm^3), Y_2 = Freezing Point (in °C), Y_3 = Iodine Value, and Y_4 = Saponification. There is one multi-valued variable, Y_5 = Fatty Acids. This variable identifies nine major fatty acids contained in the oils and takes values from the list of acids $\mathcal{Y}_5 = \{A, C, L, Ln, Lu, M, O, P, S\} \equiv$ {arachic, capric, linoleic, linolenic, lauric, myristic, oleic, palmitic, stearic}.

Table 7.8 Ichino's oils data.

w_u	Oil	Y_1 Specific Gravity	Y_2 Freezing Point	Y_3 Iodine Value	Y_4 Saponification	Y_5 Fatty Acids
w_1	linseed	[0.930, 0.935]	[−27.0, −8.0]	[170.0, 204.0]	[118.0, 196.0]	$\{L, Ln, M, O, P\}$
w_2	perilla	[0.930, 0.937]	[−5.0, −4.0]	[192.0, 208.0]	[188.0, 197.0]	$\{L, Ln, O, P, S\}$
w_3	cotton seed	[0.916, 0.918]	[−6.0, −1.0]	[99.0, 113.0]	[189.0, 198.0]	$\{L, M, O, P, S\}$
w_4	sesame	[0.920, 0.926]	[−6.0, −4.0]	[104.0, 116.0]	[187.0, 193.0]	$\{A, L, O, P, S\}$
w_5	camellia	[0.916, 0.917]	[−21.0, −15.0]	[80.0, 82.0]	[189.0, 193.0]	$\{L, O\}$
w_6	olive	[0.914, 0.919]	[0.0, 6.0]	[79.0, 90.0]	[187.0, 196.0]	$\{L, O, P, S\}$
w_7	beef	[0.860, 0.870]	[30.0, 38.0]	[40.0, 48.0]	[190.0, 199.0]	$\{C, M, O, P, S\}$
w_8	hog	[0.858, 0.864]	[22.0, 32.0]	[53.0, 77.0]	[190.0, 202.0]	$\{L, Lu, M, O, P, S\}$

Then, taking equal weights $c_j = 1/p = 1/5$ and $\gamma = 0$, we can obtain Ichino's city block distances shown in Table 7.9. These are calculated by substitution into Equation (7.29) for $q = 1$, where for $j = 1, \ldots, 4$, the entries $\phi_j(,)$ are obtained from Equation (7.27), and for $j = 5$, the entries $\phi_5(,)$ are obtained from Equation (7.17).

For example, consider the pair (linseed, lard) = W (say). Then,

$$\phi_1(W) = [(0.935 - 0.858) - (0.864 - 0.858)]/[(0.937 - 0.858)] = 0.899,$$

$$\phi_2(W) = [(32 - (-27)) - 0]/[(38 - (-27))] = 0.908,$$

$$\phi_3(W) = [(204 - 33) - 0]/[(208 - 40)] = 1.018,$$

$$\phi_4(W) = [(202 - 118) - (196 - 190)]/[(202 - 118)] = 0.929,$$

$$\phi_5(W) = (6 - 5)/9 = 0.111.$$

Hence,

$$d(W) = \frac{1}{5}(0.899 + \ldots + 0.111) = 0.773. \qquad \square$$

Table 7.9 City block distances – oils, $(Y_1, \ldots, Y_5)$.

w_u	Oils	w_1	w_2	w_3	w_4	w_5	w_6	w_7	w_8
w_1	linseed	0	0.320	0.471	0.488	0.466	0.534	0.853	0.809
w_2	perilla	0.320	0	0.244	0.226	0.336	0.273	0.626	0.582
w_3	cottonseed	0.471	0.244	0	0.105	0.182	0.117	0.418	0.374
w_4	sesame	0.488	0.226	0.105	0	0.192	0.141	0.503	0.459
w_5	camellia	0.466	0.336	0.182	0.192	0	0.160	0.504	0.460
w_6	olive	0.534	0.273	0.117	0.141	0.160	0	0.407	0.363
w_7	beef	0.853	0.626	0.418	0.503	0.504	0.407	0	0.181
w_8	lard	0.809	0.582	0.374	0.459	0.460	0.363	0.181	0

7.2 Clustering Structures

7.2.1 Types of clusters: definitions

We have p random variables $\{Y_j, j = 1, \ldots, p\}$ with actual realizations $\xi_u = (\xi_1, \ldots, \xi_p)$ on observations $w_u, u = 1, \ldots, m$, $w_u \in E = \{w_1, \ldots, w_m\}$ (if symbolic data) or realizations $x_i = (x_1, \ldots, x_p)$ for observations $i \in \Omega$, $i = \{1, \ldots, n\}$ (if classical data). We give definitions in terms of the symbolic observations in E; those for classical observations in Ω are special cases.

Definition 7.24: A **partition** of E is a set of subsets $\{C_1, \ldots, C_r\}$ such that

(i) $C_i \cap C_j = \phi$, for all $i \neq j = 1, \ldots, r$ and

(ii) $\bigcup_i C_i = E$, for $i = 1, \ldots, r$.

That is, the subsets are disjoint, but are exhaustive of the entire dataset E. The subsets of a partition are sometimes called classes or **clusters**. □

Example 7.16. The data of Table 7.5 represent values for Y_1 = Weight and Y_2 = Height of 10 animal breeds. These interval-valued observations were obtained by aggregating a larger set of data detailing animals handled at a certain veterinary clinic over a one-year period (available at http://www.ceremade.dauphine.fr/%7Etouati/clinique.htm). In this case, the concept involved is breed × gender. The plotted values of the $w_u, u = 1, \ldots, 8$, observations in Figure 7.1 suggest a partitioning which puts observations w_u with $u = 1, 2, 3$ (i.e., the male (M) and female (F) horses and the male bears) into class C_1^1 say, and observations w_u for $u = 4, 5, 6, 7, 8$ into class C_2^1.

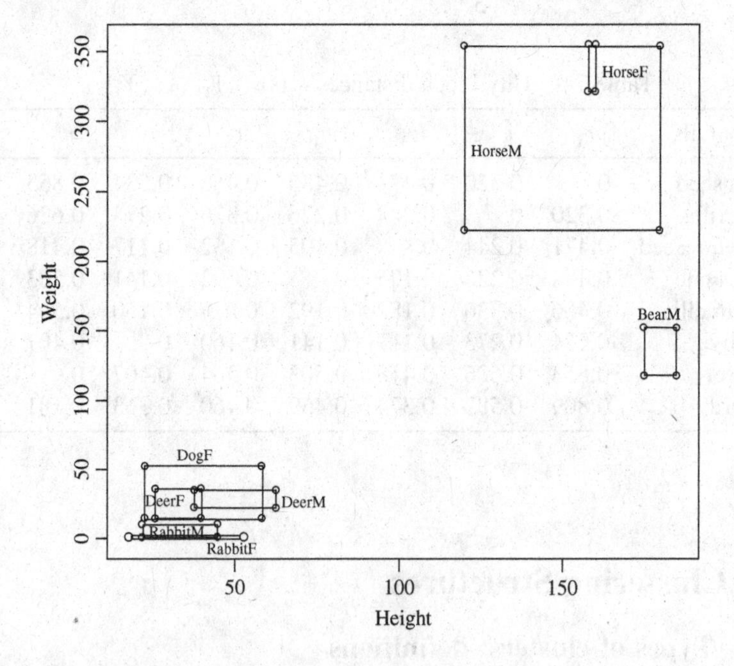

Figure 7.1 Partitions: veterinary data, u = 1, . . . , 8.

Another partitioning may have placed the horses, i.e., observations w_1 and w_2, all into a single cluster C_1^2, say, and the bear, i.e., observation w_3, into its own cluster C_2^2, with a third cluster $C_3^2 = C_2^1$. Still another partitioning might produce the four clusters C_i^3, $i = 1, \ldots, 4$, with $C_1^3 \equiv C_1^2$ and $C_2^3 \equiv C_2^2$, and with C_3^3 containing the observations w_7 and w_8 for the rabbits, and C_4^3 containing the observations w_4, w_5, w_6 corresponding to the male and female deer and the female dogs. This subpartitioning of C_2^1 is more evident from Figure 7.2 which plots the w_u, $u = 4, \ldots, 8$, observations of C_2^1 plus those for the male and female cats of w_9, w_{10}. Whether the first partitioning $P_1 = \{C_1^1, C_2^1\}$ or the second partitioning $P_2 = \{C_1^2, C_2^2, C_3^2\}$ or the third partitioning $P_3 = \{C_1^3, C_2^3, C_3^3, C_4^3\}$ is chosen in any particular analysis depends on the selection criteria associated with any clustering technique. Such criteria usually fall under the rubric of similarity–dissimilarity measures and distance measures; these were discussed in Section 7.1. □

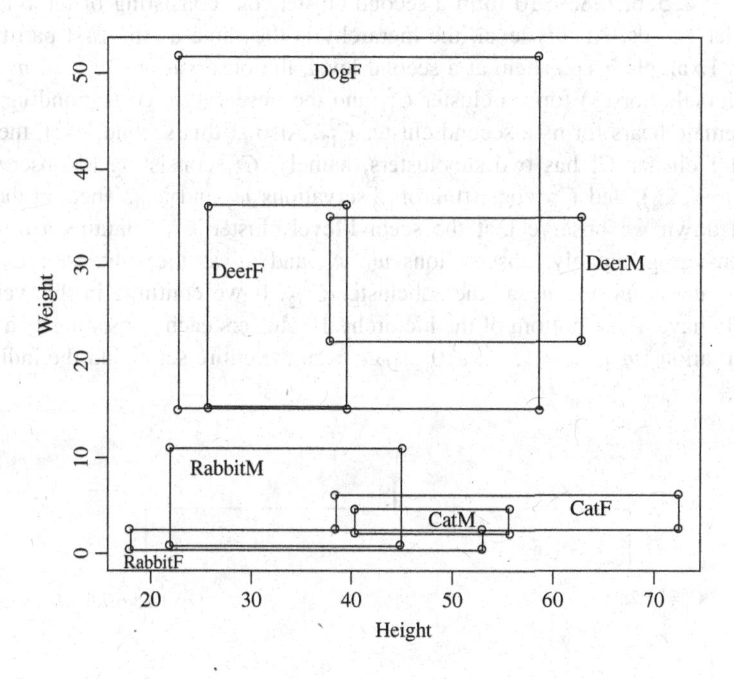

Figure 7.2 Veterinary partitions, $u = 4, \ldots, 10$.

Definition 7.25: A **hierarchy** on E is a set of subsets $H = \{C_1, \ldots, C_r\}$ such that

(i) $E \in H$;

(ii) for all single observations w_u in E, $C_u = \{w_u\} \in H$; and

(iii) $C_i \cap C_j \in \{\phi, C_i, C_j\}$ for all $i \neq j = 1, \ldots, m$. □

That is, condition (iii) tells us that either any two clusters C_i and C_j are disjoint, or one cluster is contained entirely inside the other, and every individual in E is contained in at least one cluster larger than itself.

Note that if $C_i \cap C_j = \phi$ for all $i \neq j$, then the hierarchy becomes a partitioning. Henceforth, reference to a hierarchy implies that $C_i \cap C_j \neq \phi$ for at least one set of (i, j) values.

Example 7.17. Consider all observations w_u, $u = 1, \ldots, 10$, in the veterinary data of Table 7.5. The hierarchy of Figure 7.3 may apply. Here, we start with all observations in the subset $E = C^1$. Then, at the first level of the hierarchy, the observations w_u with $u = 1, 2, 3$ (which correspond to the large animals, i.e., the male and female horses and female bears) form one cluster C_1^1 and the w_u for $u = 4, 5, 6, 7, 8, 9, 10$ form a second cluster, C_2^1, consisting of the relatively smaller breeds. At this level, the hierarchy is the same as the first partitioning P_1 of Example 7.16. Then, at a second level, the observations w_1 and w_2 (male and female horses) form a cluster C_{11}^2 and the observation corresponding to the w_3 female bears forms a second cluster C_{12}^2. Also at this second level, the other level 1 cluster C_2^1 has two subclusters, namely, C_{21}^2 consisting of observations $(w_4, \ldots, w_8)$, and C_{22}^2 consisting of observations w_9 and w_{10}. Then, at the third level down we observe that the second-level cluster C_{21}^2 contains a third tier of clustering, namely, observations w_4, w_5, and w_6 as the subcluster C_{211}^3 and observations w_7 and w_8 as the subcluster C_{212}^3. If we continue in this vein, we finally have at the bottom of the hierarchy 10 clusters each consisting of a single observation $\{w_u\}$, $u = 1, \ldots, 10$. Apart from the entire set E and the individual

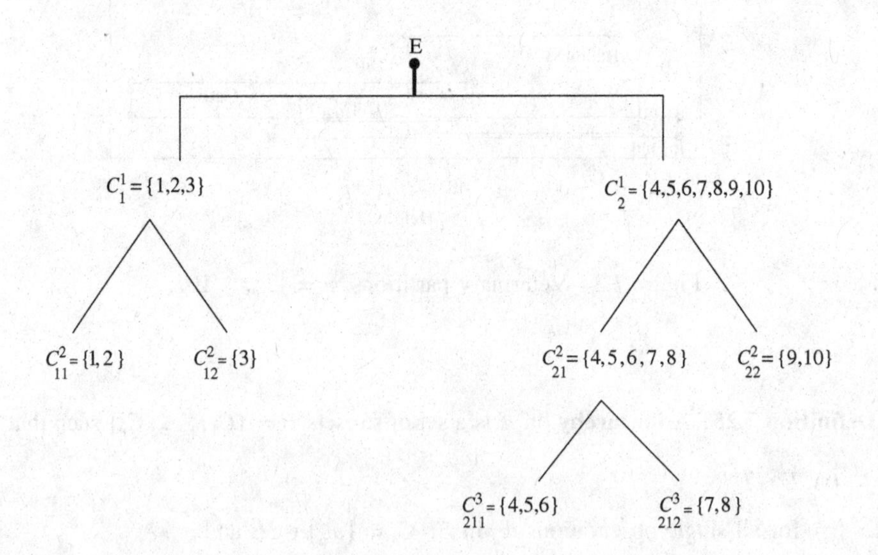

Figure 7.3 Hierarchy clusters – veterinary data.

clusters $C_u = \{w_u\}$, there are eight clusters (and relabeling the clusters) shown in Figure 7.3:

$$C_1 \equiv C_{11}^2 = \{w_1, w_2\},\ C_2 \equiv C_{12}^2 = \{w_3\},\ C_3 \equiv C_{211}^3 = \{w_4, w_5, w_6\},$$

$$C_4 \equiv C_{212}^3 = \{w_7, w_8\},\ C_5 \equiv C_{22}^2 = \{w_9, w_{10}\},\ C_6 \equiv C_{21}^2 = \{w_4, \ldots, w_8\},$$

and

$$C_7 \equiv C_1^1 = \{w_1, w_2, w_3\},\ C_8 \equiv C_2^1 = \{w_4, \ldots, w_{10}\}.$$

The hierarchy is $H = \{C_1, \ldots, C_8, \ldots, \{w_1\}, \ldots, \{w_{10}\}, E\}$. □

In Example 7.17, we started with the entire set of observations and split then into two classes, and then proceeded to split each of these classes into a second tier of classes, and so on. This 'top-down' process is an example of divisive hierarchical clustering.

Definition 7.26: Divisive clustering is a top-down clustering process which starts with the entire dataset as one class and then proceeds downward through as many levels as necessary to produce the hierarchy $H = \{C_1, \ldots, C_r\}$.

Agglomerative clustering is a bottom-up clustering process which starts with each observation being a class of size 1, and then forms unions of classes at each level for as many levels as necessary to produce the hierarchy $H = \{C_1, \ldots, C_r\}$ or a pyramid $P = \{\pi_1, \ldots, \pi_m\}$. □

In the agglomerative process, each observation is aggregated at most once for a hierarchy or at most twice for a pyramid. These two clustering processes are discussed in more detail later in Sections 7.4 and 7.5, respectively.

Example 7.17 (continued): The same clusters $C_1, \ldots, C_8$ could have been developed by agglomerative clustering. In this case, we start with the hierarchy consisting of the 10 classes $H = \{\{w_1\}, \{w_2\}, \ldots, \{w_{10}\}\}$. Then, at the bottom level, the clustering criteria form the unions $\{w_5\} \cup \{w_6\} = \{w_5, w_6\}$, i.e., the female deer and dogs with all other clusters remaining as individuals.

At the second level, the criteria produce the unions

$$C_{11}^2 = \{w_1\} \cup \{w_2\} = \{w_1, w_2\},\ C_{21}^2 = \{w_4\} \cup \{w_5, w_6\} = \{w_4, w_5, w_6\},$$

$$C_{22}^2 = \{w_7\} \cup \{w_8\} = \{w_7, w_8\},\ C_{12}^2 = \{w_3\};$$

while at the third level, we have the two clusters

$$C_1^1 = C_{11}^2 \cup C_{12}^2 = \{w_1, w_2, w_3\},$$

$$C_2^1 = C_{21}^2 \cup C_{22}^2 = \{w_4, \ldots, w_{10}\};$$

hence we reach the complete set $E = C_1^1 \cup C_2^1 = \{w_1, \ldots, w_{10}\}$. □

Notice that the top two levels of this hierarchy tree correspond to the partitioning P_3 of Example 7.16. A hierarchy has additional levels as necessary to reach single units at its base.

Definition 7.27: A set of clusters $H = \{C_1, \ldots, C_r\}$ where at least one pair of classes overlap, i.e., where there exists at least one set (C_i, C_j), $i \neq j$, such that $C_i \cap C_j \neq \phi$ and where there exists an order on the observations such that each cluster defines an interval for this order, is called a **non-hierarchical pyramid.** □

Definition 7.32 gives a more general definition of a pyramid. That definition includes the notion of ordered sets and allows a hierarchy to be a special case of a pyramid.

Example 7.18. Suppose for the animal breeds data of Table 7.5 we start with the 10 clusters $H = \{\{1\}, \ldots, \{10\}\}$ corresponding to the observations w_u, $u = 1, \ldots, 10$. Suppose the clustering criteria are such that the five clusters

$$C_1 = \{w_1, w_2\},\ C_2 = \{w_3\},\ C_3 = \{w_4, w_5, w_6\},\ C_4 = \{w_7, w_8\},\ C_5 = \{w_8, w_9, w_{10}\}$$

emerge; see Figure 7.4. Notice that the w_8 (female rabbits) observation is in both the C_4 and C_5 clusters. □

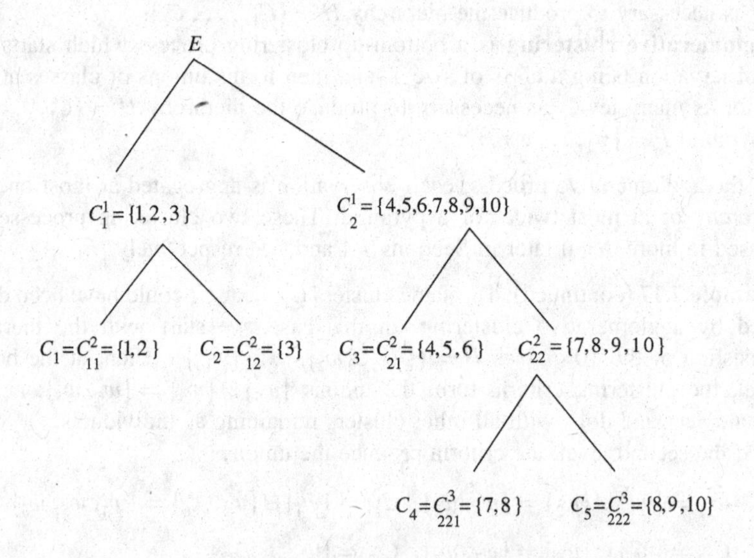

Figure 7.4 Pyramid clusters – veterinary data.

This overlap of clusters in Figure 7.4 is what distinguishes a pyramid tree from a hierarchy tree. The clusters $C_1, \ldots, C_4$ of Example 7.18 are identical, respectively, to the clusters C^2_{11}, C^2_{12}, C^3_{211}, and C^3_{212} in the hierarchy of Example 7.17 and Figure 7.3. The difference is in the fifth cluster. Here, in the pyramid tree, the observation w_8 corresponding to the female rabbit is in two clusters, i.e., C_4 consisting of male and female rabbits, and C_5 consisting of the male and female cats plus the female rabbits. We notice from Figure 7.2 that the data are highly suggestive that the cluster C_5 should contain all three (w_8, w_9, w_{10}) observations, and that the rabbits together form their own cluster C_4 as well.

A pictorial comparison of the three clustering types can be observed by viewing Figures 7.1–7.4.

Definition 7.28: Let $P = (C_1, \ldots, C_r)$ be a hierarchy of E. A class C_j is said to be a **predecessor** of a class C_i if C_i is completely contained in but is not entirely C_j, and if there is no class C_k contained in C_j that contains C_i; that is, there is no k such that if $C_i \subset C_j$ then $C_i \subset C_k \subset C_j$. Equivalently, we can define C_i as a **successor** of C_j. □

In effect, a predecessor cluster of a particular cluster refers to the cluster immediately above that cluster in the hierarchy tree. Any cluster in a hierarchy has at most one predecessor, whereas it may have up to two predecessors in a pyramid tree.

Example 7.19. Let $H = (C_1, C_2, C_3, \ldots, C_9)$ of Figure 7.5, where

$$C_1 = E = \{a, b, c, d, e\}$$
$$C_2 = \{a, b, c\},\ \ C_3 = \{b, c\},\ \ C_4 = \{d, e\},$$
$$C_5 = \{a\}, \ldots, C_9 = \{e\}.$$

Then, C_1 is a predecessor of C_2. However, C_1 is not a predecessor of C_3 since, although $C_3 \subset C_1$, there does exist another class, here C_2, which is contained in C_1 and which also contains C_3. □

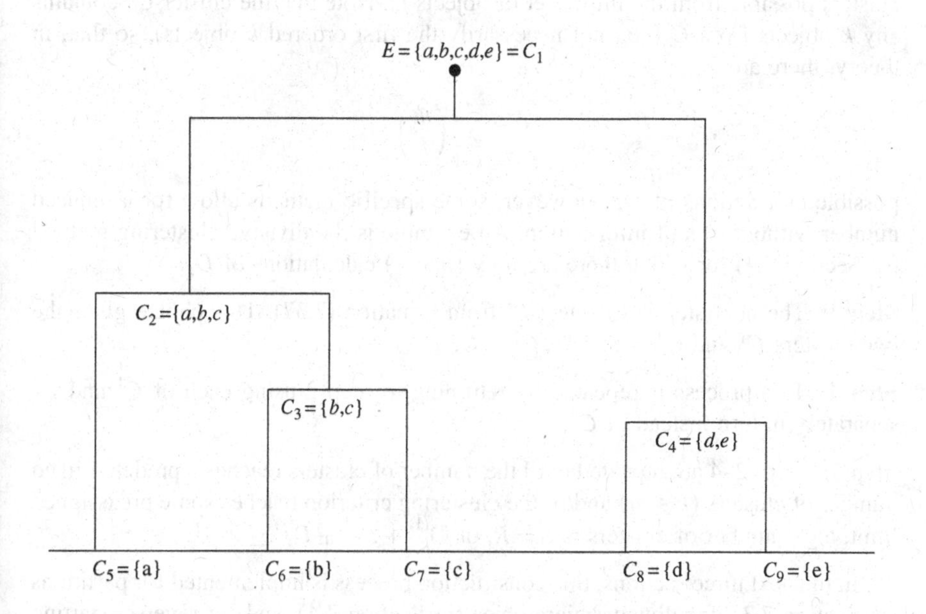

Figure 7.5 Predecessors.

7.2.2 Construction of clusters: building algorithms

As noted in Section 7.2.1 and Definition 7.26, hierarchies can be built from the top down, or from the bottom up. In this section, we discuss the divisive top-down clustering process; the pyramid bottom-up clustering process is covered in Section 7.5. Regardless of the type of cluster, an algorithm is necessary to build its structure. These algorithms will typically be iterative in nature.

Step 1: The first step is to establish the criterion (or criteria) by which one particular clustering structure would be selected over another. This criterion would be based on dissimilarity and distance measures such as those described in Section 7.1. Therefore, if we have a set of objects $C = \{a_1, \ldots, a_m\}$ (say), and if we take one cluster $C^1 = \{a_1, \ldots, a_k\}$ and a second cluster $C^2 = \{a_{k+1}, \ldots, a_m\}$, then a clustering criterion could be that value of $k = k^*$ for which

$$D_k = \sum_{(i,j)\in C^1} d(a_i, a_j) + \sum_{(i,j)\in C^2} d(a_i, a_j) = D_k^{(1)} + D_k^{(2)}, \tag{7.36}$$

say, is minimized, where $d(a, b)$ is the selected dissimilarity or distance measure. That is,

$$D_{k^*} = \min_k D_k. \tag{7.37}$$

The aim is to find clusters that are internally as homogeneous as possible.

Step 2: The second step therefore is to calculate D_k from Equation (7.36) for all the clusters possible from the initial set of objects C. Note that the cluster C^1 contains any k objects from C (i.e., not necessarily the first ordered k objects), so that, in theory, there are

$$N = \sum_{k=1}^{m} \binom{m}{k}$$

possible calculations of D_k. However, some specific methods allow for a reduced number without loss of information. An example is the divisive clustering method (of Section 7.4) for which there are only $(m-1)$ calculations of D_k.

Step 3: The next step is to select k^* from Equation (7.37). This choice gives the two clusters C^1 and C^2 with $C^1 \cup C^2 = C$.

Step 4: This process is repeated by returning to Step 2 using each of C^1 and C^2 separately in turn instead of C.

Step 5: Steps 2–4 are repeated until the number of clusters reaches a predetermined number of clusters ($r \leq m$) and/or the clustering criterion reaches some preassigned limit, e.g., number of clusters is $r = R$, or $D_k^{(1)} + \ldots + D_k^{(r)} \leq \delta$.

In the next three sections, this construction process is implemented for partitions (in Section 7.3), for divisive hierarchies (in Section 7.4), and (in reverse, starting at the bottom) for pyramids (in Section 7.5).

7.3 Partitions

In Example 7.16 and Figure 7.1 we were able to partition visually the veterinary data of Table 7.5 into two classes or clusters. The building algorithm of Section 7.2.2 can be used to construct this partition in a more rigorous way.

Example 7.20. Take the veterinary data of Table 7.5. Let us suppose (in Step 1) that we select our partitioning criterion to be based on the Ichino–Yaguchi distance measures of Equation (7.27) of Definition 7.18. The pairwise Ichino–Yaguchi distances between the male and female horses and male bears were calculated in Example 7.11, for each variable Y_1 = Height and Y_2 = Weight. Calculation of the other distances is left as an exercise. Let us take these distances to determine the normalized Euclidean distance *d(a,b)* of Equation (7.30) for use in the partitioning criterion of Equation (7.36). For example,

$$d(\text{HorseM, BearM}) = \left\{\frac{1}{2}\left[(30)^2/167 + (153.5)^2/354.6\right]\right\}^{1/2}$$
$$= 5.993$$

where we have taken $\gamma = 1/2$, and used the weight $w_j^* = 1/|\mathcal{Y}_j|, j = 1, 2$, where $|\mathcal{Y}_j|$ is as defined in Equation (7.26).

Then, the normalized Euclidean distance matrix for all $m = 10$ animals is

$$D = \begin{pmatrix} 0 & 2.47 & 5.99 & 11.16 & 11.76 & 11.28 & 12.37 & 12.45 & 12.06 & 11.85 \\ 2.47 & 0 & 7.74 & 13.07 & 13.62 & 13.16 & 14.25 & 14.35 & 13.97 & 13.77 \\ 5.99 & 7.74 & 0 & 8.13 & 9.04 & 8.52 & 9.36 & 9.35 & 8.74 & 8.39 \\ 11.16 & 13.07 & 8.13 & 0 & 0.98 & 0.70 & 1.26 & 1.31 & 0.98 & 0.95 \\ 11.76 & 13.62 & 9.04 & 0.98 & 0 & 0.67 & 0.78 & 1.08 & 1.19 & 1.48 \\ 11.28 & 13.16 & 8.52 & 0.70 & 0.67 & 0 & 1.11 & 1.23 & 1.26 & 1.36 \\ 12.37 & 14.25 & 9.35 & 1.26 & 0.78 & 1.11 & 0 & 0.37 & 0.81 & 1.21 \\ 12.45 & 14.35 & 9.35 & 1.31 & 1.08 & 1.23 & 0.37 & 0 & 0.69 & 1.09 \\ 12.06 & 13.97 & 8.74 & 0.98 & 1.19 & 1.26 & 0.81 & 0.69 & 0 & 0.51 \\ 11.85 & 13.77 & 8.39 & 0.95 & 1.48 & 1.36 & 1.21 & 1.09 & 0.51 & 0 \end{pmatrix}.$$

We take as our criterion in Equation (7.36) the sum of the pairwise normalized Euclidean distances for all animals in the given subsets (C^1, C^2).

Step 2 then entails calculating these total distances $D_k^{(1)}$ and $D_k^{(2)}$ for respective partitions. For example, suppose we set C^1 = {HorseM, HorseF, BearM} and therefore C^2 consists of the other $(m - k = 10 - 3 = 7)$ animals, i.e., C^2 = {DeerM, DeerF, DogF, RabbitM, RabbitF, CatM, CatF}. Then, the respective distance matrices are

$$D^{(1)} = \begin{pmatrix} 0 & 2.47 & 5.99 \\ 2.47 & 0 & 7.74 \\ 5.99 & 7.74 & 0 \end{pmatrix}$$

$$D^{(2)} = \begin{pmatrix} 0 & 0.98 & 0.70 & 1.26 & 1.31 & 0.98 & 0.95 \\ 0.98 & 0 & 0.67 & 0.78 & 1.08 & 1.19 & 1.48 \\ 0.70 & 0.67 & 0 & 1.11 & 1.23 & 1.26 & 1.36 \\ 1.26 & 0.78 & 1.11 & 0 & 0.37 & 0.81 & 1.21 \\ 1.31 & 1.08 & 1.23 & 0.37 & 0 & 0.69 & 1.09 \\ 0.98 & 1.19 & 1.26 & 0.81 & 0.69 & 0 & 0.51 \\ 0.95 & 1.48 & 1.36 & 1.21 & 1.09 & 0.51 & 0 \end{pmatrix}$$

Then, the sum of the Euclidean distances inside the cluster C^1 is the sum of the upper diagonal elements of $\boldsymbol{D}^{(1)}$ (or equivalently, half the sum of all the elements of $\boldsymbol{D}^{(1)}$), and likewise for the total of the Euclidean distances inside the cluster C^2. In this case, from Equation (7.36),

$$D_3^{(1)} = 16.20, D_3^{(2)} = 21.02$$

and hence

$$D_3 = 37.22.$$

Table 7.10 provides some partitions and their respective cluster sums of normalized Euclidean distances $D_k^{(1)}$ and $D_k^{(2)}$ and the total distance D_k. This is by no means an exhaustive set. At Step 3, we select the partition which minimizes D_k. In this case, we obtain

$$C^1 = \{\text{HorseM, HorseF, BearM}\}$$

$$C^2 = \{\text{DeerM, DeerF, DogF, RabbitM, RabbitF, CatM, CatF}\}.$$

We now have our $R = 2$ clusters, as previously depicted visually in Figure 7.1. If $R > 2$ clusters are required, then in Step 5 we repeat the process (Steps 2–4), as needed on C^1 and C^2 first, and then on their clusters. □

Table 7.10 Two partitions – veterinary data.

k	C_1	$D_k^{(1)}$	$D_k^{(2)}$	D_k
2	{HorseM HorseF},	2.47	82.54	85.01
2	{HorseF, BearM},	7.74	103.95	111.69
3	{HorseM, HorseF, BearM}	16.20	21.02	**37.22**
3	{BearM, DogF, DeerM}	17.35	127.24	144.59
4	{HorseM, HorseF, BearM, DeerM}	48.56	14.84	63.40
5	{HorseM, HorseF, BearM, DeerM, DeerF}	83.96	7.64	93.60
6	{HorseM, HorseF, BearM, DeerM, DeerF, DogF}	118.29	4.68	122.97
7	{HorseM, HorseF, BearM, DeerM, DeerF, DogF, RabbitM}	157.42	2.29	159.71
8	{HorseM, HorseF, BearM, DeerM, DeerF, DogF, RabbitM, RabbitF}	197.56	0.51	198.07

7.4 Hierarchy–Divisive Clustering

As noted in Section 7.2 and Definition 7.26, hierarchies can be built from the top down, or from the bottom up. In this section, we discuss the divisive top-down clustering process; the pyramid bottom-up clustering process is covered in Section 7.5. The divisive clustering methodology for symbolic data was developed by Chavent (1997, 1998, 2000). We describe the principles in Section 7.4.1 and then apply them to multi-valued and interval-valued variables in Sections 7.4.2 and 7.4.3, respectively.

7.4.1 Some basics

We start with all observations in the one class $C_1 \equiv E$, and proceed by bisecting this class into two classes (C_1^1, C_1^2). The actual choice for C_1^1 and C_1^2 will depend on an optimality criterion for the bisection. Then, $C_1^b (b = 1, 2)$ is bisected into $C_1^b = (C_{11}^b, C_{12}^b)$ which will be denoted by (C_2^1, C_2^2), and so on. Whether C_1^1 or C_1^2 is bisected depends on the underlying partitioning criteria being used. Suppose that at the rth stage there are the clusters $(C_1, \ldots, C_r) = P_r$, with the cluster C_k containing m_k observations. Recall that $E = \cup C_k$. One of the clusters, C_k say, is to be bisected into two subclusters, $C_k = (C_k^1, C_k^2)$, to give a new partitioning

$$P_{r+1} = (C_1, \ldots, C_k^1, C_k^2, \ldots, C_r) \equiv (C_1, \ldots, C_{r+1}).$$

The particular choice of C_k is determined by a selection criterion (described in Equation (7.42) below) which when optimally bisected will minimize the total within-cluster variance for the partition across all optimally bisected $\{C_k\}$.

The class C_k itself is bisected into C_k^1 and C_k^2 by determining which observations in C_k satisfy, or do not satisfy, some criteria $q(\cdot)$ based on the actual values of the variables involved. Suppose the class C_k contains observations identified by $u = 1, \ldots, m_k$, say. Then,

$$C_k^1 = \{u | q(u) \text{ is true, or } q(u) = 1\}$$

$$C_k^2 = \{u | q(u) \text{ is false, or } q(u) = 0\}.$$

The divisive clustering technique for symbolic data assumes that the question $q(\cdot)$ relates to one Y_j variable at a time. However, at each bipartition, all Y_j, $j = 1, \ldots, p$, variables can be considered.

Example 7.21. The data of Table 7.14 below provide values for six random variables relating to eight different breeds of horses (see Section 7.4.3 where the data are described more fully). A bipartition query on the random variable Y_1 = Minimum Height of the horses would be

$$q_1 : Y_1 \leq 130.$$

Then, comparing the interval midpoints to $q_1(\cdot)$, we have, for example $q_1(u = 1) = 0$ since for the first horse the midpoint is $(145 + 155)/2 = 150 > 130$.

Likewise, $q_1(2) = 0,\ q_1(3) = 1,\ q_1(4) = 0,\ q_1(5) = 0,\ q_1(6) = 0,\ q_1(7) = 1,$ $q_1(8) = 1$. Therefore, were the bisection to be based on this $q_1(\cdot)$ alone, the new classes would contain the breeds w_u according to

$$C_1^1 = \{w_3, w_7, w_8\},\ C_1^2 = \{w_1, w_2, w_4, w_5, w_6\}.$$

□

If there are z_j different bipartitions for the variable Y_j, then there are $z_k = z_1 + \ldots + z_p$ possible bipartitions of C_k into (C_k^1, C_k^2). The actual bipartition chosen is that which satisfies the given selection criteria.

At each stage there are two choices:

(i) For the selected cluster C_k, how is the split into two subclusters C_k^1 and C_k^2 determined?

(ii) Which cluster C_k is to be split?

The answers involve minimizing within-cluster variation and maximizing between-cluster variation for (i) and (ii), respectively.

Definition 7.29: For a cluster $C_k = \{w_1, \ldots, w_{m_k}\}$, the **within-cluster variance** $I(C_k)$ is

$$I(C_k) = \frac{1}{2\lambda} \sum_{u_1=1}^{m_k} \sum_{u_1=1}^{m_k} p_{u_1} p_{u_2} d^2(u_1, u_2) \tag{7.38}$$

where $d^2(u_1, u_2)$ is a distance measure between the observations w_{u_1}, w_{u_2} in C_k, and p_u is the weight of the observation w_u with $\lambda = \sum_{u=1}^{m_k} p_u$. □

The distances $d(u_1, u_2)$ take any of the forms presented in Section 7.1. When $p_u = 1/m$, where m is the total number of observations in E,

$$I(C_k) = \frac{1}{2mm_k} \sum_{u_1=1}^{m_k} \sum_{u_2=1}^{m_k} d^2(u_1, u_2)$$

or, equivalently, since $d(u_1, u_2) = d(u_2, u_1)$,

$$I(C_k) = \frac{1}{mm_k} \sum_{u_1=1}^{m_k} \sum_{u_2>u_1=1}^{m_k} d^2(u_1, u_2). \tag{7.39}$$

Definition 7.30: For a partition $P_r = (C_1, \ldots, C_r)$, the **total within-cluster variance** is

$$W(P_r) = \sum_{k=1}^{r} I(C_k) \tag{7.40}$$

where $I(C_k)$ is the within-cluster variance for C_k as defined in Definition 7.29. □

Note that a special case of Equation (7.40) is $W(E)$ where the complete set of observations E consists of $r = m$ clusters C_k with $C_k = \{w_{u_k}\}$ containing a single observation.

Definition 7.31: For the partition $P_r = (C_1, \ldots, C_r)$, the **between-cluster variance** is

$$B(P_r) = W(E) - W(P_r). \tag{7.41}$$

□

The divisive clustering methodology is to take in turn each cluster C_k in $P_r = (C_1, \ldots, C_r)$. Let $P_r(k) = (C_k^1, C_k^2)$ be a bipartition of this C_k. There are $n_k \leq (m_k - 1)$ distinct possible bipartitions. When all observations w_u in C_k are distinct, $n_k = m_k - 1$. The particular bipartition $P_r(k)$ chosen is that one which satisfies

$$Min_{L_k}[W(P_r(k))] = Min_{L_k}[I(C_k^1) + I(C_k^2)] \tag{7.42}$$

where L_k is the set of all bipartitions of C_k. From Equation (7.42), we obtain the best subpartition of $C_k = (C_k^1, C_k^2)$.

The second question is which value of k, i.e., which $C_k \subset P_r$, is to be selected? For a given choice C_k, the new partition is $P_{r+1} = (C_1, \ldots, C_k^1, C_k^2, \ldots, C_r)$. Therefore, the total within-cluster variation of Equation (7.40) can be written as

$$W(P_r + 1) = \sum_{k'=1}^{r} I(C_{k'}) - I(C_k) + I(C_k^1) + I(C_k^2).$$

The choice C_k is chosen so as to minimize this total within-cluster variation. This is equivalent to finding k' such that

$$I_{k'} = Max_k[I(C_k) - I(C_k^1) - I(C_k^2)] = Max_k \Delta W(C_k). \tag{7.43}$$

The third and final issue of the divisive clustering process is when to stop. That is, what value of r is enough? Clearly, by definition of a hierarchy it is possible to have $r = m$, at which point each class/cluster consists of a single observation. In practice, however, the partitioning process stops at an earlier stage. The precise question as to what is the optimal value of r is still an open one. Typically, a maximum value for r is prespecified, R say, and then the divisive clustering process is continued consecutively from $r = 1$ up to $r = R$.

This procedure is an extension in part of the Breiman *et al.* (1984) approach for classical data. At the kth bipartition of C_k into (C_k^1, C_k^2) there will be at most $p(m_k - 1)$ bipartitions possible. This procedure therefore involves much less computation than the corresponding $(2^{m_k - 1} - 1)$ possible bipartitions that would have to be considered were the Edwards and Cavalli-Sforza (1965) procedure used.

7.4.2 Multi-valued variables

The divisive clustering methodology applied to non-modal multi-valued variables will be illustrated through its application to the dataset of Table 7.11. These relate to computer capacity in $m = 7$ different offices in a given company. The random variable Y_1 = Brand identifies the manufacturer of the computer and takes coded values from $\mathcal{Y}_1 = \{B1, \ldots, B4\}$. The random variable Y_2 = Type of computer and takes values from $\mathcal{Y}_2 = \{\text{laptop (T1)}, \text{desktop (T2)}\}$. We perform the divisive clustering procedure using the Y_1 variable only. Then, we leave the procedure basing the partitioning on both variables Y_1 and Y_2 as an exercise. We will use the categorical distance measures of Definition 7.14 and Equation (7.13). In order to do this, we first transform the data according to Equation (7.12). The resulting relative frequencies $p_{u_j k_j}$ are as shown in Table 7.12.

Table 7.11 Computers.

w_u		Y_1 = Brand	Y_2 = Type
w_1	Office1	{B1, B2, B3}	{T1, T2}
w_2	Office2	{B1, B4}	{T1}
w_3	Office3	{B2, B3}	{T2}
w_4	Office4	{B2, B3, B4}	{T2}
w_5	Office5	{B2, B3, B4}	{T1, T2}
w_6	Office6	{B1, B4}	{T1}
w_7	Office7	{B2, B4}	{T2}

Table 7.12 Computer frequencies.

	Brand				Type		
	B1	B2	B3	B4	T1	T2	
Office1	1/3	1/3	1/3	0	1/2	1/2	2
Office2	1/2	0	0	1/2	1	0	2
Office3	0	1/2	1/2	0	0	1	2
Office4	0	1/3	1/3	1/3	0	1	2
Office5	0	1/3	1/3	1/3	1/2	1/2	2
Office6	1/2	0	0	1/2	1	0	2
Office7	0	1/2	0	1/2	0	1	2
Frequency	4/3	2	3/2	13/6	3.0	4.0	14.0

Looking at all the possible partitions of a cluster C into two subclusters can be a lengthy task, especially if the size of C is large. However, whenever the multi-valued random variable is such that there is a natural order to the entities in $\mathcal{Y}$, or when it is a modal-valued multi-valued variable, it is only necessary to consider those cutpoints c for which

$$\sum_{k_j<c_j} p_{jk_j} > 1/2 \tag{7.44}$$

for each Y_j, i.e., $c_j \geq Y_{jk_j}$, which ensures Equation (7.44) holds.

Example 7.22. For the computer dataset, from Table 7.11, the cutpoint for the Y_1 = Brand variable is between the possible brands B2 and B3 when comparing most offices; for the potential subcluster involving offices 4 and 5, this cutpoint lies between B3 and B4. □

Example 7.23. For the random variable Y_1 = Brand in the computer dataset of Table 7.11, we substitute the frequencies p_{uk} of Table 7.12 into Equation (7.13) where $p = 1$ here, to obtain the categorical distance matrix as

$$\boldsymbol{D} = \begin{pmatrix} 0.00 & 0.52 & 0.34 & 0.37 & 0.37 & 0.52 & 0.54 \\ 0.52 & 0.00 & 0.77 & 0.57 & 0.57 & 0.00 & 0.56 \\ 0.34 & 0.77 & 0.00 & 0.29 & 0.29 & 0.77 & 0.53 \\ 0.37 & 0.57 & 0.29 & 0.00 & 0.00 & 0.57 & 0.32 \\ 0.37 & 0.57 & 0.29 & 0.00 & 0.00 & 0.57 & 0.32 \\ 0.52 & 0.00 & 0.77 & 0.57 & 0.57 & 0.00 & 0.56 \\ 0.54 & 0.56 & 0.53 & 0.32 & 0.32 & 0.56 & 0.00 \end{pmatrix}.$$

For example,

$$d^2(w_1, w_4) = \left[\frac{(1/3-0)^2}{4/3} + \frac{(1/3-1/3)^2}{2} + \frac{(1/3-1/3)^2}{3/2} + \frac{(0-1/3)^2}{13/6}\right]^{1/2}$$
$$= 0.367.$$

Let us reorder the observations so that cutpoints change progressively from brands B1 to B4. The ordering is as shown in part (a) of Table 7.13, along with the corresponding cutpoints. Thus, for example, the cutpoint for Office3 lies between the brands B2 and B3. Under this ordering the matrix distance becomes

$$\boldsymbol{D} = \left(\begin{array}{ccc|cccc} 0.00 & 0.00 & 0.77 & 0.56 & 0.52 & 0.57 & 0.57 \\ 0.00 & 0.00 & 0.77 & 0.56 & 0.52 & 0.57 & 0.57 \\ 0.77 & 0.77 & 0.00 & 0.53 & 0.34 & 0.29 & 0.29 \\ \hline 0.56 & 0.56 & 0.53 & 0.00 & 0.54 & 0.32 & 0.32 \\ 0.52 & 0.52 & 0.34 & 0.54 & 0.00 & 0.37 & 0.37 \\ 0.57 & 0.57 & 0.29 & 0.32 & 0.37 & 0.00 & 0.00 \\ 0.57 & 0.57 & 0.29 & 0.32 & 0.37 & 0.00 & 0.00 \end{array}\right).$$

Table 7.13 Computer clusters.

	w_k	s	Cutpoint	$I(C_i^1)$	$I(C_i^2)$	$W(C_i)$
(a)	Office2	1	B1, B2	0.000	0.151	0.151
	Office6	2	B1, B2	0.000	0.096	**0.096**
	Office3	3	B2, B3	0.073	0.069	0.142
$i = 1$	Office7	4	B2, B3	0.114	0.035	0.149
	Office1	5	B2, B3	0.146	0.000	0.146
	Office4	6	B3, B4	0.172	0.000	0.172
	Office5	7	B3, B4	0.191	—	0.191
(b)	Office3	1	B2, B3	0.000	0.069	0.069
	Office7	2	B2, B3	0.038	0.035	0.073
$i = 2$	Office1	3	B2, B3	0.067	0.000	**0.067**
	Office4	4	B3, B4	0.085	0.000	0.085
	Office5	5	B3, B4	0.096	—	0.096

Suppose we want $R = 2$ clusters. With this ordering, there are $m - 1 = 6$ possible partitions, with s offices in C^1 and $m - s$ offices in C^2. Take $s = 3$. Then, $C = \{C_1^1, C_1^2\}$ with

$$C_1^1 = \{\text{Office2, Office6, Office3}\}$$

$$C_1^2 = \{\text{Office7, Office1, Office4, Office5}\}.$$

The dashed lines in the matrix $\boldsymbol{D}$ relate to this partition with the upper left-hand portion being the distance matrix associated with the cluster C_1^1 and the lower right-hand matrix being that for the cluster C_1^2. From Equation (7.39),

$$I(C_1^1) = \frac{1}{7 \times 3}(0 + 0.77 + 0.77) = 0.073$$

and

$$I(C_1^2) = \frac{1}{7 \times 4}(0.54 + 0.32 + \ldots + 0.00) = 0.069.$$

Hence,

$$W(C_1) = I(C_1^1) + I(C_1^2) = 0.073 + 0.069 = 0.142.$$

The within-cluster variance $I(C_1^1)$ and $I(C_1^2)$ along with the total within-cluster variance $W(C_1)$ for each partition is shown in part (a) of Table 7.13. Comparing the $W(C_1)$ values, we select the optimal bipartition to be

$$C_1 = \{\text{Office2, Office6}\}$$

$$C_2 = \{\text{Office1, Office3, Office4, Office5, Office7}\}$$

since this $(s = 2)$ partition produces the minimum $W(C_1)$ value; here $W(C_1) = 0.96$.

To bisect C_2 into two clusters $C_2 = \{C_2^1, C_2^2\}$, this process is repeated on the $\boldsymbol{D}$ matrix associated with C_2. The resulting within-cluster variances and total within-cluster variances are shown in part (b) of Table 7.13. From this, we observe that the optimal $R = 3$ cluster partition is

$$C_1 = \{\text{Office2, Office6}\}$$

$$C_2 = \{\text{Office1, Office3, Office7}\}$$

$$C_3 = \{\text{Office4, Office5}\}$$

since the $C_2 = \{C_2^1, C_2^2\}$ cluster when cut at $s = 3$ gives the minimum $W(C_2)$ value at 0.067.

The total within-cluster variation is

$$W(P_3) = 0.000 + (0.067 + 0.000) = 0.067 < W(E) = 0.191.$$

Therefore, a reduction in variation has been achieved. This hierarchy is displayed in Figure 7.6. □

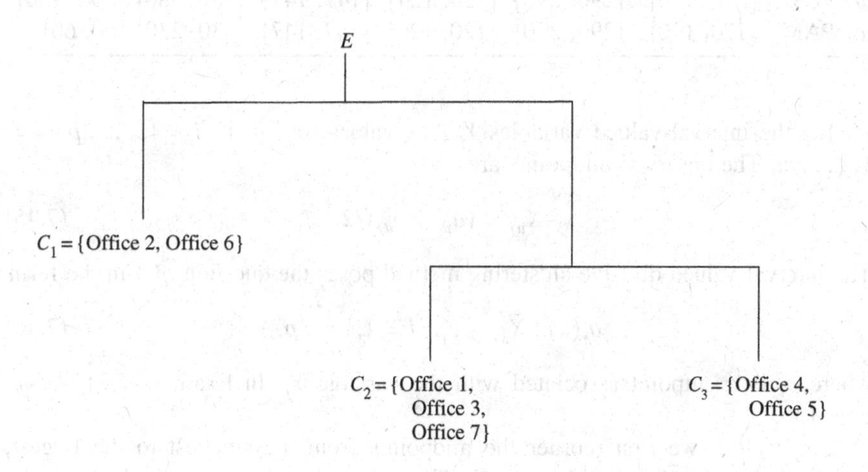

Figure 7.6 Computer hierarchy.

7.4.3 Interval-valued variables

How the divisive clustering methodology is applied to interval-valued random variables will be described through its illustration on the horses dataset of Table 7.14. This dataset consists of eight breeds of horses E ={European, Arabian European

Pony, French, North European, South European, French Pony, American Pony}, coded as E ={CES, CMA, PEN, TES, CEN, LES, PES, PAM}, respectively. There are $p = 6$ random variables: Y_1 = Minimum Weight, Y_2 = Maximum Weight, Y_3 = Minimum Height, Y_4 = Maximum Height, Y_5 = (coded) Cost of Mares, and Y_6 = (coded) Cost of Fillies. At the end of this section, the complete hierarchy on all eight breeds using all six random variables will be presented, though often portions only of the dataset may be used to demonstrate the methodology as we proceed through the various stages involved.

Table 7.14 Horses interval-valued dataset.

w_u	Breed	Minimum Weight	Maximum Weight	Minimum Height	Maximum Height	Mares Cost	Fillies Cost
w_1	CES	[410, 460]	[550, 630]	[145, 155]	[158, 175]	[150, 480]	[40, 130]
w_2	CMA	[390, 430]	[570, 580]	[130, 158]	[150, 167]	[0, 200]	[0, 50]
w_3	PEN	[130, 190]	[210, 310]	[90, 135]	[107, 153]	[0, 100]	[0, 30]
w_4	TES	[410, 630]	[560, 890]	[135, 165]	[147, 172]	[100, 350]	[30, 90]
w_5	CEN	[410, 610]	[540, 880]	[145, 162]	[155, 172]	[70, 600]	[20, 160]
w_6	LES	[170, 450]	[290, 650]	[118, 158]	[147, 170]	[0, 350]	[0, 90]
w_7	PES	[170, 170]	[290, 290]	[124, 124]	[147, 147]	[380, 380]	[100, 100]
w_8	PAM	[170, 170]	[290, 290]	[120, 120]	[147, 147]	[230, 230]	[60, 60]

Let the interval-valued variables Y_j take values $[a_{uj}, b_{uj}]$, $j = 1, \ldots, p$, $u = 1, \ldots, m$. The observed midpoints are

$$\bar{X}_{uj} = (a_{uj} + b_{uj})/2. \tag{7.45}$$

The interval-valued divisive clustering method poses the question $q(\cdot)$ in the form

$$q_j(u) : \bar{X}_{uj} \leq c_j, \ j = 1, \ldots, p, \tag{7.46}$$

where c_j is a cutpoint associated with the variable Y_j. In Example 7.21 above, $c_1 = 130$.

For each j, we first reorder the midpoints from the smallest to the largest; let us label these $\bar{X}_{sj}$, $s = 1, \ldots, m_k$. Then, the selected cutpoint values are the midpoints of consecutive $\bar{X}_{uj}$ values (for those observations u in the cluster C_k being bipartitioned). That is, we select

$$c_{sjk} = (\bar{X}_{sj} + \bar{X}_{s+1,j})/2, \ s = 1, \ldots, m_k - 1, \tag{7.47}$$

for variable Y_j in cluster C_k. Therefore, there are at most $z_j = m_k - 1$ different cutpoints and therefore at most z_j different bipartitions of C_k possible when based on Y_j alone. If two or more $\bar{X}_{uj}$ have the same value, then $z_j < m_k - 1$. It is easy to verify that for any value c in $\bar{X}_{s,j} < c < \bar{X}_{s+1,j}$ the same bipartitioning pertains.

Example 7.24. Consider the horse breeds data of Table 7.14, and the random variable $Y_1 =$ Minimum Weight $\equiv Y$, say.

The first partition P_1 divides the entire dataset C_1 into (C_1^1, C_1^2). Table 7.15 provides the midpoints $\bar{X}_{sj} \equiv \bar{X}_s$, $s = 1, \ldots, 7$, where the observations have been rearranged from the lowest to highest $\bar{X}$ value. The cutpoints $c_s (\equiv c_{jsk}, j = 1, k = 1)$ are the midpoints of these $\bar{X}_s$. For example, from the observations w_3 and w_7 which, when reordered, correspond, respectively, to $s = 1, 2$, we have

$$\bar{X}_1 = (130 + 190)/2 = 160$$

$$\bar{X}_2 = (170 + 170)/2 = 170;$$

hence,

$$c_1 = (160 + 170)/2 = 165.$$

There are $z_j = 7$ different cutpoints c_s. □

Having selected the cutpoints, we now develop an appropriate distance matrix. We use the Hausdorff distances of Equation (7.31) and the normalized Hausdorff distances of Equation (7.33) in Examples 7.25–7.31 and Examples 7.32–7.33, respectively.

Example 7.25. Consider the first four observations of the horse breeds data of Table 7.14. For the present, we assume that the breeds are in the original sequence. If we consider $Y_1 =$ Minimum Weight only, then the Hausdorff distance matrix $\boldsymbol{D}_1 = (d_j(u_1, u_2))$, $u_1, u_2 = 1, \ldots, m = 4$, is, from Equation (7.31),

$$\boldsymbol{D}_1 = \begin{pmatrix} 0 & 30 & 280 & 170 \\ 30 & 0 & 260 & 200 \\ 280 & 260 & 0 & 440 \\ 170 & 200 & 440 & 0 \end{pmatrix}. \tag{7.48}$$

For example, for $u_1 = 1$, $u_2 = 3$, we have

$$|a_{u_1 1} - a_{u_2 1}| = |410 - 130| = 280, \quad |b_{u_1 1} - b_{u_2 1}| = |460 - 190| = 270;$$

hence,

$$d_1(u_1, u_3) = Max(280, 270) = 280.$$ □

Example 7.26. We continue with the data of Table 7.14 but take $p = 2$ with $Y_1 =$ Minimum Weight and $Y_2 =$ Maximum Weight. Again, let us use the first four observations only. The Hausdorff distance matrix $\boldsymbol{D}_1$ was given in Equation (7.48) in Example 7.25. The Hausdorff distance matrix for Y_2 is

$$\boldsymbol{D}_2 = \begin{pmatrix} 0 & 50 & 340 & 260 \\ 50 & 0 & 360 & 310 \\ 340 & 260 & 0 & 580 \\ 260 & 310 & 580 & 0 \end{pmatrix}. \tag{7.49}$$

Then, the Euclidean Hausdorff distance matrix, from Equation (7.32), is $\boldsymbol{D}$ where

$$\boldsymbol{D}\#\boldsymbol{D} = (\boldsymbol{D}_1\#\boldsymbol{D}_1 + \boldsymbol{D}_2\#\boldsymbol{D}_2) \tag{7.50}$$

where if A and B have elements $A = (a_{ij})$ and $B = (b_{ij})$, then $C = A\#B$ has elements $c_{ij} = (a_{ij} \times b_{ij})$. Therefore, from Equations (7.48) and (7.49), we obtain

$$\boldsymbol{D}\#\boldsymbol{D} = \begin{pmatrix} 0 & 900 & 78400 & 28900 \\ 900 & 0 & 67600 & 40000 \\ 78400 & 67600 & 0 & 193600 \\ 28900 & 40000 & 193600 & 0 \end{pmatrix} + \begin{pmatrix} 0 & 2500 & 115600 & 67600 \\ 2500 & 0 & 129600 & 96100 \\ 115600 & 129600 & 0 & 336400 \\ 67600 & 96100 & 336400 & 0 \end{pmatrix}$$

$$= \begin{pmatrix} 0 & 3400 & 194000 & 96500 \\ 3400 & 0 & 197200 & 136100 \\ 194000 & 197200 & 0 & 530000 \\ 96500 & 136100 & 530000 & 0 \end{pmatrix}.$$

Hence,

$$\boldsymbol{D} = \begin{pmatrix} 0 & 58.31 & 440.45 & 310.64 \\ 58.31 & 0 & 444.07 & 368.92 \\ 440.45 & 444.07 & 0 & 728.01 \\ 310.64 & 368.92 & 728.01 & 0 \end{pmatrix}. \tag{7.51}$$

Thus, for example, the Hausdorff distance between the w_1 and w_3 observations is

$$d(w_1, w_3) = (280^2 + 340^2)^{1/2}$$
$$= (78400 + 115600)^{1/2} = 440.45.$$

From these $\boldsymbol{D}$ values, we observe that the observations corresponding to $u = 1, 2$ are close with $d(w_1, w_2) = 58.31$, as far as the (Y_1, Y_2) weight variables go, while the observations for $u = 3, 4$ are far apart from $d(w_3, w_4) = 728.01$. (These distances reflect the data in that the observation w_3 represents a breed of pony, while those for $w_u (u = 1, 2, 4)$ are for horses.) □

Example 7.27. Let us find the dispersion normalized Euclidean Hausdorff distances for the first four horse breeds, for Y_1 = Minimum Weight and Y_2 = Maximum Weight. The distance matrix $\boldsymbol{D}_1$ of Equation (7.48) is the non-normalized matrix for Y_1. Using these values in Equation (7.34), we obtain the normalization factor as

$$H_1 = \left\{ \frac{1}{2 \times 4^2} [0^2 + 30^2 + 280^2 + \ldots + 440^2 + 0^2] \right\}^{1/2}$$
$$= 159.96. \tag{7.52}$$

For example, for (w_1, w_3),

$$d_1(w_1, w_3) = 280/159.96 = 1.750.$$

Then the complete normalized Euclidean Hausdorff distance matrix is

$$\boldsymbol{D}_1(H_1) = \begin{pmatrix} 0 & 0.187 & 1.750 & 1.063 \\ 0.187 & 0 & 1.625 & 1.250 \\ 1.750 & 1.625 & 0 & 2.751 \\ 1.063 & 1.250 & 2.751 & 0 \end{pmatrix}. \tag{7.53}$$

For the variable Y_2 = Maximum Weight, by applying Equation (7.34) to the entries of the non-normalized matrix $\boldsymbol{D}_2$ of Equation (7.49), we obtain

$$\begin{aligned} H_2 &= \left\{ \frac{1}{2 \times 4^2} [0^2 + 50^2 + \ldots + 580^2 + 0^2] \right\}^{1/2} \\ &= 216.19. \end{aligned} \tag{7.54}$$

Hence, we can obtain the normalized Euclidean Hausdorff distances. For example, for (w_1, w_3),

$$d_2(w_1, w_3) = 340/216.19 = 1.573.$$

Then, the complete dispersion normalized Euclidean Hausdorff distance matrix for Y_2 is

$$\boldsymbol{D}_2(H_2) = \begin{pmatrix} 0 & 0.231 & 1.573 & 1.203 \\ 0.231 & 0 & 1.665 & 1.434 \\ 1.573 & 1.665 & 0 & 2.683 \\ 1.203 & 1.434 & 2.683 & 0 \end{pmatrix}. \tag{7.55}$$

Finally, we can obtain this normalization for the two variables (Y_1, Y_2) combined. We apply Equation (7.33) using now the matrices of Equations (7.48) and (7.49). Thus, we have, e.g., for w_1, w_3,

$$d(w_1, w_3) = \left[\left[\frac{280}{159.96} \right]^2 + \left[\frac{340}{216.19} \right]^2 \right]^{1/2} = 2.353. \tag{7.56}$$

Similarly, we can obtain the distances for the other (w_1, w_2) pairs; we then obtain the complete normalized Euclidean Hausdorff distance matrix as

$$\boldsymbol{D} = \begin{pmatrix} 0 & 0.298 & 2.353 & 1.605 \\ 0.298 & 0 & 2.327 & 1.902 \\ 2.353 & 2.327 & 0 & 3.842 \\ 1.605 & 1.902 & 3.842 & 0 \end{pmatrix}. \tag{7.57}$$

□

Example 7.28. Let us now calculate the span normalized Euclidean Hausdorff distance for each of the Y_1, Y_2 and (Y_1, Y_2) data of Table 7.14 using the w_u, $u = 1, \ldots, 4$, observations. Consider Y_1 = Minimum Weight. From Table 7.14 and Equation (7.26), it is clear that

$$|\mathcal{Y}_1| = 630 - 130 = 500. \tag{7.58}$$

Hence, the span normalized Euclidean Hausdorff distance for Y_1 is, from Equation (7.35),

$$\boldsymbol{D}_1 = \begin{pmatrix} 0 & 0.600 & 0.560 & 0.340 \\ 0.600 & 0 & 0.520 & 0.400 \\ 0.560 & 0.520 & 0 & 0.880 \\ 0.340 & 0.400 & 0.880 & 0 \end{pmatrix} \tag{7.59}$$

where, e.g., for (w_1, w_2), we have, from Equation (7.35) with $p = 1, j = 1$,

$$\begin{aligned} d_1(w_1, w_3) &= [\phi_1(w_1, w_3)]/|\mathcal{Y}_1| \\ &= 280/500 = 0.56 \end{aligned}$$

where $\phi_1(w_1, w_3) = 280$ is obtained from Equation (7.31).

Similarly, for Y_2 alone, we have

$$|\mathcal{Y}_2| = 890 - 210 = 680; \tag{7.60}$$

hence the span normalized Euclidean Hausdorff distance matrix is, from Equation (7.35),

$$\boldsymbol{D}_2 = \begin{pmatrix} 0 & 0.074 & 0.500 & 0.382 \\ 0.074 & 0 & 0.529 & 0.456 \\ 0.500 & 0.529 & 0 & 0.853 \\ 0.382 & 0.456 & 0.853 & 0 \end{pmatrix}. \tag{7.61}$$

For the combined (Y_1, Y_2) variables, we have from Equation (7.35) that the span normalized Euclidean Hausdorff distance matrix, also called the maximum deviation normalized Hausdorff distance matrix, is

$$\boldsymbol{D} = \begin{pmatrix} 0 & 0.095 & 0.751 & 0.512 \\ 0.095 & 0 & 0.742 & 0.606 \\ 0.751 & 0.742 & 0 & 1.226 \\ 0.512 & 0.606 & 1.226 & 0 \end{pmatrix}. \tag{7.62}$$

This $\boldsymbol{D}$ matrix is obtained in a similar way to the elementwise products of Equation (7.50) used in obtaining Equation (7.51). Thus, e.g., for (w_1, w_2), we obtain from Equations (7.56) and (7.58),

$$d(w_1, w_3) = \left[\left[\frac{280}{500}\right]^2 + \left[\frac{340}{680}\right]^2 \right]^{1/2} = 0.751. \qquad \square$$

To partition this cluster C into two subclusters, to avoid considering all possible splits (here, $\binom{4}{2} = 6$ in total) we first reorder the classes according to decreasing (or equivalently increasing) values of the midpoints c of Equation (7.47).

Example 7.29. We again consider the first four breeds and Y_1 = Minimum Weight of Table 7.14. When reordered by the cutpoint values, we have $P_1 = C$ ={TES, CES, CMA, PEN}. The corresponding Hausdorff distance matrix is

$$\boldsymbol{D} = \begin{pmatrix} 0 & 170 & 200 & 440 \\ 170 & 0 & 30 & 280 \\ 200 & 30 & 0 & 260 \\ 440 & 280 & 260 & 0 \end{pmatrix}.$$

Then, for this cluster $C = \{w_1, \ldots, w_4\}$, the within-cluster variation $I(C)$ is, from Equation (7.39),

$$I(C) = \frac{1}{4 \times 4}[170^2 + 200^2 + \ldots + 260^2] = 25587.5.$$

Suppose now that this cluster C is partitioned into the subclusters C_1^1 ={TES} and C_1^2 ={CES, CMA, PEN}. Then, from Equation (7.39),

$$I(C_1^1) = \frac{1}{4 \times 1}[0] = 0,$$

$$I(C_1^2) = \frac{1}{4 \times 3}[200^2 + 440^2 + 280^2] = 13000.$$

Therefore, from Equation (7.43), we have

$$I_1 = 25587.5 - 0 - 13000 = 12587.5.$$

Similarly, for the partition of C into C_2^1 ={TES, CES}, C_2^2 ={CMA, PEN}, we find

$$I_2 = 13525;$$

and for the partition of C into C_3^1 ={TES, CES, CMA} and C_3^2 ={PEN}, we find

$$I_3 = 19770.83.$$

Hence, the optimal bipartition of this C is that k' for which

$$I_{k'} = \max_k \{I_k\} = I_3 = 19770.83.$$

Therefore, the optimal partition of these four breeds is to create the partition $P_2 = \{C^1, C^2\}$ with subclusters C^1 ={TES, CES, CMA} and C^2 ={PEN}. The total within-clusters variation of P_2 is, from Equation (7.40),

$$I(P_2) = 5816.67 + 0 = 5816.67 < I(C) = 25587.5.$$

The desired reduction in total within-cluster variation has therefore been achieved. □

We can now proceed to determine the complete hierarchy. We do this for the Y_1 random variable only in Example 7.30, and then in Examples 7.31 and 7.32 the procedure is developed with both Y_1 and Y_2 for non-normalized and normalized distances, respectively. Finally, in Example 7.33, the solution for all breeds and all six random variables is presented with the calculation details left to the reader as an exercise. We let the stopping criterion for the number of partitions be $R = 4$.

Example 7.30. Consider all horse breeds and Y_1 = Minimum Weight. We first reorder the breeds so that the midpoints $\bar{X}_u$ obtained from Equation (7.45) are in decreasing order. These are shown in Table 7.15 along with the corresponding s and cutpoint values c_s obtained from Equation (7.47). The Hausdorff distance matrix for this reordered sequence is

$$\boldsymbol{D} = \begin{pmatrix} 0 & 40 & 40 & 260 & 260 & 280 & 420 & 440 \\ 40 & 0 & 0 & 280 & 260 & 290 & 440 & 460 \\ 40 & 0 & 0 & 280 & 260 & 290 & 440 & 460 \\ 260 & 280 & 280 & 0 & 220 & 240 & 240 & 240 \\ 260 & 260 & 260 & 220 & 0 & 30 & 180 & 200 \\ 280 & 290 & 290 & 240 & 30 & 0 & 150 & 170 \\ 420 & 440 & 440 & 240 & 170 & 150 & 0 & 20 \\ 440 & 460 & 460 & 240 & 200 & 170 & 20 & 0 \end{pmatrix}. \quad (7.63)$$

Table 7.15 Divisive clustering – horses first bipartition, Y_1.

Breed w_u	Minimum Weight	$\bar{X}_u$	s	c_s	$I(C_s^1)$	$I(C_s^2)$	$W(C_s)$	$\Delta W(C_s)$
w_3	(130, 190)	160	1	165.0	0	28873.2	28873.2	5559.6
w_7	(170, 170)	170	2	240.0	100.0	20450.0	20550.0	13882.8
w_8	(170, 170)	170	3	240.0	133.3	8657.5	**8790.8**	**25642.0**
w_6	(170, 450)	310	4	360.0	7112.5	3909.4	11021.9	23410.9
w_2	(390, 430)	410	5	422.5	11970.0	2158.3	14128.3	20304.5
w_1	(410, 460)	435	6	472.5	16331.3	25.0	16356.3	18076.6
w_5	(410, 610)	510	7	515.0	26071.4	0	26071.4	8361.4
w_4	(410, 630)	520					34432.8	

Then, from Equation (7.39), the within-cluster variance $I(C) = I(E)$ is

$$I(E) = \frac{1}{8 \times 8}[40^2 + \ldots + 20^2] = 34432.8.$$

There are seven possible cuts of C into $P_1 = \{C^1, C^2\}$, one for each $s = 1, \ldots, 7$. Let us denote the sth possible bipartition as $P(s) = \{C_s^1, C_s^2\}$. The first partition of E at $s = 1$ is such that $C_1 = \{C_1^1, C_1^2\}$ where $C_1^1 = \{w_3\}$ and $C_1^2 = \{w_u, u \neq 3\}$.

The within-cluster variations for each of C_1^1 and C_1^2 are calculated from Equation (7.39) with the relevant entries from the full distance matrix of Equation (7.63). Therefore,

$$I(C_1^1) = 0$$

and

$$I(C_1^2) = \frac{1}{8 \times 7}[40^2 + 260^2 + \ldots + 170^2] = 28873.2.$$

Hence, from Equation (7.41),

$$W(C_1) = I(C_1^1) + I(C_1^2) = 28873.2,$$

and, from Equation (7.43),

$$I_1 = \Delta W(C_1) = I(E) - W(C_1) = 5559.6.$$

When $s = 3$, say, the bipartition becomes $P_3 = \{C_3^1, C_3^2\}$ with $C_3^1 = \{w_1, w_2, w_3\}$, and $C_3^2 = \{w_4, \ldots, w_8\}$. In this case,

$$I(C_3^1) = \frac{1}{8 \times 3}[40^2 + 40^2 + 0] = 133.3,$$

$$I(C_3^2) = \frac{1}{8 \times 5}[220^2 + \ldots + 20^2] = 8657.5,$$

and hence $W(C_3) = 8790.8$ and $\Delta W(C_3) = 25642.0$. The cluster variations $I(C_s^1)$, $I(C_s^2)$, $W(C_s)$, and $I_s = \Delta W(C_s)$ are given in Table 7.15 for each bipartition $s = 1, \ldots, 7$.

From the results in Table 7.15, it is seen that the cut s which maximizes I_s (or, equivalently, minimizes $W(C_s)$) is $s = 3$. Therefore, the two clusters for the $r = 2$ partition are

$$P_2 = (C_1 = \{w_3, w_7, w_8\}, C_2 = \{w_1, w_2, w_4, w_5, w_6\}).$$

The question on which this partition was based is

$$q_1 : \text{Is } Y = \text{Minimum Weight} \leq 240? \tag{7.64}$$

If 'Yes', then that observation falls into the cluster C_1; if 'No', then the observation falls into C_2.

To find the best ($r = 3$) three-cluster partition, we first take each of these C_1 and C_2 clusters and find the best subpartitioning for each in the manner described above. That is, we find the optimal

$$C_1 = (C_1^1, C_1^2) \text{ and } C_2 = (C_2^1, C_2^2).$$

Thus, for example, if the first cluster $C_1 = \{w_3, w_7, w_8\}$ is split at $s = 1$, to give

$$C_1^1 = \{w_3\}, \; C_1^2 = \{w_7, w_8\},$$

we can calculate the within-cluster variations as

$$I(C_1^1) = 0, \; I(C_1^2) = 0;$$

hence

$$I(C_1^1) + I(C_1^2) = 0.$$

Similarly, for the split

$$C_1^1 = \{w_3, w_7\}, \; C_1^2 = \{w_8\},$$

we have

$$I(C_1^1) + I(C_1^2) = 100 + 0 = 100.$$

Also the total within-cluster variation for C_1 is

$$I(C_1) = \frac{1}{8 \times 3}[40^2 + 40^2] = 133.3.$$

The $I(C_s^1)$, $I(C_s^2)$, $W(C_s)$, and $I_s = \Delta W(C_s)$ values are summarized in part (a) of Table 7.16. Thus, if this cluster is selected to form the actual partition, it is the partition $C_1 = (C_1^1 = \{w_3\}, C_1^2 = \{w_7, w_8\})$ which is optimal since this bipartition has the minimum value of $W(C_s)$, or equivalently the maximum value of I_s.

Table 7.16 Second and third bipartition – horses, Y_1.

Breed w_u	Y_1	$\bar{X}_u$	s	c_s	$I(C_s^1)$	$I(C_s^2)$	$W(C_s)$	$\Delta W(C_s)$
				(a) Second partition				
w_3	(130, 190)	160	1	165.0	0	0	**0**	**133.3**
w_7	(170, 170)	170	2	240.0	100	0	100	33.3
w_8	(170, 170)	170	—	240.0			133.3	
w_6	(170, 450)	310	1	360.0	0	3909.4	***3909.4***	***4748.1***
w_2	(390, 430)	410	2	422.5	3025.0	2158.3	5183.3	3474.2
w_1	(410, 460)	435	3	472.5	4454.2	25.0	4479.2	4178.3
w_5	(410, 610)	510	4	515.0	6856.3	0	6856.3	1801.3
w_4	(410, 630)	520	—				8657.5	
				(b) Third partition				
w_3	(130, 190)	160	1	165.0	0	0	133.3	133.3
w_7	(170, 170)	170	2	240.0	100	0	**100.0**	**33.3**
w_8	(170, 170)	170	—	240.0			133.3	
w_6	(170, 450)	310	—	360.0	0	0	0	0
w_2	(390, 430)	410	1	422.5	0	2158.3	2158.3	1751.0
w_1	(410, 460)	435	2	472.5	56.3	25.0	***81.3***	***3828.1***
w_5	(410, 610)	510	3	515.0	2325.0	0	2325.0	1584.4
w_4	(410, 630)	520	—				3909.4	

Similarly, if the second cluster of five observations $C_2 = \{w_1, w_2, w_4, w_5, w_6\}$ is split into the two subclusters

$$C_2^1 = \{w_6\}, C_2^2 = \{w_1, w_2, w_4, w_5\},$$

say, we obtain the within-cluster variation from Equation (7.39) as

$$I(C_2^1) = 0, \ I(C_2^2) = 3909.4,$$

and hence

$$W(C_1) = I(C_2^1) + I(C_2^2) = 3909.4.$$

Therefore,

$$I_1 = \Delta W(C_1) = 8657.5 - 3909.4 = 4748.1.$$

The within-cluster variations for all possible partitions of C_2 are shown in part (a) of Table 7.16. The total within-cluster variation for the whole cluster C_2 is

$$I(C_2) = 8657.5.$$

Comparing the $W(C_s) = [I(C_2^1) + I(C_2^2)]$ for each s, we observe that the $s = 1$ cut gives the optimal partition, within C_2. That is, if C_2 is to be split, we obtain optimally

$$C_2 = (C_2^1 = \{w_6\}, \ C_2^2 = \{w_1, w_2, w_4, w_5\}).$$

The cluster to be partitioned at this (second) stage is chosen to be the C_k which maximizes I_s. In this case,

$$I_1(C_1) = Max_s I_s(C_1) = 133.3$$

and

$$I_1(C_2) = Max_s I_s(C_2) = 4748.1.$$

Since

$$I_1(C_2) > I_1(C_1),$$

it is the optimal partitioning of C_2, and not that of C_1, which is chosen.

Hence, the best $r = 3$ cluster partition of these data is

$$P_3 = (C_1 = \{w_3, w_7, w_8\}, \ C_2 = \{w_6\}, \ C_3 = \{w_1, w_2, w_4, w_5\}).$$

The question being asked now is the pair of questions

$$q_1 : \text{ Is } Y = \text{Minimum Weight} \leq 240?$$

If 'No', then we also ask

$$q_2 : \text{ Is } Y = \text{Minimum Weight} \leq 360?$$

The process continues in a similar manner to obtain the $R = r = 4$ partition $P_4 = (C_1, C_2, C_3, C_4)$. The variations $I(C_k), I(C_k^2), I(C_k^2), W(C_k)$, and $\Delta W(C_k)$ for the three clusters $k = 1, 2, 3$ in the $(r = 3)$ P_3 partition are displayed in part (b) of Table 7.16. It is seen that the optimal P_4 partition is

$$P_4 = (C_1 = \{w_3, w_7, w_8\},\ C_2 = \{w_6\},\ C_3 = \{w_1, w_2\},\ C_4 = \{w_4, w_5\}).$$

The partitioning criteria are the following three questions:

$$q_1: \text{ Is } Y = \text{Minimum Weight} \leq 240?$$

If 'Yes', then the observations fall into cluster C_1. If 'No', then the next question is

$$q_2: \text{ Is } Y = \text{Minimum Weight} \leq 360?$$

If 'No', then the observations fall into cluster C_2. If 'Yes', then the third question is

$$q_3: \text{ Is } Y = \text{Minimum Weight} \leq 472.5?$$

If 'No', then the observations fall into C_3; otherwise, they fall into cluster C_4.

While this partitioning was based on one variable Y_1 = Minimum Weight only, the bi-plot of (Y_1, Y_2), where Y_2 = Maximum Weight, shown in Figure 7.7 suggests that this partitioning is indeed appropriate.

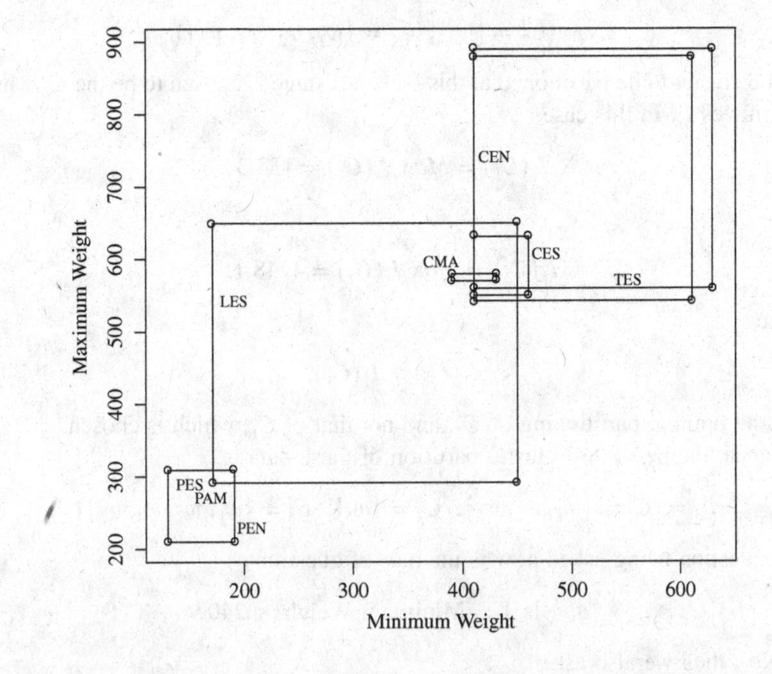

Figure 7.7 Bi-plot horses (Y_1 = Minimum Weight, Y_2 = Maximum Weight). □

Example 7.31. We want to find $R = 3$ clusters from all $m = 8$ breeds of horses from Table 7.14 based on the two variables Y_1 = Minimum Weight and Y_4 = Maximum Height.

The cut sequence based on the Y_1 variable is as shown in part (a) of Table 7.17 (and is the same as needed in Example 7.30). The corresponding cut sequence based on the Y_4 variable is in part (b) of Table 7.17. Note in particular that the ordering is slightly different (with the breeds w_1 and w_5 corresponding to $s = 6, 7$ reversed). This will impact the row and column order for the relevant distance matrices. Even though each possible cut of C into (C^1, C^2) involves one variable at a time, the distance measures used in this procedure are based on all the variables involved. We use the non-normalized Euclidean Hausdorff distances of Equation (7.32). These are displayed in Table 7.18 for each of the cuts on Y_1 and on Y_4.

Table 7.17 First partition, (Y_1, Y_4) horses.

	Breed w_u	Y_{uj}	$\bar{Y}_{uj}$	c_{js}	s	$I(C_s^1)$	$I(C_s^2)$	$W(C_s)$	$\Delta W(C_s)$
(a)	w_3	[130, 190]	160.0	165.0	1	0	28988.1	28988.1	5750.8
Cut	w_7	[170, 170]	170.0	170.0	2	200.0	20522.3	20722.2	14016.6
on	w_8	[170, 170]	170.0	240.0	3	266.7	8670.2	***8936.9***	***25801.9***
Y_1	w_6	[170, 450]	3100	360.0	4	7295.6	3119.0	11214.6	23524.3
	w_2	[390, 430]	410.0	422.5	5	12182.9	2166.4	14349.3	20389.5
	w_1	[410, 460]	435.0	472.5	6	16599.4	30.8	16630.2	18108.7
	w_5	[410, 610]	510.0	515.0	7	26366.5	0	26366.5	8372.4
	w_4	[410, 630]	520.0	—	—			34738.8	
(b)	w_3	[107, 153]	130.0	138.50	1	0	28988.1	28988.1	5750.8
Cut	w_7	[147, 147]	147.0	147.00	2	200.0	20522.3	20722.3	14016.6
on	w_8	[147, 147]	147.0	152.75	3	266.7	8670.2	**8936.9**	**25801.9**
Y_4	w_6	[145, 170]	158.5	158.50	4	7295.6	3919.0	11214.6	23524.3
	w_2	[150, 167]	158.5	161.00	5	12182.9	2166.4	14349.3	20389.5
	w_5	[155, 172]	163.5	165.00	6	23845.0	923.2	24768.2	9980.7
	w_1	[158, 175]	166.5	168.00	7	26366.5	0	26366.5	8372.4
	w_4	[147, 172]	169.5	—	—			34738.8	

For each s value for each of the Y_1 and Y_4 cuts, we calculate the within-cluster variations $I(C_s^1)$ and $I(C_s^2)$, $s = 1, \ldots, 7$. These are shown in part (a) and (b) of Table 7.17 for cuts on the Y_1 and Y_4 orderings, respectively. For example, if we bipartition C at $s = 3$ on Y_1, we have the two subclusters

$$C = (C_3^1 = \{w_3, w_7, w_8\},\ C_3^2 = \{w_6, w_2, w_1, w_5, w_4\}.$$

The calculations for $I(C_3^1)$ take the upper left-hand 3×3 submatrix of $\boldsymbol{D}(Y_1)$ and those for $I(C_3^2)$ take the lower right-hand 5×5 submatrix of $\boldsymbol{D}(Y_1)$. Then, from Equation (7.39),

$$I(C_3^1) = \frac{1}{8 \times 3}[(56.57)^2 + (56.57)^2 + 0^2] = 266.67,$$

Table 7.18 Euclidean Hausdorff distance matrix.

				$D(Y_1)$ Cut on Y_1				
u	3	7	8	6	2	1	5	4
3	0.000	56.569	56.569	263.059	263.532	284.607	422.734	441.814
7	56.569	0.000	0.000	280.943	260.768	291.349	440.710	460.679
8	56.569	0.000	0.000	280.943	260.768	291.349	440.710	460.679
6	263.059	280.943	280.943	0.000	220.021	240.252	240.133	240.008
2	263.532	260.768	260.768	220.021	0.000	31.048	180.069	200.063
1	284.607	291.349	291.349	240.252	31.048	0.000	150.030	170.356
5	422.734	440.710	440.710	240.133	180.069	150.030	0.000	21.541
4	441.814	460.679	460.679	240.008	200.063	170.188	22.825	0.000

				$D(Y_4)$ Cut on Y_4				
u	3	7	8	6	2	5	1	4
3	0.000	56.569	56.569	263.059	263.532	422.734	284.607	441.814
7	56.569	0.000	0.000	280.943	260.768	440.710	291.349	460.679
8	56.569	0.000	0.000	280.943	260.768	440.710	291.349	460.679
6	263.059	280.943	280.943	0.000	220.021	240.133	240.252	240.008
2	263.532	260.768	260.768	220.021	0.000	180.069	31.048	200.063
5	422.734	440.710	440.710	240.133	180.069	0.000	150.030	21.541
1	284.607	291.349	291.349	240.252	31.048	150.030	0.000	170.356
4	441.814	460.679	460.679	240.008	200.063	170.188	22.825	0.000

and also from Equation (7.39),

$$I(C_3^2) = \frac{1}{8 \times 5}[(220.02)^2 + (240.25)^2 + \ldots + (21.54)^2]$$
$$= 8670.15.$$

Hence, from Equation (7.40),

$$W(C_3) = I(C_3^1) + I(C_3^2) = 266.7 + 8670.2 = 8936.9.$$

The total within-cluster variation for P_1 (before its bisection) is, from Equation (7.39) and Table 7.18,

$$I(C) = \frac{1}{8 \times 8}[(56.57)^2 + \ldots + (21.54)^2] = 34738.8.$$

Therefore,

$$\Delta W(C_3) = 34738.8 - 8936.9 = 25801.9.$$

Since the cut orders on Y_1 and Y_4 are the same for $s = 1, \ldots, 5, 8$, these $W(C_3)$ values are the same in both cases, as Table 7.17 reveals. They differ, however, for the $s = 6$ and $s = 7$ cuts since, when cutting on Y_1, $C_7^2 = \{w_5, w_4\}$, but when cutting on Y_4, $C_7^2 = \{w_1, w_4\}$.

The optimal bipartition is that for which $I(C_s)$ is minimum. In this case, this occurs at $s = 3$. Therefore, the second partition is

$$P_2 = (C_1 = \{w_3, w_7, w_8\},\ C_2 = \{w_6, w_2, w_1, w_5, w_4\}).$$

The question that separates C into C_1 and C_2 is

$$\text{Is } Y_1 = \text{Minimum Weight} \leq 240? \tag{7.65}$$

If 'Yes', then the observation goes into cluster C_1, but if 'No', then it goes into cluster C_2. Since the cut on the Y_4 order gives the same minimum $I(C_3)$ value, the question might have been phrased in terms of Y_4 as

$$\text{Is } Y_4 = \text{Maximum Height} \leq 152.75? \tag{7.66}$$

However, the distance between the breeds w_8 (PAM) and w_6 (LES) for the Y_1 variable is $d_1(w_8, w_6) = 280$, while the distance between them for the Y_4 variable is $d_2(w_8, w_6) = 23$. Therefore, the criterion selected is that based on Y_1, i.e., Equation (7.65) rather than Equation (7.66).

By partitioning P_1 into P_2, the new total within-cluster variation, from Equation (7.40), is

$$W(P_2) = W(C_3) = I(C_3^1) + I(C_3^2) = 8936.9.$$

The proportional reduction in variation is

$$[I(P_1) - W(C_3)]/I(P_1) = (34738.8 - 8936.9)/34738.8 = 0.743.$$

That is, 74.3% of the total variation is explained by this particular partitioning.

The second partitioning stage is to bipartition $P_2 = (C_1, C_2)$ into three subclusters. We take each of C_k; $k = 1, 2$, and repeat the previous process by taking cuts s on each of C_1 and C_2 for each of Y_1 and Y_4 using the relevant distance submatrices extracted from the full distance matrices in Table 7.18. The resulting $I(C_s^1)$, $I(C_s^2)$, $W(C_s)$, and I_s values in each case are displayed in Table 7.19. We note again that these values differ for the Y_1 and Y_4 variables at the $s = 3$ and $s = 4$ cuts on C_2 because of the different orders for the breeds in C_2.

Then, if the cluster to be partitioned is C_1, its optimal bisection is into

$$C_1 = (C_1^1 = \{w_3\},\ C_1^2 = \{w_7, w_8\})$$

since $W(C_1) = Min_s[W(C_s)] = 0$, and $I_1^1 = Max_s I_s(C_s^1) = 266.7$.

If the cluster to be partitioned is C_2, its optimal bisection is

$$C_2 = (C_2^1 = \{w_6\},\ C_2^2 = \{w_2, w_1, w_5, w_4\})$$

since for subclusters within C_2,

$$W(C_1) = Min_s[W(C_s)] = 3919.0$$

Table 7.19 Second partition, (Y_1, Y_4) horses.

	Breed w_u	Y_{uj}	$\bar{Y}_{uj}$	c_{js}	s	$I(C_s^1)$	$I(C_s^2)$	$W(C_s)$	$\Delta W(C_s)$
(a)	w_3	[130, 190]	160.0	165.0	1	0	0	**0**	**266.7**
Cut	w_7	[170, 170]	170.0	170.0	2	200.0	0	200.0	66.7
on	w_8	[170, 170]	170.0	240.0	3			266.7	
Y_1	w_6	[170, 450]	3100	360.0	4	0	3919.0	***3919.0***	***4751.2***
	w_2	[390, 430]	410.0	422.5	5	3025.6	2166.4	5192.0	3478.2
	w_1	[410, 460]	435.0	472.5	6	4462.3	29.0	4491.3	4178.9
	w_5	[410, 610]	510.0	515.0	7	6865.4	0	6865.4	1804.8
	w_4	[410, 630]	520.0	—	—			8670.2	
(b)	w_3	[107, 153]	130.0	138.50	1	0	0	**0**	**266.7**
Cut	w_7	[147, 147]	147.0	147.00	2	200.0	0	200.0	66.7
on	w_8	[147, 147]	147.0	152.75	3			266.7	
Y_4	w_6	[145, 170]	158.5	158.50	4	0	3919.0	**3919.0**	**4751.2**
	w_2	[150, 167]	158.5	161.00	5	3025.6	2166.4	5192.0	3478.2
	w_5	[155, 172]	163.5	165.00	6	5770.8	1813.8	7584.6	1085.6
	w_1	[158, 175]	166.5	168.00	7	6865.4	0	6865.4	1804.8
	w_4	[147, 172]	169.5	—	—			8670.2	

and $I_1^2 = Max_s I_s(C_s^2) = 4751.2$. The choice of whether to bisect C_1 or C_2 is made by taking the optimal bisection within C_1 or C_2 for which I_s is maximized. Here,

$$I_1^2 = Max\{I_1^1, I_1^2\} = Max_k[266.7, 4751.2] = 4751.2.$$

Therefore, it is the C_2 cluster that is bisected. This gives us the new partitioning as

$$P_3 = (C_1 = \{w_3, w_7, w_8\},\ C_2 = \{w_6\},\ C_3 = \{w_2, w_1, w_5, w_4\}).$$

The criterion now becomes the two questions:

$$q_1 : \text{Is } Y_1 = \text{Minimum Weight} \leq 240?$$

If 'No', then we ask

$$q_2 : \text{Is } Y_1 = \text{Minimum Weight} \leq 360? \tag{7.67}$$

Again, the criterion of Equation (7.67) is based on the Y_1 variable since the distance between w_6 (LES) and w_2 (CMA) is greater for the Y_1 variable than it is for the Y_4 variable (at 160 and 5, respectively).

The third partitioning stage to give $P_4 = (C_1, C_2, C_3, C_4)$ proceeds in a similar manner. The relevant within-cluster variations are shown in Table 7.20. Note that, this time, the optimal partitioning of C_3 is different depending on whether the cut is based on Y_1 or Y_4, with the Y_1 cut producing

$$C_3 = (C_3^1 = \{w_2, w_1\},\ C_3^2 = \{w_5, w_4\})$$

Table 7.20 Third partition, (Y_1, Y_4) horses.

	Breed w_u	Y_{uj}	$\bar{Y}_{uj}$	c_{js}	s	$I(C_s^1)$	$I(C_s^2)$	$W(C_s)$	$\Delta W(C_s)$
(a)	w_3	[130, 190]	160.0	165.0	1	0	0	**0**	**266.7**
Cut	w_7	[170, 170]	170.0	170.0	2	200.0	0	200.0	66.7
on	w_8	[170, 170]	170.0	240.0	3			266.7	
Y_1	w_6	[170, 450]	3100	360.0	4	0	0	**0**	**0**
	w_2	[390, 430]	410.0	422.5	5	0	2166.4	2166.4	1752.6
	w_1	[410, 460]	435.0	472.5	6	60.2	29.0	***89.2***	***3829.7***
	w_5	[410, 610]	510.0	515.0	7	2329.1	0	2329.1	1589.9
	w_4	[410, 630]	520.0	—	—			3919.0	
(b)	w_3	[107, 153]	130.0	138.50	1	0	0	**0**	**266.7**
Cut	w_7	[147, 147]	147.0	147.00	2	200.0	0	200.0	66.7
on	w_8	[147, 147]	147.0	152.75	3			266.7	
Y_4	w_6	[145, 170]	158.5	158.50	4	0	0	**0**	**0**
	w_2	[150, 167]	158.5	161.00	5	0	2166.4	**2166.4**	**1752.6**
	w_5	[155, 172]	163.5	165.00	6	2026.6	1813.8	3840.4	78.6
	w_1	[158, 175]	166.5	168.00	7	2329.1	0	2329.1	1589.9
	w_4	[147, 172]	169.5	—	—			3919.0	

and the Y_4 cut producing

$$C_3 = (C_3^1 = \{w_2\},\ C_3^2 = \{w_5, w_1, w_4\}).$$

However, the Y_1 cut is chosen as it has a higher $I_s = W(C_s)$ value (3829.7 for the Y_1 cut compared to 1752.6 for the Y_4 cut).

From these results we conclude that the best partitioning of E into four clusters gives

$$P_4 = (C_1 = \{w_3, w_7, w_8\},\ C_2 = \{w_6\},\ C_3 = \{w_1, w_2\},\ C_4 = \{w_4, w_5\}).$$

The progressive schemata are displayed in Figure 7.8.

The complete criteria on which this bipartitioning is made are

$$q_1: \text{ Is } Y_1 = \text{Minimum Weight} \leq 240?$$

If 'Yes', then the breed goes into C_1. If 'No', then we ask

$$q_2: \text{ Is } Y_1 \leq 360?$$

If 'Yes', then the breed goes into C_2. If 'No', then we ask

$$q_3: \text{ Is } Y_1 \leq 472.5?$$

If 'Yes', then the breed goes into C_3. If 'No', it goes into C_4.

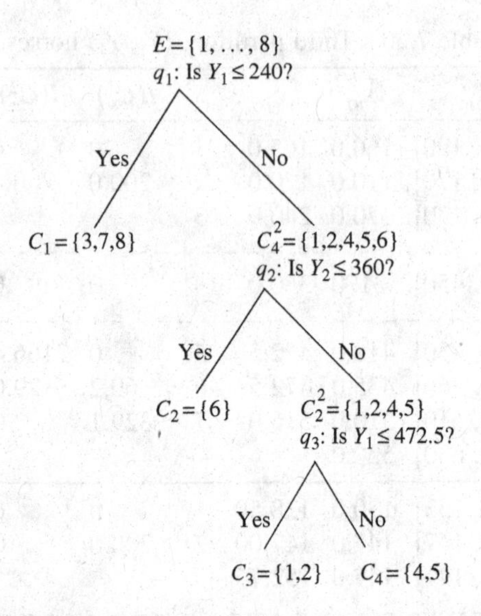

Figure 7.8 Hierarchy based on $(Y_1,\ Y_4)$ – horses, by breed #.

Finally, the total within-cluster variation for the partition P_4 is

$$I(P_4) = \sum_{k=1}^{4} I(C_k)$$
$$= 266.7 + 0 + 60.2 + 29.0 = 355.9$$

Therefore, the total between-cluster variation for P_4 is, from Equation (7.41),

$$B(P_4) = I(P_1) - I(P_4) = 34738.8 - 355.9 = 34382.9$$

Thus, 99.98% of the total variation is explained by the differences between clusters. □

Example 7.32. To find the hierarchy for the same horse breeds of Table 7.14 based on the random variables Y_1 = Minimum Weight and Y_4 = Maximum Height by using the dispersion Euclidean Hausdorff distances, we proceed as in Example 7.31. However, the distances are calculated by use of Equation (7.33). Therefore, at each cut s, the distances must be recalculated since, from Equation (7.34), the normalization factor H_j is based only on those observations in the (sub)cluster under consideration.

Table 7.21 gives the elements of the dispersion Euclidean distance matrix used when cutting by Y_1. Consider the $s = 3$ cut on Y_1. We need to calculate the

Table 7.21 Dispersion Euclidean distance matrix – cut on Y_1.

w_u	3	7	8	6	2	1	5	4
w_3	0.00	2.30	2.30	2.68	2.83	3.28	3.56	3.29
w_7	2.30	0.00	0.00	2.00	1.81	2.24	2.77	2.86
w_8	2.30	0.00	0.00	2.00	1.81	2.24	2.77	2.86
w_6	2.68	2.00	2.00	0.00	1.20	1.44	1.37	1.30
w_2	2.83	1.81	1.81	1.20	0.00	0.49	1.01	1.12
w_1	3.28	2.24	2.24	1.44	0.49	0.00	0.83	1.11
w_5	3.56	2.77	2.77	1.37	1.01	0.83	0.00	0.47
w_4	3.29	2.86	2.86	1.30	1.12	1.02	0.64	0.00

relevant distance matrices for each of the subclusters $C^1 = \{w_3, w_7, w_8\}$ and $C^2 = \{w_6, w_2, w_1, w_5, w_4\}$. Take C^2. Then, for the Y_1 variable, from Equation (7.34),

$$H_1^2 = \frac{1}{5^2}[220^2 + \ldots + 20^2] = 13852,$$

$$H_4^2 = \frac{1}{5^2}[3^2 + \ldots + 8^2] = 20.24.$$

Hence, from Equation (7.33), for instance,

$$d(w_6, w_2) = \left[\left(\frac{220}{117.7}\right)^2 + \left(\frac{3}{4.5}\right)^2\right]^{1/2} = 0.28.$$

Proceeding likewise for all elements in C^2, we obtain

$$\boldsymbol{D}(C^2) = \begin{pmatrix} 0 & 1.98 & 2.71 & 3.18 & 2.09 \\ 1.98 & 0 & 1.89 & 1.80 & 2.03 \\ 2.71 & 1.89 & 0 & 1.44 & 1.79 \\ 3.18 & 1.80 & 1.44 & 0 & 2.84 \\ 2.09 & 2.03 & 1.79 & 2.84 & 0 \end{pmatrix}.$$

Similarly,

$$\boldsymbol{D}(C^1) = \begin{pmatrix} 0 & 3 & 3 \\ 3 & 0 & 0 \\ 3 & 0 & 0 \end{pmatrix}.$$

We proceed as in Example 7.31 using at each s cut the recalculated distance matrix. Then, we calculate the respective inertias $I(C_s^1), I(C_s^2), W(C_2)$, and $\Delta W(C_s) \equiv I_s$. Hence, for each $r = 2, 3, 4$, we obtain the optimal cut to give the best $P_r = (C_1, \ldots, C_r)$. The final hierarchy of $R = 4$ clusters is found to be

$$P_4 = (C_1 = \{w_3, w_7, w_8\},\ C_2 = \{w_6\},\ C_3 = \{w_1, w_2\},\ C_4 = \{w_4, w_5\}).$$

The details are left to the reader as an exercise. □

Example 7.33. The hierarchy based on all horse breeds and all six random variables of the data of Table 7.14, when using the normalized Hausdorff distances of Equation (7.33) for $R = 4$ is obtained as

$$P_4 = (C_1 = \{w_3\},\ C_2 = \{w_2, w_4, w_6\},\ C_3 = \{w_7, w_8\},\ C_4 = \{w_1, w_5\}).$$

This is displayed in Figure 7.9. The numbers indicated at the nodes in this figure tell us the sequence of the partitioning. Thus, after partitioning E into

$$P_1 = (C_1^1 = \{3, 7, 8\},\ C_1^2 = \{1, 2, 4, 5, 6\})$$

at the first stage, it is the C_1^1 subcluster which is partitioned at the second stage. The criteria on which this bipartitioning proceeds are the questions q_1, q_2, q_3:

$$q_1 : \text{Is } Y_1 = \text{Minimum Weight} \leq 240? \tag{7.68}$$

$$q_2 : \text{Is } Y_4 = \text{Maximum Height} \leq 138.5? \tag{7.69}$$

$$q_3 : \text{Is } Y_4 = \text{Maximum Height} \leq 161.5? \tag{7.70}$$

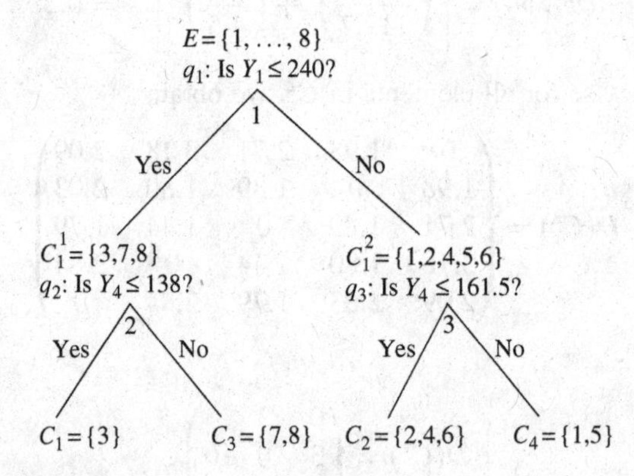

Figure 7.9 Hierarchy based on $(Y_1, \ldots, Y_6)$ – horses, by breed # (# at node indicates bipartition sequence).

If the answer to q_1 is 'Yes' and to q_2 is 'Yes', then the breeds go into C_1. If the answer to q_1 is 'Yes' and to q_2 is 'No', then the breeds go into C_3. If the answer to q_1 is 'No' and to q_3 is 'Yes', then the breeds go into C_2. If the answer to q_1 is 'No' and to q_3 is 'No', then they go into C_4. The details are left as an exercise. □

7.5 Hierarchy–Pyramid Clusters

7.5.1 Some basics

Pyramid clustering was introduced by Diday (1984, 1986) for classical data. Bertrand (1986) and Bertrand and Diday (1990) extended this concept to symbolic data, followed by a series of developments in, for example, Brito (1991, 1994) and Brito and Diday (1990). The underlying structures have received a lot of attention since then. We restrict our treatment to the basic principles. Recall from Section 7.2 that pyramid clusters are built from the bottom up, starting with each observation as a cluster of size 1. Also, clusters can overlap. This overlapping of clusters distinguishes a pyramid structure from the pure hierarchical structures of Section 7.4. See Section 7.5.2 below for an example of this comparison.

Example 7.34. Figure 7.10 represents a pyramid constructed on the five observations in $\Omega = \{a, b, c, d, e\}$. The 11 clusters

$$\begin{aligned} &C_1 = \{a\},\ C_2 = \{b\},\ C_3 = \{c\},\ C_4 = \{d\},\ C_5 = \{e\},\\ &C_6 = \{b, c\},\ C_7 = \{c, d\},\ C_8 = \{a, b, c\},\ C_9 = \{b, c, d\},\\ &C_{10} = \{a, b, c, d\},\ C_{11} = \{a, b, c, d, e\} \end{aligned}$$

together form a pyramid, written as

$$P = (C_1, \ldots, C_{11}).$$

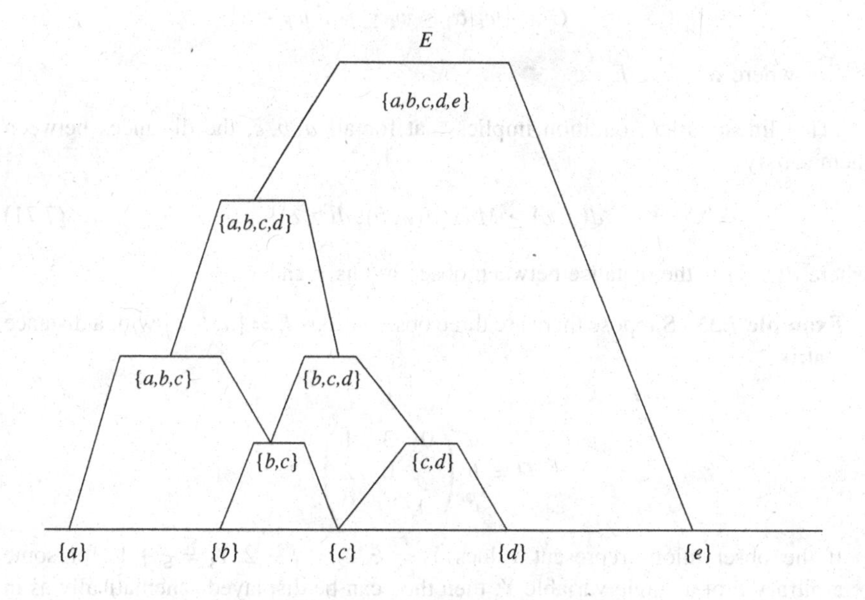

Figure 7.10 Pyramid clustering.

For example, the clusters C_6 and C_7 contain the observations {b, c} and {c, d}, respectively. Notice that the observation {c} appears in both the C_6 and C_7 clusters, i.e., they overlap. □

We have the following formal definitions; these are expansions of the earlier Definition 7.27 for pyramids.

Definition 7.32: A **pyramidal** classification of the observations in $E = \{w_1, \ldots, w_m\}$ is a family of non-empty subsets $(C_1, C_2, \ldots) = P$ such that

(i) $E \in P$;

(ii) for all $w_u \in E$, all single observations $\{w_u\} \in P$;

(iii) for any two subsets C_i, C_j in P,

$$C_i \cap C_j = \emptyset \text{ or } C_i \cap C_j \in P,$$

for all i, j; and

(iv) there exists a linear order such that every C_i in P spans a range of observations in E, i.e., if the subset C_i covers the interval

$$C_i = [\alpha_i,\ \beta_i]$$

then C_i contains all (but only those) observations w_k for which $\alpha_i \leq w_k \leq \beta_i$, or

$$C_i = \{w_k | \alpha_i \leq w_k \leq \beta_i,\ w_k \in E\}$$

where $\alpha_i,\ \beta_i \in E$. □

This linear order condition implies that for all a, b, c, the distances between them satisfy

$$d(a, c) \geq Max\{d(a, b), d(b, c)\} \tag{7.71}$$

where $d(x, y)$ is the distance between observations x and y.

Example 7.35. Suppose there are three observations $E = \{a, b, c\}$ with a distance matrix

$$\boldsymbol{D} = \begin{matrix} \\ a \\ b \\ c \end{matrix} \begin{matrix} a & b & c \\ \left(\begin{matrix} 0 & 3 & 1 \\ & 0 & 2 \\ & & 0 \end{matrix}\right) \end{matrix}.$$

If the observations represent values $Y_1 = \xi$, $Y_2 = \xi + 2$, $Y_3 = \xi + 1$, for some arbitrary ξ of a single variable Y, then they can be displayed schematically as in Figure 7.11.

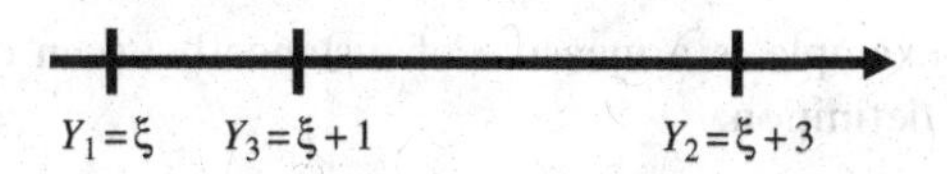

Figure 7.11 Observations for Example 7.35.

For condition (iv) to hold, it is necessary to relabel (a, b, c), so to speak, into the linear order $a' = a$, $b' = c$, and $c' = b$, with the new distance matrix

$$\boldsymbol{D}' = \begin{matrix} a' \\ b' \\ c' \end{matrix} \overset{\begin{matrix} a' & b' & c' \end{matrix}}{\begin{pmatrix} 0 & 1 & 3 \\ & 0 & 2 \\ & & 0 \end{pmatrix}}.$$

Note that this new distance matrix $\boldsymbol{D}'$ is a Robinson matrix. □

A pyramidal clustering can reflect a measure of the proximity of clusters by the 'height', h, of the pyramid branches. Thus, we see in Figure 7.12 two different pyramids, each with the same observations in the respective clusters, but with different proximity measure. Definition 7.32 gives us both pyramids of Figure 7.12. The two pyramids are seen as two different pyramids when their indexes are taken into account. For example, consider the cluster $\{a, b, c\}$. In pyramid (a), this cluster has height $h = 2.5$, whereas in pyramid (b), the height is $h = 2$. Indexes can take

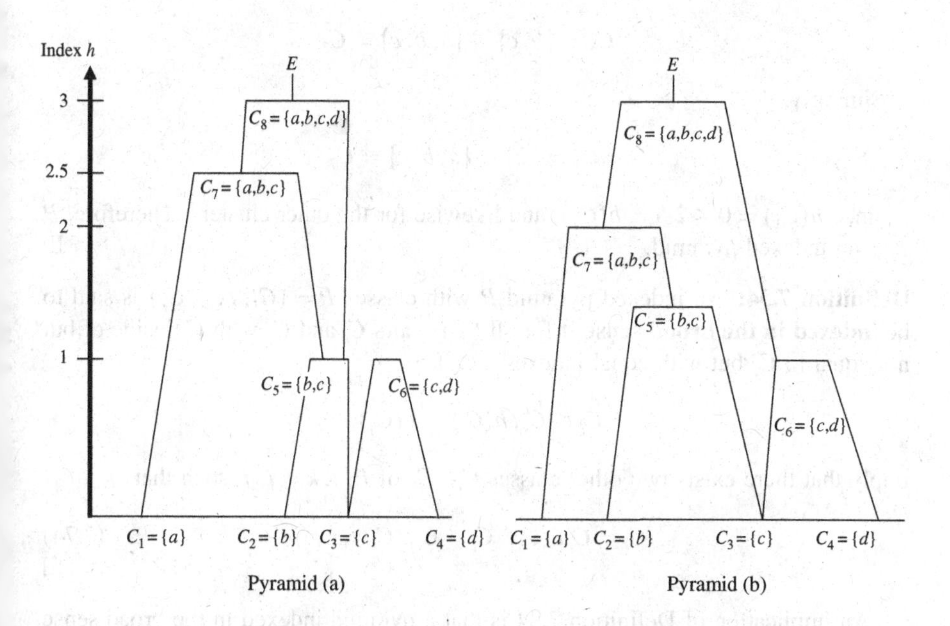

Figure 7.12 Two pyramids, different proximity.

various forms. An example is a measure of distance between clusters. This leads us to the following definition.

Definition 7.33: Let P be a pyramid with the properties (i)–(iv) of Definition 7.32, and let $h(C_i) \geq 0$ be an **index** associated with the clusters C_i in $P, i = 1, \dots, r$, where r is the number of clusters in P. If, for all (i, j) pairs in E with C_i contained in but not equal to C_j, there exists an index h that is isotonic on P, i.e.,

$$h(C_i) \leq h(C_j) \quad \text{if} \quad C_i \subset C_j, \tag{7.72}$$

and

$$h(C_i) = 0 \quad \text{if } C_i \text{ is a singleton}, \tag{7.73}$$

then (P, h) is defined to be an **indexed pyramid**. □

Example 7.36. Let E contain the observations $E = \{a, b, c, d\}$. Let the pyramid be $P = (C_1 = \{a\}, C_2 = \{b\}, C_3 = \{c\}, C_4 = \{d\}, C_5 = \{b, c\}, C_6 = \{c, d\}, C_7 = \{a, b, c\}, C_8 = \{a, b, c, d\})$, and let $h(C_5) = 1.0$ and $h(C_7) = 2.5$; see Figure 7.12, pyramid(a).

Then,

$$h(C_5) = 1.0 \leq 2.5 = h(C_7)$$

and

$$C_5 = \{b, c\} \subset \{a, b, c\} = C_7.$$

Similarly,

$$C_1 = \{a\} \subset \{a, b, c\} = C_7$$

since $h(C_1) = 0 \leq 2.5 = h(C_7)$, and likewise for the other clusters. Therefore, P is an indexed pyramid. □

Definition 7.34: An indexed pyramid P with classes $P = (C_1, \dots, C_r)$ is said to be **indexed in the broad sense** if for all (i, j) pairs C_i and C_j with C_i inside of but not equal to C_j but with equal indexes, i.e., for

$$C_i \subset C_j, h(C_i) = h(C_j),$$

imply that there exists two other classes C_l, C_k of P, $l, k \neq i, j$, such that

$$C_i = C_l \cap C_k, \ C_i \neq C_l, \ C_i \neq C_k. \tag{7.74}$$

□

An implication of Definition 7.34 is that a pyramid indexed in the broad sense has clusters C_i each with two predecessors.

Example 7.37. To illustrate a pyramid, indexed in the broad sense consider the structure of Figure 7.13 which defines the pyramid of $E = \{a, b, c, d\}$ by $P = (C_1, \ldots, C_{10})$ where

$$C_1 = \{a, b, c, d\},\ C_2 = \{a, b, c\},\ C_3 = \{b, c, d\},$$
$$C_4 = \{a, b\},\ C_5 = \{b, c\},\ C_6 = \{b, c, d\},$$
$$C_7 = \{a\},\ C_8 = \{b\},\ C_9 = \{c\},\ C_{10} = \{d\},$$

and where the index heights are

$$h(C_4) = 0.75,\ h(C_5) = h(C_6) = 1.0,\ h(C_2) = h(C_3) = 2.0,\ h(E) = 3.0.$$

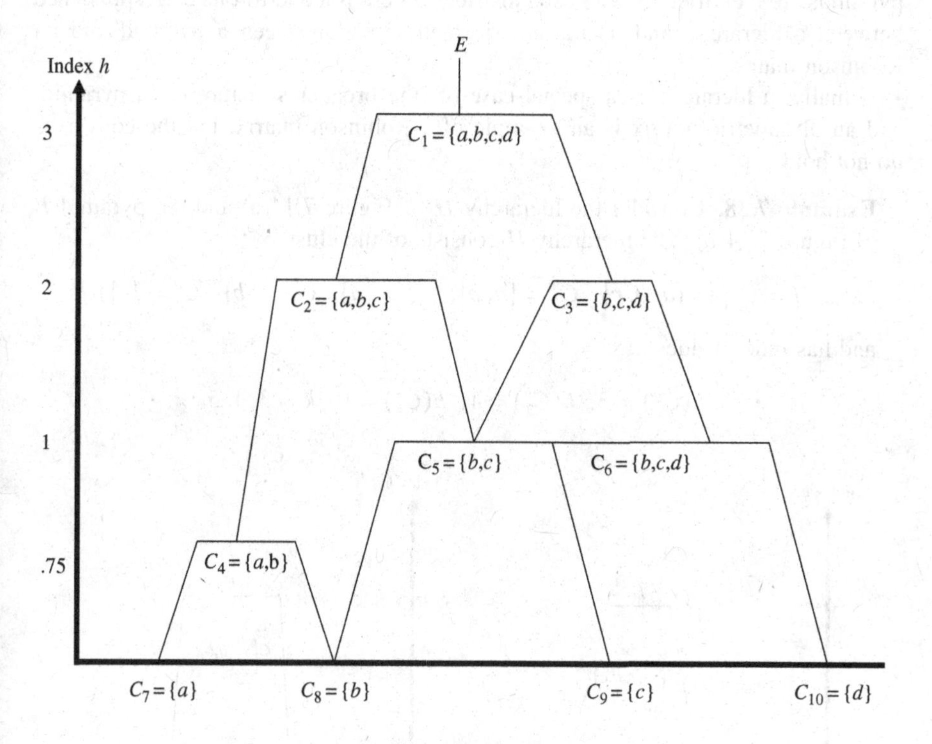

Figure 7.13 Pyramid indexed in broad sense.

In particular, consider the clusters C_5 and C_6. Here $h(C_5) = h(C_6)$ and $C_5 \subset C_6$ completely. Then, by Definition 7.34, there exist two other clusters whose intersection is C_5. In this case, we see that the clusters C_2 and C_3 satisfy this condition. Note too that since C_5 is contained completely within C_6, and since they have the same index value, this pyramid indexed in the broad sense can be arbitrarily approximated by an indexed pyramid. □

7.5.2 Comparison of hierarchy and pyramid structures

By 'hierarchy' in this subsection, we refer to Definition 7.25. Our hierarchy can be created by a top-down (divisive) process or a bottom-up (agglomerative) process, but for the present purposes the hierarchy has no overlapping clusters. Both hierarchy and pyramid structures consist of nested clusters. However, a pyramid can contain overlapping clusters while a pure hierarchy structure does not. An implication of this statement is that any one cluster in a hierarchy has at most one predecessor, while for a pyramid any one cluster has at most two predecessors. Thirdly, though index values may be comparable, they may induce different distance matrices depending on whether the structure is a hierarchy or a pyramid.

Indeed, hierarchies are identified by ultrametric dissimilarity matrices, while pyramids are identified by Robinson matrices. There is a one-to-one correspondence between a hierarchy and its ultrametric matrix, and between a pyramid and its Robinson matrix.

Finally, a hierarchy is a special case of (the broader structure of) a pyramid, and an ultrametric matrix is an example of a Robinson matrix, but the converses do not hold.

Example 7.38. Consider the hierarchy H of Figure 7.14(a) and the pyramid P of Figure 7.14(b). The hierarchy H consists of the clusters

$$H = (C_1^h = \{a, b, c\},\ C_2^h = \{a, b\},\ C_3^h = \{a\},\ C_4^h = \{b\},\ C_5^h = \{c\})$$

and has index values of

$$h(C_1^h) = 3,\ h(C_2^h) = 1,\ h(C_k^h) = 0,\ k = 3, 4, 5.$$

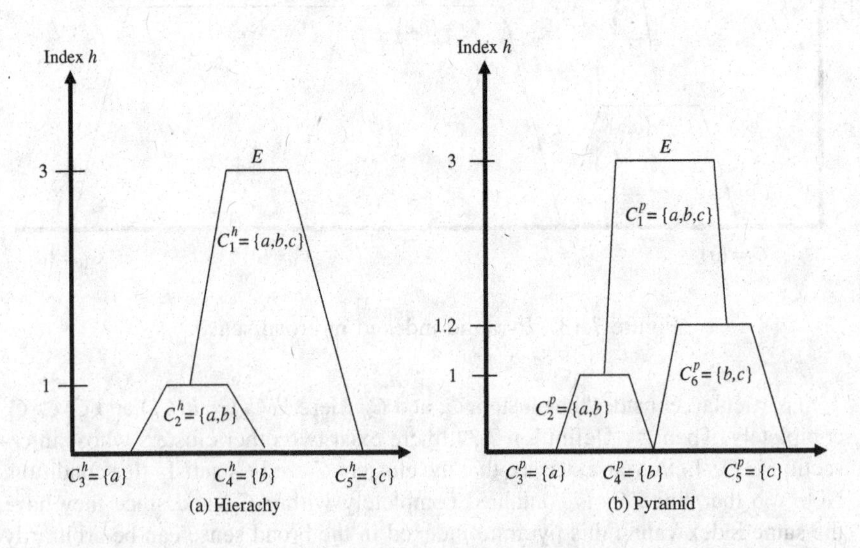

Figure 7.14 Comparison of hierarchy (a) and pyramid (b).

The pyramid P consists of the clusters

$$P = (C_1^p \equiv C_1^h,\ C_2^p \equiv C_2^h,\ C_3^p \equiv C_3^h,\ C_4^p \equiv C_4^h,\ C_5^p \equiv C_5^h,\ C_6^p = \{b, c\}),$$

with the index values

$$h(C_k^p) \equiv h(C_k^h),\ k = 1, \ldots, 5,\ h(C_6^p) = 1.2,$$

i.e., the structures are similar except that the pyramid contains the additional cluster $C_6^p = \{b, c\}$. Notice that the two pyramid clusters C_2^p and C_6^p overlap, each containing the observation $\{b\}$. Therefore, this implies that the cluster $C_4^p = \{b\}$ has two predecessors in P, namely, C_2^p and C_6^p. In contrast, this cluster has only one predecessor in H, namely, $C_2^h = C_2^p$.

Further, let us suppose that the index function $h(\cdot)$ is the distance $d(x, y)$ induced from the hierarchy, or the pyramid, as the lowest height/distance between the observations $\{x\}$ and $\{y\}$ in H, or P, where we start at $h = 0$ for the individuals themselves. Therefore, in this case, we see from Figure 7.14(a) that the distance matrix for the hierarchy H is

$$\boldsymbol{D}^h = \begin{array}{c} a \\ b \\ c \end{array} \overset{\begin{array}{ccc} a & b & c \end{array}}{\begin{pmatrix} 0 & 1 & 3 \\ & 0 & 3 \\ & & 0 \end{pmatrix}}$$

where, for example, the distance between the observations a and b is $d(a, b) = 1$ since the lowest $C_i \in H$ containing both a and b is $C_2^h = \{a, b\}$, and not $C_1^h = \{a, b, c\}$. That is, C_2^h is lower down the tree than C_1^h, and so is chosen. Rather than 'lower' down the tree, we can think of C_2^h as being the smallest C_k^h which contains both $\{a\}$ and $\{b\}$. The distance $d(a, c)$ is the index value between the clusters $C_1^h = \{a, b, c\}$ and $C_5^h = \{c\}$, i.e., $d(a, c) = 3$; likewise, $d(b, c) = 3$.

In contrast, the corresponding distance matrix for the pyramid P in Figure 7.14(b) is

$$\boldsymbol{D}^p = \begin{array}{c} a \\ b \\ c \end{array} \overset{\begin{array}{ccc} a & b & c \end{array}}{\begin{pmatrix} 0 & 1 & 3 \\ & 0 & 1.2 \\ & & 0 \end{pmatrix}}.$$

For example, $d(b, c) = 1.2$, since in P the lowest C_k^p that contains $\{b\}$ and $\{c\}$ is $C_6^p = \{b, c\}$ at an index of $h = 1.2$ above $\{c\}$ and not the $h = 3$ between $C_1^p = \{a, b, c\}$ and $C_5^p = \{c\}$.

It is observed that $\boldsymbol{D}^h$ is an ultrametric similarity matrix, while $\boldsymbol{D}^p$ is a Robinson matrix. That $\boldsymbol{D}^h$ is ultrametric is readily shown from the fact that, from Definition 7.3 in Section 7.1,

$$\begin{aligned} d(a, b) = 1 &\leq Max\{d(a, c), d(c, b)\} \\ &= Max\{3, 3\} = 3, \\ d(a, c) = 3 &\leq Max\{d(a, b), d(b, c)\} \\ &= Max\{1, 3\} = 3, \\ d(b, c) = 3 &\leq Max\{d(a, b), d(a, c)\} \\ &= Max\{1, 3\} = 3. \end{aligned}$$

In addition, (a, b, c) forms an isosceles triangle with base (a, b) having a length $d(a, b) = 1$, which is shorter than the length of the other two sides, $d(a, c) = d(b, c) = 3$.

Notice also that $\boldsymbol{D}^h$ is a Robinson matrix, as is $\boldsymbol{D}^p$, since its elements do not decrease in value as they move away from the diagonal. An application of Definition 7.5 verifies this. However, the Robinson matrix $\boldsymbol{D}^p$ is not an ultrametric matrix. To see this, it is only necessary to note that

$$\begin{aligned} d(a, c) = 3 &\not\leq Max\{d(a, b), d(b, c)\} \\ &= Max\{1, 1.2\} = 1.2. \end{aligned}$$

□

7.5.3 Construction of pyramids

We construct our pyramid using the methodology of Brito (1991, 1994). Since pyramid clustering is an agglomerative bottom-up method, the principles of cluster construction outlined in Section 7.2.2 apply in reverse. That is, instead of starting with the complete set E and partitioning clusters C into two subclusters at each stage (C^1 and C^2), we start with m clusters, one for each observation singly, and merge two subclusters (C^1 and C^2, say) into one new cluster C. The other features of the construction process remain, i.e., selection of a criterion for choosing which two subclusters C^1 and C^2 prevail, and selection of a stopping rule. For these we need Definitions 7.35 and 7.36.

Definition 7.35: Let E be a set of m observations and let $C \subseteq E$ be a subset (or cluster) consisting of elements of E. Suppose that the observations are represented by the realizations $\boldsymbol{Y}(w_1), \ldots, \boldsymbol{Y}(w_m)$ of the random variable $\boldsymbol{Y} = (Y_1, \ldots, Y_p)$. Let s be the symbolic assertion object

$$s = f(C) = \bigwedge_{j=1}^{p} [Y_j \subseteq \underset{w_u \in C}{} Y_j(w_u)] \tag{7.75}$$

where f is the mapping with $f(s) = Ext(s|C)$; likewise let $g = g(E)$ be the corresponding mapping with $g(s) = Ext(s|E)$. Suppose h is an iterated mapping satisfying

$$h = f \; o \; g \text{ gives } h(s) = f(g(s))$$
$$h' = g \; o \; f \text{ gives } h'(C) = g(f(C)).$$

Then a symbolic object s is **complete** if and only if $h(s) = s$. □

This is clearly a formal definition, and is based on definitions presented and illustrated in Sections 2.2.2 and 2.2.3. Brito (1991) relates this to the notion of a concept (see Definition 2.22) through the result that, for a complete object s for which $C = g(s) = Ext(s|E)$, the pair (C, s) is a concept, where $C \subseteq E$ is a subset (cluster) of E. In a less formal way, completeness corresponds to the 'tightest' condition which describes the extension of a set C.

Example 7.39. Consider the data of Table 7.22 (extracted from Table 2.1a) for the random variables Y_3 = Age, Y_4 = Gender, Y_{10} = Pulse rate and Y_{15} = LDL cholesterol level. Suppose we have the two assertion objects

$$s_1 = (\text{Age} \subseteq [20, 60]) \wedge (\text{Gender} \subseteq \{F\}) \wedge (\text{Pulse} \subseteq [75, 80]),$$
$$s_2 = (\text{Age} \subseteq [27, 56]) \wedge (\text{Gender} \subseteq \{F\}) \wedge (\text{Pulse} \subseteq [76, 77]) \wedge (\text{LDL} \subseteq [71, 80]). \tag{7.76}$$

The extensions are therefore $Ext(s_1) = \{\text{Carly, Ellie}\} = Ext(s_2)$.

Table 7.22 Health measurements.

w_u	Y_3 Age	Y_4 Gender	Y_{10} Pulse	Y_{15} LDL
Amy	70	F	72	124
Carly	27	F	77	80
Ellie	56	F	76	71
Matt	64	M	81	124
Sami	87	F	88	70

Although the extensions of s_1 and s_2 contain the same two individuals, the description of the set $C = \{\text{Carly, Ellie}\}$ is more precise in s_2 than it is in s_1. That is, s_1 is not a complete object, whereas s_2 is a complete object. □

From Definition 7.35, it follows that a cluster C as a single observation is complete. Also, the union of two complete objects is itself a complete object.

Example 7.40. To continue with Example 7.39 and the data of Table 7.22, suppose we have the clusters

$$C_1 = \{\text{Amy}\}$$
$$C_2 = \{\text{Carly, Ellie}\}.$$

Then, the single observation C_1 with

$$s_1 = s_1(C_1)$$
$$= (\text{Age} \subseteq [70, 70]) \wedge (\text{Gender} \subseteq \{F\}) \wedge (\text{Pulse} \subseteq [72, 72]) \wedge (\text{LDL} \subseteq [124, 124])$$

is a complete object. Also, $s_2 = s_2(C_2)$ where s_2 is given in Equation (7.76) is a complete object. Merging C_1 and C_2, we obtain the new cluster

$$C = C_1 \cup C_2 = \{\text{Amy, Carly, Ellie}\}$$

with extension

$$s = s(C) = (\text{Age} \subseteq [27, 70]) \wedge (\text{Gender} \subseteq \{F\}) \wedge (\text{Pulse} \subseteq [72, 80]) \wedge (LDL \subseteq [71, 124]).$$

This $s = s(C)$ is a complete object. □

Definition 7.36: Let $Y = (Y_1, \ldots, Y_p)$ be a random variable with Y_j taking values in the domain $\mathcal{Y}_j$, where $\mathcal{Y}_j$ is a finite set of categories (or list of possible values for a multi-valued Y_j) or bounded intervals on $\Re$ (for an interval-valued Y_j). Let the object s have description

$$s = \bigwedge_{j=1}^{p} [Y_j \in D_j], \ D_j \subseteq \mathcal{Y}_j, \ j = 1, \ldots, p.$$

Then, the **generality degree of s** is

$$G(s) = \prod_{j=1}^{p} c(D_j)/c(\mathcal{Y}_j) \tag{7.77}$$

where $c(A)$ is the cardinality of A if Y_j is a multi-valued variable and $c(A)$ is the length of the interval A if Y_j is an interval-valued variable. □

Example 7.41. Consider the first four horse breeds in Table 7.14. Take the random variable Y_3 = Minimum Height. Suppose we are interested in the cluster C = {CES, CMA}. Then, the descriptions of $\mathcal{Y}$ and C are

$$\mathcal{Y} = [90, 165], \ C = [130, 158].$$

Hence, the generality degree of C for the random variable Y_3 is

$$G(C) = (158 - 130)/(165 - 90) = 0.373.$$

If our interest is on both Y_3 and Y_4 = Maximum Height, the descriptions of $\mathcal{Y}$ and C become

$$\mathcal{Y} = ([90, 165], \ [107, 175])$$
$$C = ([130, 158], \ [150, 175]).$$

Hence, the generality degree of C for both Y_3 and Y_4 is

$$G(C) = \left(\frac{158-130}{165-90}\right)\left(\frac{175-150}{175-107}\right) = 0.137. \qquad \square$$

Example 7.42. Consider the color and habitat data of birds shown in Table 7.2. Take the object or cluster $C = \{\text{species 1, species 2}\}$. Then, using both variables, we have

$$\mathcal{Y} = (\{\text{red, black, blue}\},\ \{\text{rural, urban}\})$$

and

$$C = (\{\text{red, black}\},\ \{\text{rural, urban}\}).$$

Hence, the generality degree of C is

$$G(C) = \left(\frac{2}{3}\right)\left(\frac{2}{2}\right) = 0.667. \qquad \square$$

The stopping rule for a pyramid is that the 'current' cluster is the entire set E itself or that the conditions, for any two subclusters C_1 and C_2, are such that:

(i) neither C_1 nor C_2 has already been merged twice in earlier steps and there is a total linear order (as defined in Definition 7.32) ≤ 1 on E such that there is an interval with respect to ≤ 1 on the merged clusters ($C = C_1 \cup C_2$);

(ii) the merged cluster C is complete (as defined in Definition 7.35).

The selection criterion for choosing which particular pair (C_1, C_2) to merge is to take that pair which minimizes the generality degree $G(s)$ defined in Definition 7.36 and Equation (7.77).

Example 7.43. We construct a pyramid on the first four observations in Table 7.14, i.e., for the horse breeds CES ($u = 1$), CMA ($u = 2$), PEN ($u = 3$), and TES ($u = 4$). We take Y_1 = Minimum Weight only. At the initial step $t = 0$, we identify the four singleton clusters.

Let us measure the distance between any two observations by the generality degree given above in Equation (7.77), for $j = p = 1$. Clearly, the length of the entire set $\mathcal{Y}$ is

$$d(\mathcal{Y}) = \max_u(b_u) - \min_u(a_u) = 630 - 130 = 500.$$

Hence, the distance matrix is

$$D' = \begin{pmatrix} 0 & 0.14 & 0.66 & 0.44 \\ 0.14 & 0 & 0.60 & 0.48 \\ 0.66 & 0.60 & 0 & 1.00 \\ 0.44 & 0.48 & 1.00 & 0 \end{pmatrix} \begin{matrix} \text{CES} \\ \text{CMA} \\ \text{PEN} \\ \text{TES} \end{matrix}$$

with columns CES, CMA, PEN, TES, where, for example, from Equation (7.77),

$$\begin{aligned} d(1,3) &= c(\text{CES, PEN})/c(\mathcal{Y}) \\ &= \{Max(b_1, b_3) - Min(a_1, a_3)\}/d(\mathcal{Y}) \\ &= (460 - 130)/500 = 0.66. \end{aligned}$$

Transforming these observations so that a linear order applies (recall Example 7.35), we obtain the Robinson distance matrix

$$D_1 = \begin{pmatrix} 0 & 0.60 & 0.66 & 1.00 \\ 0.60 & 0 & 0.14 & 0.48 \\ 0.66 & 0.14 & 0 & 0.44 \\ 1.00 & 0.48 & 0.44 & 0 \end{pmatrix} \begin{matrix} \text{PEN} \\ \text{CMA} \\ \text{CES} \\ \text{TES} \end{matrix}. \quad (7.78)$$

with columns PEN, CMA, CES, TES.

It is easily checked from Definition 7.32(iv) that this gives a linear order, at least of the initial clusters denoted as

$$C_1 = \{\text{PEN}\}, \ C_2 = \{\text{CMA}\}, \ C_3 = \{\text{CES}\}, \ C_4 = \{\text{TES}\}.$$

Then since the smallest distance is

$$d(2,3) = d(\text{CMA, CES}) = 0.14,$$

the cluster formed at this first ($t = 1$) stage is

$$C_5 = C_2 \cup C_3 = \{\text{CMA, CES}\}.$$

The second ($t = 2$) step is to form a new cluster by joining two of the clusters $C_1, \ldots, C_5$. Since $C_1, \ldots, C_4$ and hence $C_1, \ldots, C_5$ have an internal linear order, it is necessary to consider only the four possibilities:

$$C_1 \cup C_2, \ C_1 \cup C_5, \ C_5 \cup C_4, \ C_3 \cup C_4.$$

The distances between C_1 and C_2, and C_3 and C_4, are known from D_1 in Equation (7.78). To calculate the distances between $C_1 = \{\text{PEN}\}$ and $C_5 = \{\text{CMA, CES}\}$, we first observe that C_5 consists of the interval

$$\begin{aligned} \xi_5 &= [Min(a_2, a_3), \ Max(b_2, b_3)] \\ &= [390, \ 460]. \end{aligned}$$

Then, the generality degree for $C_1 \cup C_5$ is, from Equation (7.77),

$$d(C_1, C_5) = \{Max(460, 190) - Min(390, 130)\}/500$$
$$= (460 - 130)/500 = 0.66.$$

Similarly,

$$d(C_4, C_5) = \{Max(630, 460) - Min(410, 390)\}/500$$
$$= (630 - 390)/500 = 0.48,$$

or equivalently,

$$G(C_1 \cup C_2) = 0.60, \; G(C_1 \cup C_5) = 0.66,$$
$$G(C_4 \cup C_5) = 1.00, \; G(C_3 \cup C_4) = 0.44.$$

Since

$$Min\; G(C_i \cup C_j) = G(C_3 \cup C_4) = 0.44,$$

the two clusters C_3 and C_4 are merged to form a new cluster C_6 = {CES, TES}. This C_6 cluster consists of the new interval $\xi_6 = [410, 630]$.

For the third ($t = 3$) step, the possible merged clusters are

$$C_1 \cup C_2, \; C_1 \cup C_5, \; C_5 \cup C_6.$$

Notice in particular that C_3 = {CES} cannot be merged with another cluster at this stage, as it already has two predecessors, having been merged into the new clusters C_5 and then C_6. The possible clusters have generality degree values of

$$G(C_1 \cup C_2) = 0.60, \; G(C_1 \cup C_5) = 0.66, \; G(C_5 \cup C_6) = 0.48,$$

where we have

$$G(C_5 \cup C_6) = G(\{\text{CMA, CES, TES}\}),$$
$$= [Max(430, 460, 630) - Min(390, 410, 410)]/500$$
$$= (630 - 390)/500 = 0.48.$$

Or equivalently, the new distance matrix is

$$D_3 = \begin{pmatrix} 0 & 0.60 & 0.66 & 1.00 \\ 0.60 & 0 & - & - \\ 0.66 & - & 0 & 0.48 \\ 1.00 & - & 0.48 & 0 \end{pmatrix} \begin{matrix} \text{PEN} \\ \text{CMA} \\ \text{(CMA, CES)} \\ \text{(CES, TES)} \end{matrix}.$$

(Columns: PEN, CMA, (CMA, CES), (CES, TES).)

Since the smallest generality degree value is $G(C_5 \cup C_6) = 0.48$, the new cluster C_7 is formed from merging C_5 and C_6, i.e.,

$$C_7 = C_5 \cup C_6 = \{\text{CMA, CES, TES}\}$$

and spans the interval

$$\begin{aligned}\xi_7 &= [Min(390, 410, 410),\ Max(430, 460, 630)]\\ &= [390,\ 630].\end{aligned}$$

Similarly, at the next $t = 4$th step, we have

$$G(C_1 \cup C_2) = 0.60,\ G(C_1 \cup C_5) = 0.66,\ G(C_5 \cup C_7) = 1.00,$$

$$\boldsymbol{D}_4 = \begin{array}{c} \begin{array}{cccc} \text{PEN} & \text{CMA} & (\text{CMA, CES}) & (\text{CMA, CES, TES}) \end{array} \\ \left(\begin{array}{cccc} 0 & 0.60 & 0.66 & 1.00 \\ 0.60 & 0 & - & - \\ 0.66 & - & 0 & - \\ 1.00 & - & - & 0 \end{array}\right) \end{array} \begin{array}{c} \text{PEN} \\ \text{CMA} \\ (\text{CMA, CES}) \\ (\text{CES, TES}) \end{array}.$$

Hence, C_1 and C_2 are merged to give $C_8 = C_1 \cup C_2 = \{\text{PEN, CMA}\}$ which is the interval $\xi_8 = [130, 430]$, at a distance $d = 0.60$.

At the $t = 5$th step, we have

$$G(C_5 \cup C_8) = 0.66,\ G(C_7 \cup C_8) = 1.00,\ G(C_5 \cup C_7) = 0.48,$$

to give the distance matrix

$$\boldsymbol{D}_5 = \begin{array}{c} \begin{array}{ccc} C_5 & C_7 & C_8 \end{array} \\ \left(\begin{array}{ccc} 0 & 0.48 & 0.66 \\ 0.48 & 0 & 1.00 \\ 0.66 & 1.00 & 0 \end{array}\right) \end{array} \begin{array}{c} C_5 \\ C_7 \\ C_8 \end{array}.$$

Hence, the new cluster is $C_9 = C_5 \cup C_7 = \{\text{PEN, CMA, CES}\}$ which is the interval $\xi_9 = [130, 460]$.

Finally, the $t = 6$th step gives $C_{10} = C_7 \cup C_9 = E$. Since C_{10} contains all the observations, the pyramid is complete.

The complete pyramid $P = (C_1, \ldots, C_{10})$ showing the observations contained in each C_i is displayed in Table 7.23. Also given is the symbolic description of each cluster as well as the generality index (or minimum distance selected) when that cluster was merged. For example, the C_6 cluster consists of two breeds, CES and TES, which together have a minimum weight interval value of $\xi_6 = [410, 630]$. The generality index value which resulted when this cluster was formed is $d = 0.44$. The pictorial representation of this pyramid is displayed in Figure 7.15. □

Table 7.23 Pyramid clusters for horses data.

Cluster C_k	Elements	Interval ξ_k	Generality $G(C_k)$
C_1	{PEN}	[130, 190]	0
C_2	{CMA}	[390, 430]	0
C_3	{CES}	[410, 460]	0
C_4	{TES}	[410, 630]	0
C_5	{CMA, CES}	[390, 460]	0.14
C_6	{CES, TES}	[410, 630]	0.44
C_7	{CMA, CES, TES}	[390, 630]	0.48
C_8	{PEN, CMA}	[130, 430]	0.60
C_9	{PEN, CMA, CES}	[130, 460]	0.66
C_{10}	{PEN, CMA, CES, TES}	[130, 630]	1.00

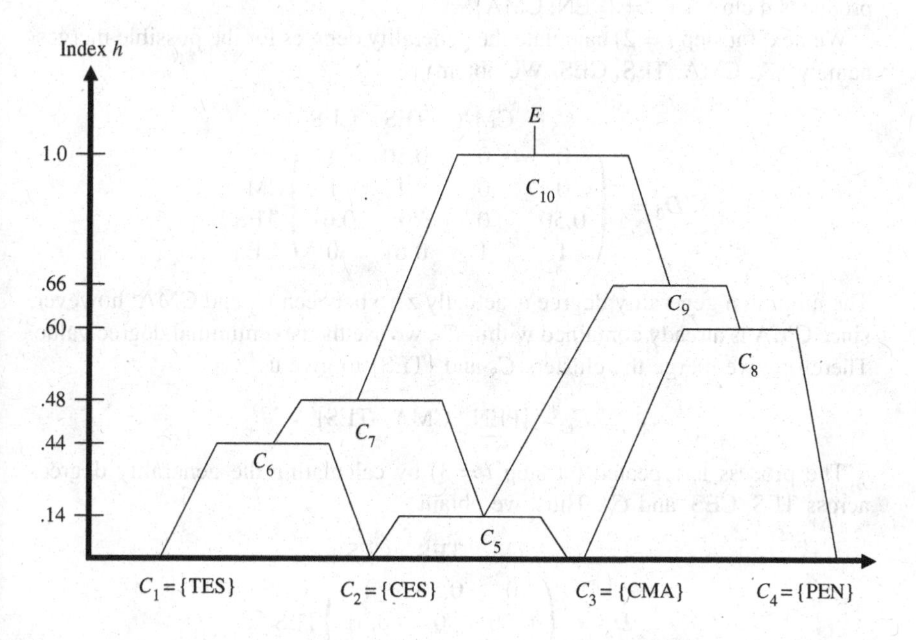

Figure 7.15 Pyramid of horses {CES, CMA, PEN, TES}, Y_1.

Example 7.44. Let us take the same four horses (CES, CMA, PEN, TEX) used in Example 7.43, but now let us build the pyramid based on the two random variables Y_5 = Mares and Y_6 = Fillies. We again use the generality degree of Equation (7.77). We calculate the relevant $\boldsymbol{D}$ matrix and after reordering to attain the desired linear order, we obtain the Robinson matrix

$$\boldsymbol{D}_1 = \begin{array}{c} \\ \end{array}\begin{matrix} \text{PEN} & \text{CMA} & \text{TES} & \text{CES} \\ \end{matrix}$$

$$\boldsymbol{D}_1 = \left(\begin{array}{cccc} 0 & 0.16 & 0.50 & 1 \\ 0.16 & 0 & 0.50 & 1 \\ 0.60 & 0.50 & 0 & 0.61 \\ 1 & 1 & 0.61 & 0 \end{array}\right)\begin{array}{l} \text{PEN} \\ \text{CMA} \\ \text{TES} \\ \text{CES} \end{array}.$$

For example,

$$G(\text{PEN, CMA}) = \left(\frac{200-0}{480-0}\right)\left(\frac{50-0}{130-0}\right) = 0.16$$

since {PEN, CMA} spans [0, 200], [0, 50] and $\mathcal{Y} = (\mathcal{Y}_1, \mathcal{Y}_2)$ spans ([0, 480], [0, 130]).

Then, at the first step ($t = 1$), the two clusters {PEN} and {CMA} are merged since of all possible mergers this has the minimum generality degree. This produces a cluster $C_5 = \{\text{PEN, CMA}\}$.

We next (at step $t = 2$) calculate the generality degrees for the possible merges, namely, C_5, CMA, TES, CES. We obtain

$$\boldsymbol{D}_2 = \begin{array}{c} \begin{matrix} C_5 & \text{CMA} & \text{TES} & \text{CES} \end{matrix} \\ \left(\begin{array}{cccc} 0 & 0 & 0.50 & 1 \\ 0 & 0 & 0 & 1 \\ 0.50 & 0 & 0 & 0.61 \\ 1 & 1 & 0.61 & 0 \end{array}\right) \end{array}\begin{array}{l} C_5 \\ \text{CMA} \\ \text{TES} \\ \text{CES} \end{array}.$$

The minimum generality degree is actually zero between C_5 and CMA; however, since CMA is already contained within C_5, we use the next minimal degree value. Therefore, we merge the clusters C_5 and {TES} to give us

$$C_6 = \{\text{PEN, CMA, TES}\}.$$

The process is repeated (at step $t = 3$) by calculating the generality degrees across TES, CES, and C_6. Thus, we obtain

$$\boldsymbol{D}_3 = \begin{array}{c} \begin{matrix} C_6 & \text{TES} & \text{CES} \end{matrix} \\ \left(\begin{array}{ccc} 0 & 0.50 & 1 \\ 0.50 & 0 & 0.61 \\ 1 & 0.61 & 0 \end{array}\right) \end{array}\begin{array}{l} C_6 \\ \text{TES.} \\ \text{CES} \end{array}$$

As in the construction of C_6, this time we ignore the 'possible' merger of TES with C_6, instead merging {TES} and {CES}, which two clusters otherwise have the minimum generality degree at 0.61. This produces the cluster $C_7 = \{\text{TES, CES}\}$.

Finally at the last step ($t = 4$) we merge C_6 and $C_7 = E$. The complete pyramid structure is shown statistically in Table 7.24 and pictorially in Figure 7.16. Notice that for these two variables (Y_5, Y_6), only two clusters, $C_6 = \{\text{TES, CMA, PEN}\}$ and $C_7 = \{\text{CES, TES}\}$, overlap, with the horse TES common to both. □

Table 7.24 Pyramid clusters for horses data – (Y_5, Y_5).

Cluster C_k	Elements	$\xi_k = (Y_5, Y_6)$	Generality degree $G(C_k)$
C_1	{CES}	([150, 480], [40, 130])	0
C_2	{TES}	([100, 350], [30, 90])	0
C_3	{CMA}	([0, 200], [0, 50])	0
C_4	{PEN}	([0, 100], [0, 30])	0
C_5	{CMA, PEN}	([0, 200], [0, 50])	0.16
C_6	{TES, CMA, PEN}	([0, 350], [0, 90])	0.50
C_7	{CES, TES}	([100, 480], [30, 130])	0.61
C_8	{CES, TES, CMA, PEN}	([0, 480], [0, 130])	1

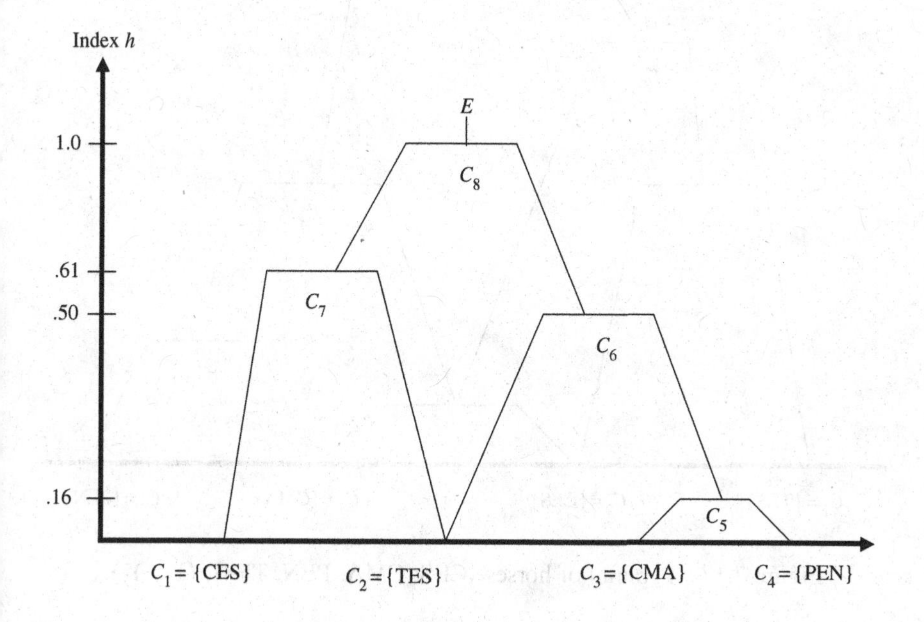

Figure 7.16 Pyramid of horses {CES, CMA, PEN, TES}, (Y_5, Y_6).

Example 7.45. Consider the two random variables Y_1 = Minimum Weight and Y_2 = Maximum Weight for the basis for constructing a pyramid on the same four horses (CES, CMA, PEN, TES) used in Examples 7.43 and 7.44. We show in Figure 7.17 the pyramid obtained when the merging criterion used is a dissimilarity matrix whose elements are those for generalized weighted Minkowski distances obtained from Equation (7.28) when $q = 1$, using the Ichino–Yaguchi measures of Equation (7.27) with $\gamma = 0$. The linearly ordered distance matrix is

$$
\boldsymbol{D}_1 = \begin{array}{c} \\ \end{array}\begin{pmatrix} 0 & 1.144 & 1.678 & 2.000 \\ 1.144 & 0 & 0.203 & 0.911 \\ 1.678 & 0.203 & 0 & 0.737 \\ 2.000 & 0.911 & 0.737 & 0 \end{pmatrix} \begin{matrix} \text{PEN} \\ \text{CMA} \\ \text{CES} \\ \text{TES} \end{matrix}
$$

(columns: PEN, CMA, CES, TES).

The details are omitted, and left as an exercise for the reader.

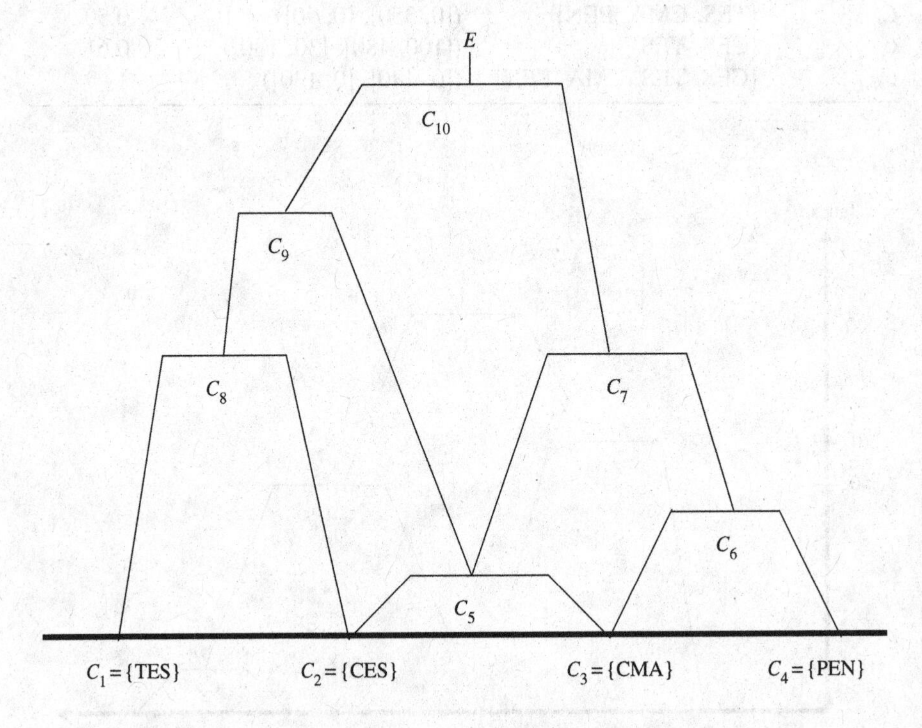

Figure 7.17 Pyramid of horses {CES, CMA, PEN, TES}, (Y_1, Y_2).

Notice that there are three overlapping clusters, specifically $C_5 = \{\text{CES, CMA}\}$ and $C_6 = \{\text{CMA, PEN}\}$ with CMA in common, C_5 and $C_8 = \{\text{TES, CES}\}$ with CES in common, and $C_7 = \{\text{CES, CMA, PEN}\}$ and $C_9 = \{\text{TES, CES, CMA}\}$ with two horses, CES and CMA, in common. □

By comparing the three examples Examples 7.43–7.45, we can identify some interesting features not uncommon in pyramid structures. We see that Figure 7.15 and Figure 7.17 are in effect almost mirror images of each other, both with the same relative linear ordering of the four horses, but with the roles of the horses TES and PEN reversed in the pyramid itself. One (Figure 7.15) was constructed using the generality degree as the optimality criterion, while the other used a dissimilarity measure. Of course, in this case, it could be a more dominant influence of the random variable Y_2(MaximumWeight) that prescribed this difference; we leave that determination as an exercise for the reader. The pyramid built on the variables Y_5 (Mares) and Y_6 (Fillies) shown in Figure 7.16 is quite different from those in Figures 7.15 and 7.17. Not only is the linear order different (with the horse TES now taking an 'interior' ordering), but there is only one pair of overlapping clusters compared to three overlapping clusters for the other two pyramids. These differing structures allow for enhanced interpretations from a now richer knowledge base generated by these analyses. Whatever the effect(s) of the inclusion or exclusion of certain variables, it remains that the impact of differing distance measures is in general still an open question.

This technique can also be extended to the case of mixed variables where some random variables are multi-valued and some are interval-valued.

Example 7.46. Take the last four oils, namely, camellia Ca, olive Ol, beef B, and hog H, of Ichino's oils dataset given in Table 7.8. Let us build a pyramid from these four oils based on the three random variables Y_1 = Specific Gravity, Y_3 = Iodine Value, and Y_5 = Fatty Acids, and let us use the generality degree of Equation (7.77) as our optimality criterion.

The Y_1 and Y_3 variables are interval-valued, taking values, respectively, over the ranges $\mathcal{Y}_1 = [0.858, 0.919]$ with cardinality $c(\mathcal{Y}_1) = 0.061$, and $\mathcal{Y}_3 = [40, 90]$ with $c(\mathcal{Y}_3) = 50$; Y_5 is a multi-valued variable taking values over $\mathcal{Y}_5 = \{C, L, Lu, M, O, P, S\}$ with $c(\mathcal{Y}_5) = 7$.

After reordering the oils so that a linear order pertains, we obtain the distance matrix

$$\boldsymbol{D}_1 = \begin{array}{c} \begin{array}{cccc} \text{Ol} & \text{Ca} & \text{H} & \text{B} \end{array} \\ \left(\begin{array}{cccc} 0 & 0.010 & 0.634 & 0.829 \\ 0.010 & 0 & 0.481 & 0.673 \\ 0.634 & 0.481 & 0 & 0.146 \\ 0.829 & 0.673 & 0.146 & 0 \end{array} \right) \begin{array}{c} \text{Ol} \\ \text{Ca} \\ \text{H} \\ \text{B} \end{array} \end{array}.$$

For example, the generality degree for the pair Ol and H is

$$G(\{\text{Ol}, \text{H}\}) = \left(\frac{0.061}{0.061}\right)\left(\frac{37}{50}\right)\left(\frac{6}{7}\right) = 0.634,$$

since the cluster {Ol, H} has description ([0.858, 0.919], [53.0, 90.0], $\{L, Lu, M, O, P, S\}$). Then, at this step $t = 1$, since the cluster {Ol, Ca} has the smallest generality degree value (at 0.010) in $\boldsymbol{D}_1$, the new cluster is C_5 = {Ol, Ca}. The description for this cluster is ([0.914, 0.919], [79.0, 90.0], $\{L, O, P, S\}$), as given in Table 7.25.

Table 7.25 Pyramid clusters – Oils (Ol, Ca, H, B).

Cluster	Oils	Description $\xi_k = (Y_1, Y_3, Y_5)$	$G(C_k)$
C_1	{Ol}	((0.914, 0.919], [79.0, 90.0], {L, O, P, S})	0
C_2	{Ca}	((0.916, 0.917], [80.0, 82.0], {L, O})	0
C_3	{H}	([0.858, 0.864], [53.0, 77.0], {L, Lu, M, O, P, S})	0
C_4	{B}	([0.860, 0.870], [40.0, 48.0], {C, M, O, P, S})	0
C_5	{Ol, Ca}	([0.914, 0.919], [79.0, 90.0], {L, O, P, S})	0.01
C_6	{H, B}	([0.858, 0.870], [40.0, 77.0], {C, L, Lu, M, O, P, S})	0.15
C_7	{Ca, H}	([0.858, 0.917], [53.0, 82.0], {L, Lu, M, O, P, S})	0.48
C_8	{Ol, Ca, H}	([0.858, 0.919], [53.0, 90.0], {L, Lu, M, O, P, S})	0.63
C_9	{Ol, Ca, H, B}	([0.858, 0.919], [40.0, 90.0], {C, L, Lu, M, O, P, S})	1.0

We repeat the process taking into account the clusters $C_2, \ldots, C_5$. We obtain the distance matrix as

$$\boldsymbol{D}_2 = \begin{pmatrix} 0 & 0 & 0.634 & 0.829 \\ 0 & 0 & 0.481 & 0.673 \\ 0.634 & 0.481 & 0 & 0.146 \\ 0.829 & 0.673 & 0.146 & 0 \end{pmatrix} \begin{matrix} C_5 \\ \text{Ca} \\ \text{H} \\ \text{B} \end{matrix}, \quad \text{columns: } C_5, \text{Ca}, \text{H}, \text{B}.$$

Since the smallest generality degree value which produces a new cluster is 0.146, we merge (at this step $t = 2$) the two oils H and B, to give $C_6 = \{\text{H, B}\}$.

Now, for step $t = 3$, we consider the four clusters C_5, Ca, H and C_6. These give the distance matrix

$$\boldsymbol{D}_3 = \begin{pmatrix} 0 & 0 & 0.634 & 1 \\ 0 & 0 & 0.481 & 0.812 \\ 0.634 & 0.481 & 0 & 0.146 \\ 1 & 0.812 & 0.146 & 0 \end{pmatrix} \begin{matrix} C_5 \\ \text{Ca} \\ \text{H} \\ C_6 \end{matrix}, \quad \text{columns: } C_5, \text{Ca}, \text{H}, C_6.$$

Then, from $\boldsymbol{D}_3$, the smallest generality degree value to give us a new cluster is 0.481 obtained when Ca and H are merged to form $C_7 = \{\text{Ca, H}\}$.

We now take the three clusters C_5, C_7, C_6 (in this linear order) and find the distance matrix to be

$$\boldsymbol{D}_4 = \begin{pmatrix} 0 & 0.634 & 1 \\ 0.634 & 0 & 0.812 \\ 1 & 0.812 & 0 \end{pmatrix} \begin{matrix} C_5 \\ C_7 \\ C_6 \end{matrix}, \quad \text{columns: } C_5, C_7, C_6.$$

Therefore, since the smallest generality degree value is 0.634, the optimal merge (at this step $t = 4$) is to form $C_8 = \{C_5, C_7) = (\text{O, Ca, H})$. The final step ($t = 5$) is to merge the two clusters C_5 and C_8, which produces the entire dataset E.

The pyramid therefore consists of the nine clusters $P = (C_1, \ldots, C_8, E)$. The oils contained in each cluster along with their statistical descriptions and the corresponding index h(= the generality degree value at their formation) are given in Table 7.25. The pictorial representation is shown in Figure 7.18. □

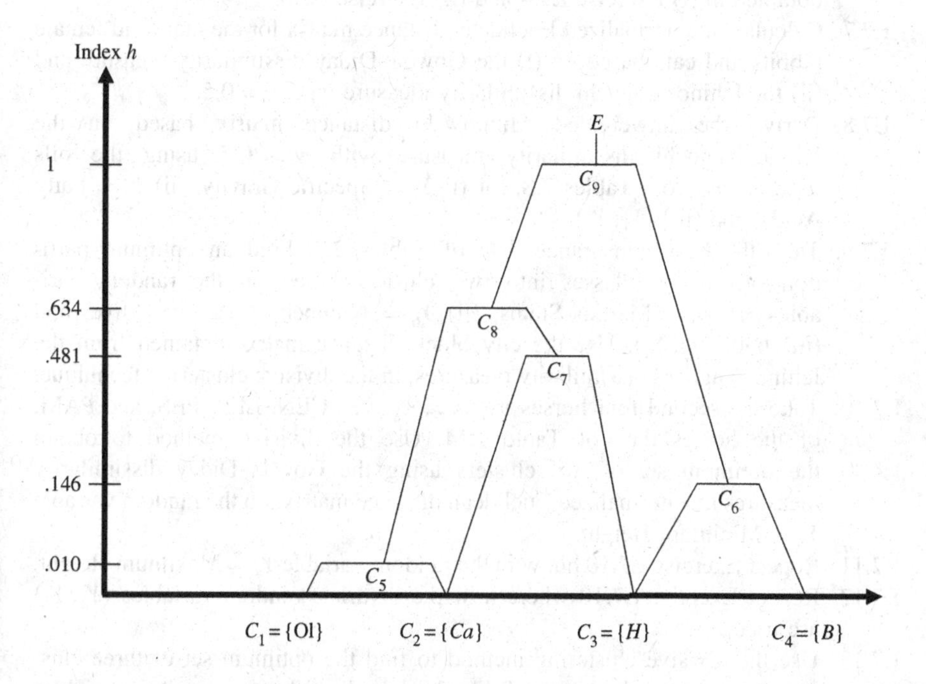

Figure 7.18 Pyramid – oils (Ol, Ca, H, B), (Y_1, Y_3, Y_5).

Exercises

E7.1. Consider the bird data of Table 7.2, and assume that listed values of the random variables Y_1 = Color and Y_2 = Habitat are equally likely. Obtain the categorical distance measures for all pairs of observations by using Equation (7.13). Hence, give the distance matrix $\boldsymbol{D}$.

E7.2. For the computer data of Table 7.11, find the normalized Euclidean distance matrix for (i) Y_1 = Brand, (ii) Y_2 = Type, and (iii) (Y_1, Y_2), respectively, based on the Gowda–Diday dissimilarity measure.

E7.3. Redo Exercise E7.2 but base the distance matrix in each case on the Ichino–Yaguchi dissimilarity measure with (a) $\gamma = 0$, (b) $\gamma = 0.5$.

E7.4. Consider the veterinary clinic data of Table 7.5. Obtain the Gowda–Diday distance matrix for the four animals: male and female rabbits and cats. Use (i) Y_1 = Height, (ii) Y_2 = Weight, and (iii) (Y_1, Y_2).

E7.5. Repeat Exercise E7.4 but this time obtain the Ichino–Yaguchi distance matrix for these four animals. Calculate these distances when (a) $\gamma = 0$, (b) $\gamma = 0.5$.

E7.6. Calculate the city block distance matrix based on the dissimilarity measures obtained in (i) Exercise E7.4 and (ii) Exercise E7.5.

E7.7. Calculate the normalized Euclidean distance matrix for the male and female rabbits and cats based on (i) the Gowda–Diday dissimilarity measure and (ii) the Ichino–Yaguchi dissimilarity measure with $\gamma = 0.5$.

E7.8. Derive the unweighted Minkowski distance matrix based on the Ichino–Yaguchi dissimilarity measure with $\gamma = 0.5$ using the oils $w_1, \ldots, w_6$ from Tables 7.8, for (i) Y_1 = Specific Gravity, (ii) Y_5 = Fatty Acids, and (iii) (Y_1, Y_5).

E7.9. Take the health insurance data of Tables 2.2. Find an optimum partition of the six classes into two clusters based on the random variables (i) Y_5 = Martial Status, (ii) Y_6 = Number of Parents Alive, and (iii) both (Y_5, Y_6). Use the city block distance matrix obtained from the Ichino–Yaguchi dissimilarity measures, in the divisive clustering technique.

E7.10. Take the second four horses $w_5, \ldots, w_8$, i.e., CEN, LES, PES, and PAM, of the horses data of Table 7.14. Use the divisive method to obtain the optimum set of two clusters using the Gowda–Diday dissimilarity measures in a normalized Euclidean distance matrix, on the random variable Y_3 = Minimum Height.

E7.11. Repeat Exercise E7.10 but with the random variable Y_4 = Maximum Height.

E7.12. Repeat Exercise E7.10 where now the bivariate random variables (Y_3, Y_4) are used.

E7.13. Use the divisive clustering method to find the optimum set of three clusters among the oils data of Table 7.8 based on the two random variables (Y_1, Y_5) and the city block distance matrix obtained via the Ichino–Yaguchi dissimilarity measures with $\gamma = 0.5$.

E7.14. Repeat Exercise E7.13 but use the Gowda–Diday dissimilarity measures when calculating the distance matrix.

E7.15. Take the health insurance data of Tables 2.2. Obtain the pyramid built from the random variables (i) Y_1 = Age, (ii) Y_{13} = Cholesterol, and (iii) both (Y_1, Y_{13}). Use the generality degree as the basis of the distance matrix.

E7.16. Repeat Exercise E7.15 where now the distance matrix is obtained from a normalized Euclidean distance based on the Ichino–Yaguchi dissimilarity measure with $\gamma = 0$.

References

Bertrand, P. (1986). Etude de la Représentation Pyramidile. Thése de 3eme Cycle, Université Paris, Dauphine.

Bertrand, P. and Diday, E. (1990). Une Généralisation des Arbres Hiérarchiques: les Représentations Pyramidales. *Revue de Statistique Appliquée* 38, 53–78.

Breiman, L., Friedman, J. H., Olshen, R. A., and Stone, C. J. (1984). *Classification and Regression Trees*. Wadsworth, Belmont, CA.

Brito, P. (1991). Analyse de Données Symboliques. Pyramides d'Héritage. Thése de Doctorat, Université Paris, Dauphine.

Brito, P. (1994). Use of Pyramids in Symbolic Data Analysis. In: *New Approaches in Classification and Data Analysis* (eds. E. Diday, Y. Lechevallier, M. Schader, P. Bertrand, and B. Burtschy). Springer-Verlag, Berlin, 378–386.

Brito, P. and Diday, E. (1990). Pyramidal Representation of Symbolic Objects. In: *Knowledge, Data and Computer-Assisted Decisions* (eds. M. Schader and W. Gaul). Springer-Verlag, Heidelberg, 3–16.

Chavent, M. (1997). Analyse de Données Symboliques. Une Méthode Divisive de Classification. Thése de Doctoral, Université Paris, Dauphine.

Chavent, M. (1998). A Monothetic Clustering Algorithm. *Pattern Recognition Letters* 19, 989–996.

Chavent, M. (2000). Criterion-based Divisive Clustering for Symbolic Data. In: *Analysis of Symbolic Data Exploratory Methods for Extracting Statistical Information from Complex Data* (eds. E. Diday and H.-H Bock). Springer-Verlag, Berlin, 299–311.

Diday, E. (1984). Use Repésentation Visuelle des Classes Empiétantes les Pyramides. *Rapport de Recherche* 291, INRIA, Rocquencourt.

Diday, E. (1986). Orders and Overlapping Clusters by Pyramids. In: *Multivariate Data Analysis* (eds. J. De Leeuw, W. J. Heiser, J. J. Meulman, and F. Critchley), DSWO Press, Leiden, 201–234.

Diday, E. (1989). *Data Analysis, Learning Symbolic and Numeric Knowledge*. Nova Science, Antibes.

Edwards, A. W. F. and Cavalli-Sforza, L. L. (1965). A Method for Cluster Analysis. *Biometrics* 21, 362–375.

Esposito, F., Malerba, D., and Tamma, V. (2000). Dissimilarity Measures for Symbolic Data. In: *Analysis of Symbolic Data: Exploratory Methods for Extracting Statistical Information from Complex Data* (eds. H.-H. Bock and E. Diday). Springer-Verlag, Berlin, 165–185.

Gordon, A. D. (1999). *Classification* (2nd ed.). Chapman and Hall, London.

Gowda, K. C. and Diday, E. (1991). Symbolic Clustering Using a New Dissimilarity Measure. *Pattern Recognition* 24, 567–578.

Gower, J. C. (1985). Measures of Similarity, Dissimilarity, and Distance. In: *Encyclopedia of Statistical Sciences, Volume 5* (eds. S. Kotz and N. L. Johnson). John Wiley & Sons, Inc., New York, 397–405.

Ichino, M. (1988). General Metrics for Mixed Features - The Cartesian Space Theory for Pattern Recognition. In: *Proceedings of the 1988 Conference on Systems, Man, and Cybernetics*. Pergamon, Oxford, 494–497.

Ichino, M. and Yaguchi, H. (1994). Generalized Minkowski Metrics for Mixed Feature Type Data Analysis. *IEEE Transactions on Systems, Man, and Cybernetics* 24, 698–708.

Data Index

Author Index

Symbolic Data Analysis: Conceptual Statistics and Data Mining L. Billard and E. Diday

Subject Index

Symbolic Data Analysis: Conceptual Statistics and Data Mining L. Billard and E. Diday